Concepts and Applications of Energy Storage

Concepts and Applications of Energy Storage

Edited by **Patricia Sachs**

CLANRYE
INTERNATIONAL

New Jersey

Published by Clanrye International,
55 Van Reypen Street,
Jersey City, NJ 07306, USA
www.clanryeinternational.com

Concepts and Applications of Energy Storage
Edited by Patricia Sachs

International Standard Book Number: 978-1-63240-114-4 (Hardback)

Contents

Permissions

List of Contributors

Preface

This book has been a concerted effort by a group of academicians, researchers and scientists, who have contributed their research works for the realization of the book. This book has materialized in the wake of emerging advancements and innovations in this field. Therefore, the need of the hour was to compile all the required researches and disseminate the knowledge to a broad spectrum of people comprising of students, researchers and specialists of the field.

The concepts and applications of energy storage techniques have been elucidated in this comprehensive and up-to-date book. Researchers and experts have contributed valuable information in this book from across the globe. It discusses a number of energy storage systems including electrochemical energy storage, compressed air energy storage, hybrid energy storage, etc. There is also a detailed description on the future of energy storage systems. The primary aim of this book is to serve as a useful source of information for researchers, scientists as well as students interested in gaining knowledge regarding energy storage systems, their technologies as well as applications.

At the end of the preface, I would like to thank the authors for their brilliant chapters and the publisher for guiding us all-through the making of the book till its final stage. Also, I would like to thank my family for providing the support and encouragement throughout my academic career and research projects.

Editor

Estimation of Energy Storage and Its Feasibility Analysis

Mohammad Taufiqul Arif, Amanullah M. T. Oo and A. B. M. Shawkat Ali

Additional information is available at the end of the chapter

1. Introduction

Storage significantly adds flexibility in Renewable Energy (RE) and improves energy management. This chapter explains the estimation procedures of required storage with grid connected RE to support for a residential load. It was considered that storage integrated RE will support all the steady state load and grid will support transient high loads. This will maximize the use of RE. Proper sized RE resources with proper sized storage is essential for best utilization of RE in a cost effective way. This chapter also explains the feasibility analysis of storage by comparing the economical and environmental indexes.

Most of the presently installed Solar PV or Wind turbines are without storage while connected to the grid. The intermittent nature of solar radiation and wind speed limits the capacity of RE to follow the load demand. The available standards described sizing and requirements of storage in standalone systems. However standards available for distributed energy resources (DER) or distributed resources (DR) to connect to the grid while considering solar photovoltaic (PV), wind turbine and storage as DR. Bearing this limitation, this chapter followed the sizing guidelines for standalone system to estimate the required storage for the grid connected RE applications.

Solar PV is unable provide electricity during night and cloudy days; similarly wind energy also unable to follow load demand. Moreover PV and/or wind application is not able to follow the load demand; when these RE generators are just in the stage to start generating energy and when these RE are in highest mode of generating stage while load demand falls to the lowest level. Therefore it can be said that RE is unable to generate energy by following the load demand which is a major limitation in energy management. Storage can play this critical role of proper energy management. Moreover storage helps in reducing the intermittent nature of RE and improve the Power Quality (PQ). This study considers regional Australia as the study area also considered residential load, solar radiation and wind speed data of that location for detailed analysis. Figure 1 shows the daily load profile (summer: January 01, 2009 and winter: July 01 2009) of Capricornia region of Rockhampton, a regional city in Australia. Ergon Energy [1] is the utility operator in that area. However load demand of the residential load in that area depends on the work time patter which is different than the load profile of Figure 1. Overall electricity demand is very high in the evening and also in the morning for the residential load however PV generates electricity mostly during the day time therefore residents need to purchase costly electricity during peak demand in the evening. Similarly wind energy also unable to follow the residential load profile. Therefore properly estimated storage needs to be integrated to overcome this situation.

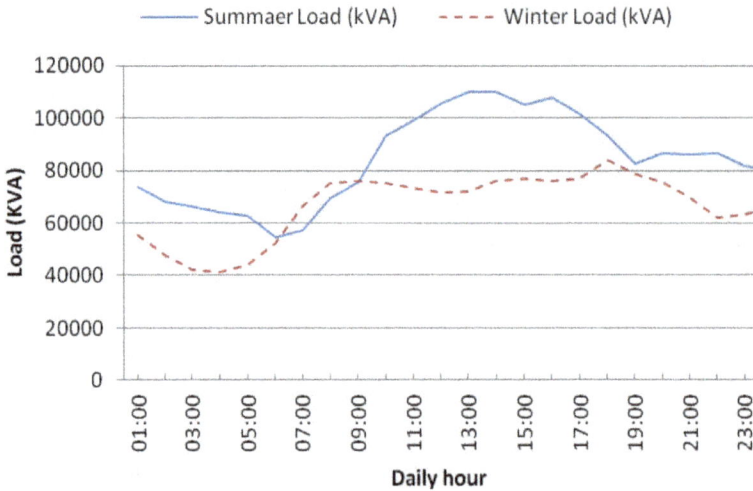

Figure 1. Daily load profile of Capricornia region in 2009 (Summer & Winter)

This chapter explores the need of storage systems to maximize the use of RE, furthermore estimates the required capacity of storage to meet the daily need which will gradually eliminate the dependency on conventional energy sources. Estimation of storage sizing is explained in section 3. This chapter also conducts the feasibility assessment of storage in terms of economic and environmental perspective which is explained in section 4.

2. Background

Solar and Wind are the two major sources of RE. Australia is one of best places for these sources. In regional areas of Australia, roof top Solar PV is installed in many residential houses either in off-grid or grid connected configurations and most residential wind turbine are for specific applications in off-grid configuration. In grid connected solar PV systems where storage is not integrated, the energy output from this system does not satisfy to the desired level. Currently installed most of the residential PV systems are designed in an unplanned way that even with battery integrated system is not able to support the load in reliable way. Figure 2 illustrates a typical situation when whole system in jeopardize as the estimation of storage system was not done correctly.

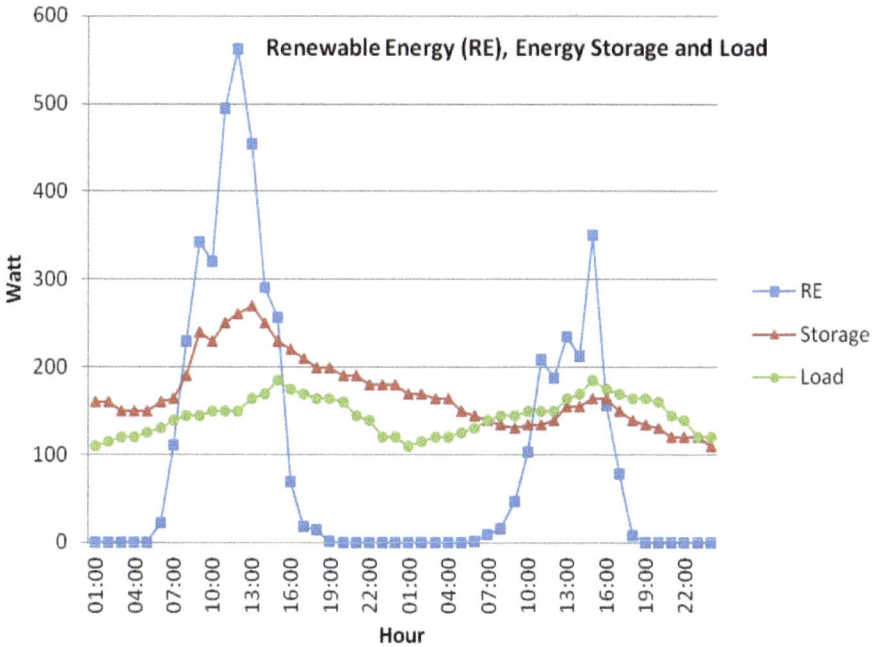

Figure 2. Typical condition of failure in storage integrated RE system

The adoption of storage with the PV system certainly incurs additional cost to the system but the benefits of adding storage has not been clearly assessed. Therefore this chapter aims to achieve two objectives. One is to estimate the required storage for the grid connected PV system or grid connected wind turbine or combination of grid connected PV and wind turbine system to achieve the maximum daily use of RE. Second objective is to identify the effects of storage on the designed system in terms of environment and economic by comparing the same system with and without storage. The feasibility of the designed system is expressed as, the Cost of Energy (COE) is closer to the present system while providing environmental benefits by reducing Greenhouse Gas (GHG) emission and improving the Renewable Fraction (RF).

Data was collected for the Capricornia region of Rockhampton city in Queensland, Australia. Load data was collected for a 3 bed room house by estimating all the electrical appliances demand and average usage period considering its ratings. Daily load profile drawn from hourly load data and total daily load was estimated by calculating the area under the daily load profile curve using trapezoidal method. Weather data was collected for the year 2009 from [2] for this location and calculated the energy output from PV array and wind turbine. Figure 3 shows the solar radiation and wind speed of Rockhampton for the year 2009. Hourly energy output curve of PV and wind was drawn and compared with the load profile to estimate the required storage. For estimation, choice was taken considering the worst month weather data and it was observed that May to July is worst period when solar and wind have lower energy density as shown in Figure 3.

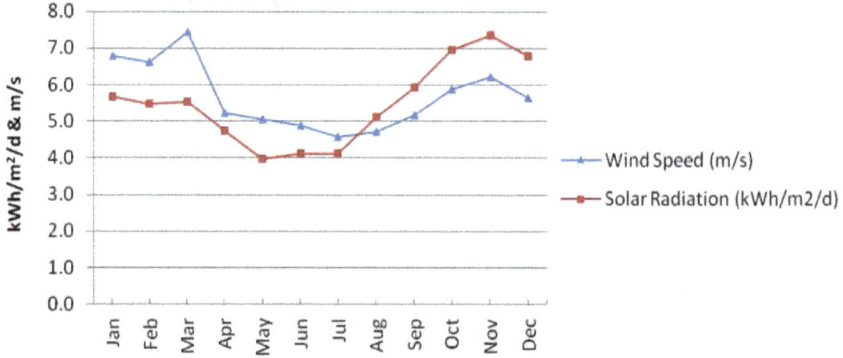

Figure 3. Solar radiation and wind speed in Rockhampton in 2009

Solar power

PV array considered as a device that produces DC electricity in direct proportion to the global solar radiation. Therefore, the power output of the PV array can be calculated by Equation 1[3-4].

$$P_{PV} = Y_{PV} f_{PV} \left(\frac{\bar{G}_T}{\bar{G}_{T,STC}} \right) [1 + \alpha_P (T_C - T_{C,STC})] \tag{1}$$

If there is no effect of temperature on the PV array, the temperature coefficient of the power is zero, thus the above equation can be simplified as Equation 2 [3-4].

$$P_{PV} = Y_{PV} f_{PV} \left(\frac{\bar{G}_T}{\bar{G}_{T,STC}} \right) \tag{2}$$

where Y_{PV} - rated capacity of PV array, meaning power output under standard test conditions [kW]; f_{PV} - PV de-rating factor [%]; G_T - solar radiation incident on PV array in current time step [kW/m²]; $G_{T,STC}$ - incident radiation under standard test conditions [1 kW/m²]; α_P - temperature coefficient of power [%/°C]; T_C - PV cell temperature in current time step [°C]; $T_{C,STC}$ -PV cell temperature under standard test conditions [25°C]. Performance of PV array depends on derating factors like temperature, dirt and mismatched modules.

Wind Power

Kinetic energy of wind can be converted into electrical energy by using wind turbine, rotor, gear box and generator. The available power of wind is the flux of kinetic energy, which the air is interacting with rotor per unit time at a cross sectional area of the rotor and that can be expressed [5] as per Equation 3:

$$P = \frac{1}{2}\rho AV^3 \qquad (3)$$

where, P is Power output from wind turbine in Watts, ρ is the air density (1.225kg/m3 at 15°C and 1-atmosphere or in sea level), A is rotor swept area in m² and V is the wind speed in m/s.

The swept area of a horizontal axis wind turbine of rotor diameter (D) in meter (or blade length = D/2) can be calculated by Equation 4.

$$A = \pi(\frac{D}{2})^2 \text{ sq.m} \qquad (4)$$

As power in the wind is proportional to the cube of the wind speed therefore increase in wind speed is very significant. One way to get more power is by increasing the tower height. Hourly wind speed at different height above ground level can be calculated by the vertical wind profile Equation 5 [6-7]:

$$V_2 = V_1(\frac{H_2}{H_1})^\alpha \qquad (5)$$

where v_1 and v_2 are the wind speeds at heights H_1 and H_2 and α is the wind shear component or power law exponent or friction coefficient. A typical value of α is 0.14 for countryside or flat plane area. Equation 5 commonly used in United States and the same is expressed in Europe by Equation 6 [7]:

$$V_2 = V_1(\frac{\ln(H_2/z)}{\ln(H_1/z)}) \qquad (6)$$

where z is the roughness length in meters. A typical value of z for open area with a few windbreaks is 0.03m.

Temperature has effect on air density which changes the output of wind turbine. Average wind speed globally at 80m height is higher during day time (4.96m/s) than night time (4.85m/s) [8].

However German physicist Albert Betz concluded in 1919 that no wind turbine can convert more than 16/27 or 59.3% of the kinetic energy of the wind into mechanical energy by turning a rotor. This is the maximum theoretical efficiency of rotor and this is known as Betz Limit or Betz' Law. This is also called "power coefficient" and the maximum value is: C_P= 0.59. Therefore Equation-3 can be written as:

$$P = \frac{1}{2}C_p\rho AV^3 \qquad (7)$$

The following subsections describe the residential load, solar & wind energy of Rockhampton area and also describe the importance of storage.

2.1. Estimation of daily residential load

Preferred method of determining load is bottom-up approach in which daily load is anticipated and summed to yield an average daily load. This can be done by multiplying the power rating of all the appliances by the number of hours it is expected to operate on an average day to obtain Watt-hour (Wh) value as shown in Table 1. The load data collected from a 3 bed room house in Kawana, Rockhampton in Australia and total land area of the house is 700m² where 210m² is the building area with available roof space. For grid connected household appliances daily average load can also be obtained from monthly utility bills.

Appliances	Rating	Daily time of use	Qty	Daily use (Wh/day)
Refrigerator	602kWh/year (300W)	Whole day	1	1650
Freezer	88W	Whole day	1	880
Electrical Stove	2100W	Morning & Evening (1-2hrs)	1	2100
Microwave Oven	1000W	Morning & Evening (30 min to 1 hr)	1	500
Rice cooker	400W	Evening (30 minutes)	1	200
Toaster	800W	Morning (10 - 30 minutes)	1	80
Ceiling Fan	65W	Summer night (4 -5 hrs) & Holidays	5	1300
Fluorescent light	16W	Night (6 - 8 hours)	20	320
Washing machine (vertical axis)	500W	Weekends (1hr/week)	1	71
Vacuum Cleaner	1400W	Weekends (1hr/week)	1	200
Air conditioner (Window)	1200W	Summer night & Holidays (1hr)	3	1200
TV 32″ LCD (Active/Standby)	150/3.5W	Morning & night (4 hrs)	1	670
DVD player (Active/Standby)	17/5.9W	Night (2 hrs)	1	50
Cordless phone	4W	Whole day	1	96
Computer (Laptop)	20W	Night (4 - 5hrs)	1	80
Clothe iron	1400W	Night & Holidays (15 - 30 minutes)	1	350
Heater (Portable)	1200W	Winter night & Holidays (30 minutes)	1	600
Hot Water System	1800W	Whole day(3- 4 hrs)	1	5400
Total:				15,747

Table 1. Daily load consumption of a house
Data source: Product catalogue and [7]

Load profile of a residential house varies according to the residents work time pattern. Working nature of the residents of Kawana suburb is such that most of the residents start for work between 7:00AM to 8:00AM and returns home between 5:00PM to 6:00PM during weekday from Monday to Friday. A 24 hour load profile of a particular day as shown in Figure 4. It was found that maximum load demand was in the evening from 6:00PM to 10:00PM and in the morning 7:00AM to 9:00AM.

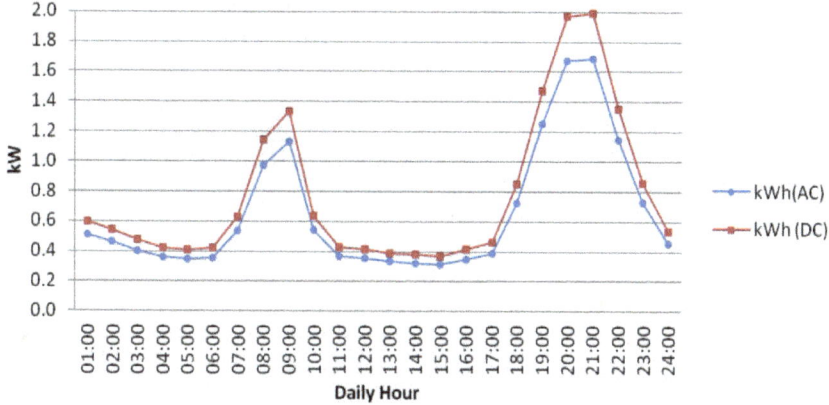

Figure 4. Daily load profile of a Residential house

Hourly load is a time series data and total daily load can be estimated by calculating the area under the load profile curve using Equation 8.

$$\text{Daily Load} = P_{\text{Load}} = \int_{t_1}^{t_{24}} f(x)dt \qquad (8)$$

Where $f(x) = \frac{1}{2}(p_{t1} + p_{t2})T_{12}$

Where p_{t1} = Load (in kW) at time t_1 =1 in hour, p_{t2} = Load (in kW) at time t_2 = 2 in hour, T_{12} = time difference b/w t_1 and t_2 in hour.

Following Equation 8 total daily AC (Alternating current) load is the area under this load curve which is 15.7kWh and the equivalent DC (Direct current) load is shown considering efficiency of the converter as 85% which is 18.47kWh.

2.2. Estimation of daily available solar energy

Solar radiation varies with time and season. For estimation of available useful solar energy, worst month solar radiation was considered to ensure that the designed system can operate year-round. In Australia yearly average sunlight hours varies from 5 to 10 hours/day and maximum area is over 8 hours/day [2]. From the collected data it was found that in Rockhampton solar radiation over 5.0kWh/m²/d varies from 08:00AM to 16:00PM i.e. sun hour is 8hrs/day.

The daily average solar radiation of Kawana suburb in the Capricornia region of Rockhampton city is as shown in the Figure 5. It was found that annual average solar radiation was 5.48kWh/m2/day. Lowest monthly average solar radiation was 4kWh/m2/day on May and highest solar radiation was from October to December in 2009. PV system designed to supply entire load considering the worst month solar radiation, which will deliver sufficient energy during rest of the year.

For estimating daily solar energy, worst month (May) solar radiation was considered and Figure 5 shows hourly solar radiation for May 07, 2009. Daily total solar energy was estimated by calculating area under the solar radiation curve using Equation 8. Therefore total solar radiation in May 07, 2009 was $1.582975kWh/m^2/d$. This energy generated by $1m^2$ PV area. Total solar radiation will increase with the increased area of the PV array.

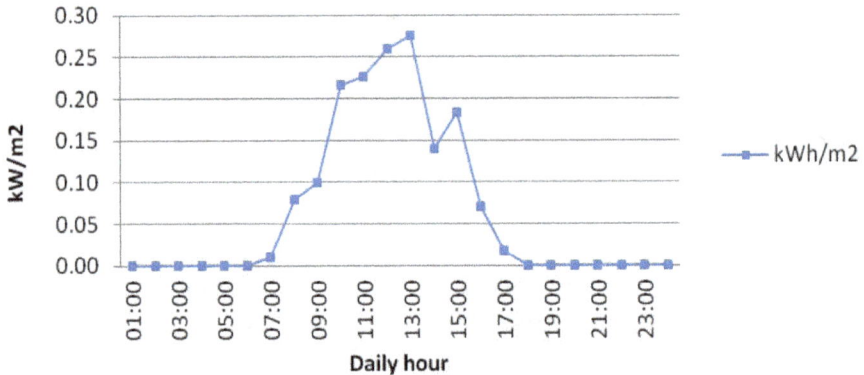

Figure 5. Daily solar radiation (May 07, 2009) in Rockhampton

2.3. Estimation of daily available wind energy

Wind speed varies with different natural factors, time and season. To estimate the available useful wind energy, worst month wind speed was considered to ensure that the designed system can operate year-round. From the collected data of Rockhampton it was found that July had the worst wind speed scenario as shown in Figure 3. It was found that in 2009, wind speed of Rockhampton was 6m/s or more for daily average duration of 10 hours. However for the month of July and August it was only 5 hours as shown in Table 2. Therefore wind speed data of July was considered for estimation of daily energy.

Three hourly wind speed data at 10.4m above sea level was collected from [2] for the year 2009, which was interpolated to get hourly data. At rotor height of 10m, 40m and 80m corresponding wind speed as shown in Figure 6. For energy estimation, July 03, 2009 wind speed data was considered and corresponding energy was calculated for $1m^2$ of rotor wind area at 40m rotor height using Equation 7 as shown in Figure 7. Betz limit, gearbox, bearing and generator efficiency was considered and overall efficiency of the wind turbine was taken 25%. Total energy output from wind turbine on July 03, 2009 is the area under the curve of Figure 7 (11:00AM to 09:00PM) which is $0.232785kWh/m^2/d$.

Month	Daily Time period	Time window (hrs)
Jan	06:00 - 20:00	14
Feb	03:00 - 17:00	14
Mar	00:00 - 15:00, 22:00 - 24:00	17
Apr	00:00 - 04:00, 20:00 - 24:00	8
May	16:00 - 24:00	8
Jun	12:00 - 19:00	7
Jul	10:00 - 15:00	5
Aug	07:00 - 12:00	5
Sep	01:00 - 10:00	9
Oct	00:00 - 06:00, 19:00 - 24:00	11
Nov	00:00 - 03:00, 15:00 - 24:00	12
Dec	13:00 - 24:00	11

Table 2. Wind speed period or window (6m/s or more)

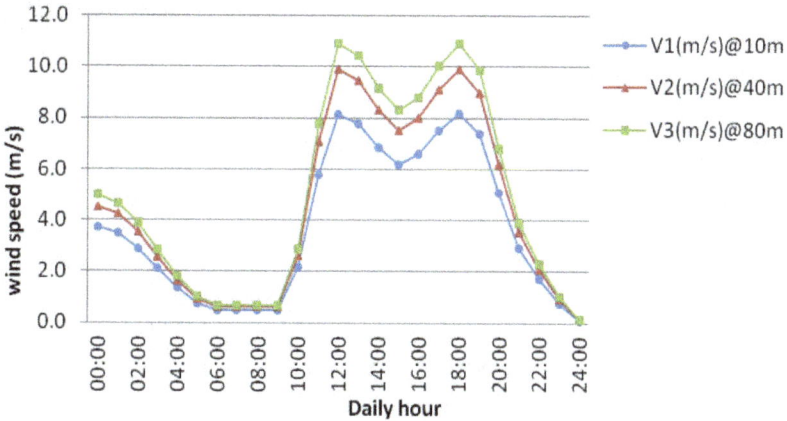

Figure 6. Wind speed at 10m, 40m and 80m height in Rockhampton

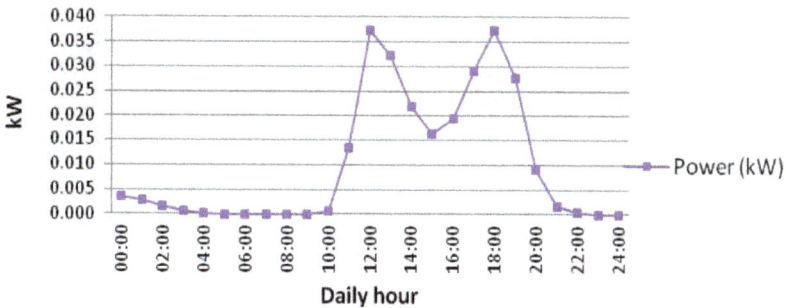

Figure 7. Energy converted per m² area at 40m height wind speed in Rockhampton

2.4. Importance of storage

The ability to store large amounts of energy would allow electrical utilities to have greater flexibility in their operation, because with this option the supply demand do not have to be matched instantaneously [9]. Table 3 shows the benefits and limitations of different storage systems.

Storage	Advantage			Limitations
	Efficiency (%)	Life time (years)	Response time (S)	
Pumped hydro	80	50	10	Location specific. Expensive to build.
Compressed air energy storage (CAES)	$^a f_{cef}=1.3$ $^b f_{fhr}=4300kJ/k$ Wh	25	360	Location specific. Expensive to build.
Flywheel	85	20	0.1	Low energy density. Large standby loss
Thermal energy storage(TES)	75	30	Tens of minutes	Storage tank is expensive
Batteries	80	10	0.01	Early stage technology. Expensive
Superconducting magnetic energy storage (SMES)	90	30	0.01	Low energy density. Expensive.
Capacitor	80	10	0.01	Low energy density. Expensive.
Hydrogen	50	25	360	Highly flammable

Table 3. advantage and limitations of few storage systems [12-14]
a. charge energy factor b. Fuel heat rate

The role of Energy Storage (ES) with Renewable Electricity generation is mentioned in[10] that the selection of ES system depends on application which is largely determined by the length of discharge. Based on the length of discharge, ES applications are often divided into three categories named power quality, bridging power and energy management applications. Although large scale storage is still expensive but research is going on for inexpensive and efficient batteries [11] suitable for large scale RE applications.

RE can be considered for different kinds of applications i.e. from small stand-alone remote systems to large scale grid-connected solar/wind power application. However development goes on to remote areas and brings the remote areas close to the grid network and eventually connected to the power grid and these RE generator are expected to operate as grid connected Distributed Generator (DG). Grid connected PV/wind with battery as storage can provide future-proof energy autonomy and allow home or office to generate clean energy and supply extra energy to the grid.

A recent study on high penetration of PV on present grid, mentioned that energy storage is the ultimate solution for allowing intermittent sources to address utility base load needs [15]. Storage integrated PV/Wind systems provides a combination of operational, financial and environmental benefits.

3. Estimation of storage sizing

Improper sized PV/Wind system is unable to meet the load requirements, sometimes electrical energy from RE wasted which neither can be used by the load nor can be stored in battery. This event occurs when the battery State of Charge (SOC) exceeds its maximum allowable value and the solar/wind power output exceeds load demand. The amount of wasted/lost energy can be avoided or reduced by proper choice of battery and PV/Wind generation sizes. G.B. Shrestha et.al. in [16] mentioned that PV panel size and the battery size have different impacts on the indices of performance and proper balance between the two is necessary. A proper match between the installed capacities with the load demand is essential to optimize such installation.

Brahmi Nabiha et. al. in [17] presents sizing of mini autonomous hybrid grid, including PV, wind, generator and battery. The performance of any battery, expressed essentially by the voltage, load capacity and SOC or the Depth of Discharge (DOD). The usable energy in a battery can be expressed by Equation 9.

$$E_{usable} = C \times V_{bat} \times DOD_{max} \qquad (9)$$

where C is battery capacity and V_{bat} is the battery cell voltage.

IEEE Std-1013-2007 [18] provides the recommendations for sizing of lead-acid batteries for stand-alone PV systems. This recommended practice provides a systematic approach for determining the appropriate energy capacity of a lead-acid battery to satisfy the energy requirements of the load for residential, commercial and industrial stand-alone PV systems. IEEE Std-1561-2007 [19]provides guideline for optimizing the performance and life of Lead-Acid batteries in remote hybrid power systems; which includes PV, wind, batteries. It also explains the battery sizing considerations for the application. IEEE Std 1547-2003 [20] provides guideline to connect Distributed Resources (DR), such as PV, wind and storage with the power grid at the distribution level. Grid connected system sizing for storage integrated PV system also explained in [7].

Considering the above sizing practices and guidelines Figure 8 shows the steps for estimation of required storage for steady state residential load. For the easy of this analysis both PV and Wind turbine are considered to produce DC power which than converted to AC by inverter, also considered battery as storage device.

The following steps are summarized for estimation.

Step 1. Determine the daily load of a residential house
Step 2. Determine the required PV or Wind turbine rating for the load
Step 3. Determine daily energy output from the PV array or Wind turbine

Step 4. Estimate PV array size and wind turbine rotor diameter

Step 5. Compare the daily energy output (from PV or wind turbine) with the daily load, find the required load that storage needs to support

Step 6. For the load on storage estimate the required Battery/Storage size in Ah.

The following sub-sections describe the estimation of required storage for grid connected PV, Wind and hybrid systems considering the residential load of Rockhampton as estimated in section 2.1.

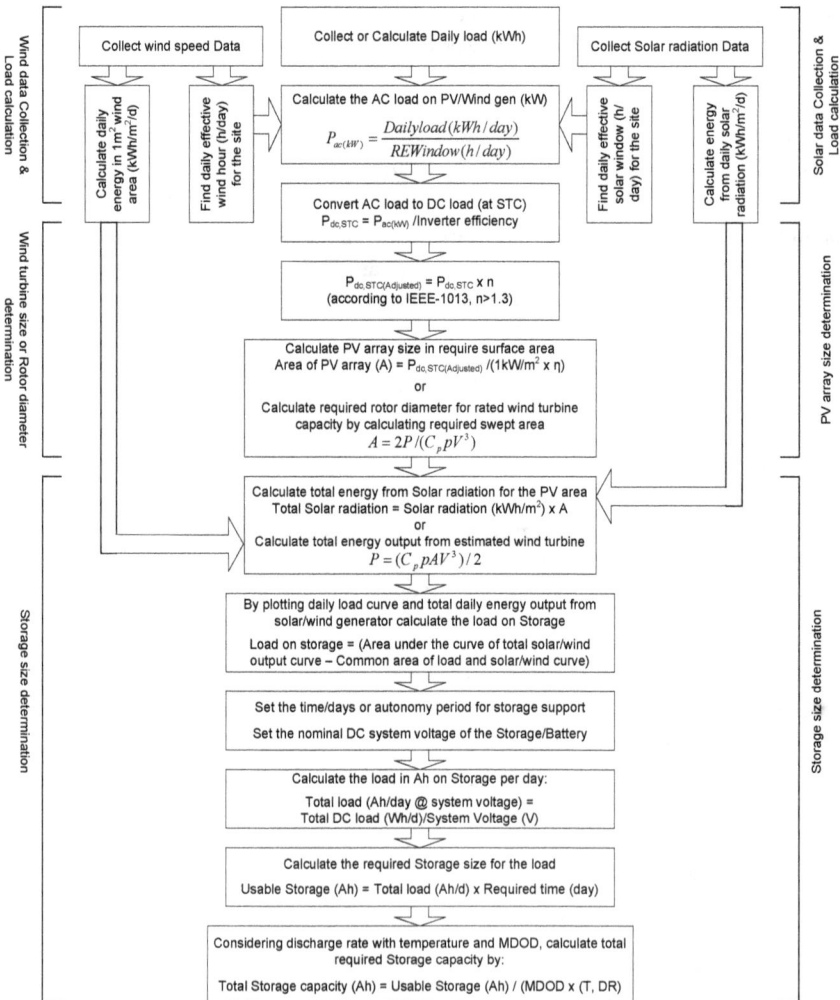

Figure 8. Storage size estimation steps

3.1. Estimation of storage for grid connected residential solar PV

The size of the PV array is determined by the daily average load divided by the available solar window or sun-hours per day. Generally, grid connected PV systems are designed to provide from 10% to 60% of energy needs with the difference being supplied from power utility[21]. However PV contribution can be increased to 100% of average steady state load. Following the steps in Figure 8, the estimation starts by calculating required PV size.

Daily extractable solar energy as calculated in section 2.2 was considered for the estimation of required PV size for the residential load. Daily load of a three bed room house as calculated in section 2.1 is 15.7kWh. Therefore the PV array should support at least 15.7kWh of load everyday at the solar energy rate of 1.582975kWh/m²/day. Solar window is 8 hours or more in Rockhampton [2], therefore the required PV array capacity for the AC load as:

$$P_{ac}(kW) = \frac{\text{Energy (kWh/day)}}{\text{Solar window (h/day)}} = \frac{15.7}{8} = 1.9625kW$$

Equivalent DC load on PV array can be found by considering the efficiency as:

η = inverter efficiency * dirty collector * mismatched modules = 85%

$$P_{dc,STC} = \frac{P_{ac}(kW)}{\eta} = \frac{1.9625}{0.85} = 2.31kW$$

To use battery as storage system, size of the PV array needs to be more than 1.3 times the load [18] in stand-alone configuration. But for the grid connected configuration 1.0 or 1.1 is good enough to avoid over design. For this designed residential load, it was considered 1.1. So the adjusted PV array size for the equivalent DC load becomes:

$$P_{dc,STC(Adjusted)} = 1.1xP_{dc,STC} = 1.1x2.31 = 2.541kW$$

Therefore, for this three bed room house 2.541kW capacity of PV array with proper sized storage required to support its load for 24 hours a day.

For known PV efficiency and for 1kW/m² rated PV module required surface area of the PV array can be calculated. The efficiency of crystal silicon PV module is 12.5% [7], however LG Polycrystalline PV module efficiency is 13.7% [22], therefore surface area becomes:

$$P_{dc,STC} = (1kW/m^2)\text{insolation} * A * \eta$$

$$A = \frac{P_{dc,STC}}{(1kW/m^2)\eta} = \frac{2.541}{1x0.125} = 20.328m^2$$

Therefore 20.328m² PV module with PV array efficiency of 12.5% will support the load with sufficient storage size. This PV area is much smaller than the area of the designed house roof area.

This PV size was considered to calculate the total energy from PV array and to estimate the required storage for the load. Batteries last longer if they are shallow cycled. The capacity of the battery bank can be calculated by multiplying the daily load on battery by the autonomy day or the number of days it should provide power continuously. The ampere-hour (Ah) rating of the battery bank can be found after dividing the battery bank capacity by the battery bank voltage (e.g. 24V or 48V). It is generally not recommended to design for more than 12 days of autonomy for off-grid system and for grid connected system one day autonomy is good to design.

Total solar energy generated by the 20.328m² PV array at the solar radiation rate of Rockhampton is plotted in Figure 9 and calculated as 32.17872kWh which is the area under the PV output curve. Now superimpose the DC load curve on the PV output curve to find the load that needs to be supported by the storage as shown in Figure 9. The common area under the curve is 6.196kWh which is the area of the load that served by the PV array during day time while charging the batteries as well. The remaining load is (18.47 - 6.196) =12.274kWh/day that needs to be served by the storage. This is the daily minimum load on storage. However the design was based on to support total load, therefore the remaining energy from the PV array should be managed by the storage system which is (32.17872 - 6.196) = 25.98272kWh/day. This is the maximum load on storage, if total energy generated by PV array needs to be managed by the storage.

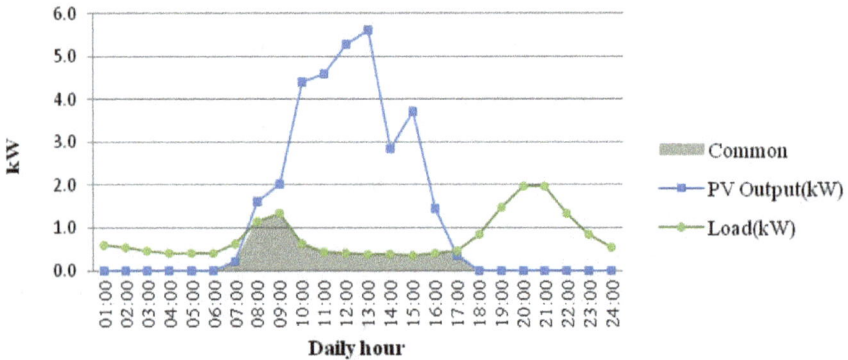

Figure 9. PV output and daily load curve shows the load on storage

Inverters are specified by their DC input voltage as well as by their AC output voltage, continuous power handling capability and the amount of surge power they can supply for brief periods of time. Inverter's DC input voltage which is the same as the voltage of the Battery bank and the PV array is called the system voltage. The system voltage usually considered as 12V, 24V or 48V. The system voltage for this designed DC system was considered 24V and this system was designed for one day. Considering inverter efficiency of 95% [23], the required battery capacity can be calculated.

$$\text{Daily minimum load in Ah @ system voltage} =$$

$$= \frac{\text{Load (Wh/day)}}{\text{System Voltage}} = \frac{12.274 \times 10^3}{24} = 511.416\text{Ah/d}$$

Daily maximum load in Ah @ system voltage =

$$= \frac{\text{Load (Wh/day)}}{\text{System Voltage}} = \frac{25.98272 \times 10^3}{24} = 1082.613 \text{Ah/d}$$

Energy storage in a battery typically given in Ah, at system voltage and at some specified discharge rate. Table 4 shows characteristics of several types of batteries.

Battery type	MDOD	Energy Density (Wh/kg)	Cycle Life (Cycles)	Calendar Life (Year)	Efficiencies Ah%	Wh%
Lead-acid, SLI	20%	50	500	1-2	90	75
Lead-acid, golf cart	80%	45	1000	3-5	90	75
Lead-acid, deep-cycle	80%	35	2000	7-10	90	75
Nickel-cadmium	100%	20	1000-2000	10-15	70	60
Nickel-metal hydride	100%	50	1000-2000	8-10	70	65

Table 4. Comparison of Battery Characteristics[7]

The Ah capacity of a battery is not only rate-dependent but also depends on temperature. The capacity under varying temperature and discharge rates to a reference condition of C/20 at 25°C is explained in [7]. Lead-acid battery capacity decreases dramatically in colder temperature conditions. However heat is also not good for batteries. In Rockhampton average temperature is above 20°C. The Maximum depth of discharge (MDOD) for Lead-acid batteries is 80%, therefore for one day discharge the batteries need to store:

$$\text{Battery storage (minimum)} = \frac{\text{Load (Ah/day)} \times \text{No of days}}{\text{MDOD}} = \frac{511.416 \times 1}{0.80} = 639.27 \text{Ah}$$

$$\text{Battery storage (maximum)} = \frac{\text{Load (Ah/day)} \times \text{No of days}}{\text{MDOD}} = \frac{1082.613 \times 1}{0.80} = 1353.26 \text{Ah}$$

The rated capacity of battery is specified at standard temperature. At 25°C, the discharge rate of C/20 type battery (i.e. discharge for 20 hours), becomes 96% [7], therefore finally required battery capacity becomes:

Required minimum Battery storage(25°C,20hour-rate)=

$$= \frac{\text{Battery storage}}{\text{Rated capacity}} = \frac{639.27}{0.96} = 665.90 \text{Ah}$$

Required maximum Battery storage(25°C,20hour-rate)=

$$= \frac{\text{Battery storage}}{\text{Rated capacity}} = \frac{1353.26}{0.96} = 1409.64 \text{Ah}$$

3.2. Estimation of storage for grid connected residential wind power

Following the similar steps in section 3.1, required wind turbine capacity was calculated and then required storage was estimated for the same load of 15.7kWh/day.

Energy generated by wind turbine at 40m height for $1m^2$ rotor wind area was calculated in section 2.3, which is 0.232785kWh/m^2/d. The output of the wind turbine needs to be improved such that at least 15.7kWh of load should be supported each day. It was found that in July, wind speed was 6m/s or above only for 5hrs/day at 10m height, however at 40m height wind speed was 6m/s or above for 10hrs/day, therefore the rotor height was considered 40m. The required wind turbine size for the load can be calculated as:

$$P_{ac}(kW) = \frac{Load\ (kWh/day)}{Windwindow\ (h/day)} = \frac{15.7}{10} = 1.57kW$$

This estimation is for required storage which is a DC component; it requires inverter to support the load. DC capacity of the wind turbine can be calculated considering inverter efficiency of 90%.

$$P_{dc,STC} = \frac{P_{ac}(kW)}{\eta} = \frac{1.57}{0.90} = 1.744kW$$

Likewise PV assumption, wind turbine capacity is considered 1.1 times the required load in grid connected configuration, to charge batteries while supporting load.

$$P_{dc,STC(Adjusted)} = 1.1xP_{dc,STC} = 1.1x1.744 = 1.92kW$$

Energy generated by wind turbine on July 03, 2009 was 0.232785kWh/m^2/d. To support total load, rotor swept area needs to be adjusted. Equation 7 shows that power output is not linear for increase in rotor diameter. It was found that at 40m rotor height, wind speed varied b/w 6.17m/s to 9.92m/s, therefore average wind speed of 8m/s was considered to calculate the rotor diameter for the rated wind turbine capacity of 1.92kW. The rotor diameter was calculated as 5.58m and calculated total energy is 26.355kWh which is the area under the wind turbine output curve as shown in Figure 10.

Daily load curve was plotted on the daily energy output curve and calculated the common area to estimate the required load on storage to support for the day. It was found that 7.736kWh of load was supported by the wind turbine while charging the storage. The remaining (18.47 - 7.736) = 10.734kWh of load needs to be supported by the storage each day. This is the minimum load on storage. However the design was considering to manage 100% load therefore remaining (26.355 - 7.736) = 18.619kWh of energy must be managed by the storage. This is the maximum load on storage.

Considering the DC system voltage as 24V, load on battery in Ah can be calculated for one day as:

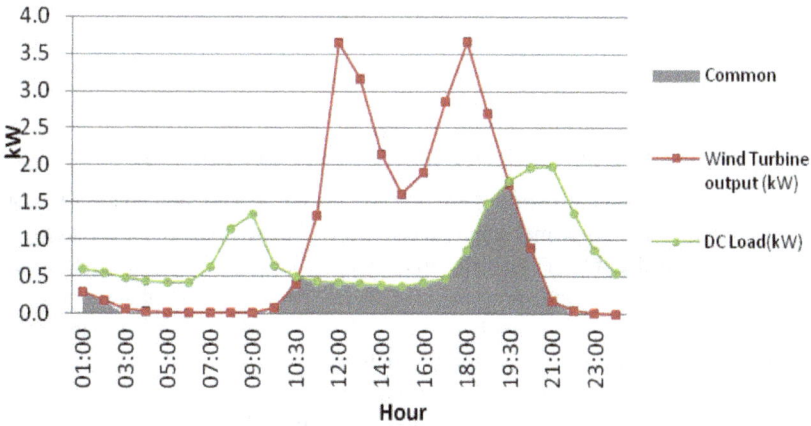

Figure 10. Wind turbine output and daily load curve shows the load on storage

Daily minimum load in Ah @ system voltage =

$$= \frac{\text{Load (Wh/day)}}{\text{System Voltage}} = \frac{10.734 \times 10^3}{24} = 447.25 \text{Ah/d}$$

Daily maximum load in Ah @ system voltage =

$$= \frac{\text{Load (Wh/day)}}{\text{System Voltage}} = \frac{18.619 \times 10^3}{24} = 775.79 \text{Ah/d}$$

Energy storage in a battery typically given in Ah, at system voltage and at some specified discharge rate. Consider MDOD for Lead-Acid batteries is 80%, therefore for one day discharge the battery needs to store the energy as:

Battery storage (minimum) =

$$= \frac{\text{Load (Ah/day) x No of days}}{\text{MDOD}} = \frac{447.25 \times 1}{0.80} = 559.0625 \text{Ah}$$

Battery storage (maximum) =

$$= \frac{\text{Load (Ah/day) x No of days}}{\text{MDOD}} = \frac{775.79 \times 1}{0.80} = 969.7375 \text{Ah}$$

The rated capacity of battery is specified at standard temperature. At 25°C, the discharge rate of C/20 (i.e. discharge for 20 hours), becomes 96% [7], therefore finally required battery capacity becomes:

Required minimum Battery storage(25°C,20hour-rate)=

$$= \frac{\text{Battery storage}}{\text{Rated capacity}} = \frac{559.0625}{0.96} = 582.356 \, Ah$$

Required maximum Battery storage($25°$C,20hour-rate)=

$$= \frac{\text{Battery storage}}{\text{Rated capacity}} = \frac{969.7375}{0.96} = 1010.143\,Ah$$

3.3. Estimation of storage for grid connected residential hybrid system

Many studies indicated that hybrid system is always better than any single RE system. However the practical implementation depends on the availability of adequate solar radiation, wind speed and their seasonal variation. Other critical point is adequate space for hybrid system installation and moreover the overall cost of the installation. The study location of this analysis is suitable for both solar and wind energy. It was found that for little variation of wind speed, convertible energy variation is much higher therefore wind energy fluctuation is higher than solar energy. Considering all the scenarios and for the easy of analysis it was considered that 50% of load to be supported by solar and 50% by wind energy.

Following the steps in Figure 8 and earlier sections, required storage is estimated.

For Solar PV: 50% AC Load is (15.7/2) = 7.85kWh/d

Required PV array capacity becomes:

$$P_{ac}(kW) = \frac{\text{Energy (kWh/day)}}{\text{Solar window (h/day)}} = \frac{7.85}{8} = 0.98125kW$$

Equivalent DC load can be found by considering the efficiency of the PV system as:

η = inverter efficiency * dirty collector * mismatched modules = 85%

$$P_{dc,STC} = \frac{P_{ac}(kW)}{\eta} = \frac{0.98125}{0.85} = 1.1544kW$$

For this designed house load, PV capacity considered 1.1 times the load. So the adjusted PV array size for the equivalent DC load becomes:

$$P_{dc,STC(Adjusted)} = 1.1 x P_{dc,STC} = 1.1 x 1.1544 = 1.27kW$$

Therefore it requires 1.27kW capacity of PV array with proper sized storage to support 50% load for 24 hours a day.

Considering the crystal silicon PV module whose efficiency is 12.5% [7], therefore the surface area of PV module becomes:

$$P_{dc,STC} = (1kW/m^2)\text{insolation} * A * \eta$$

$$A = \frac{P_{dc,STC}}{(1kW/m^2)\eta} = \frac{1.27}{1 x 0.125} = 10.16m^2$$

Therefore 10.16m² of PV area required for this hybrid system. The output energy from this PV module is plotted in Figure 11. For the remaining load the required wind turbine is estimated as:

For Wind turbine: 50% AC Load is (15.7/2) = 7.85kWh/d

Required wind turbine capacity becomes:

$$P_{ac}(kW) = \frac{Load\ (kWh/day)}{Windwindow\ (h/day)} = \frac{7.85}{10} = 0.785kW$$

The inverter considered with this wind turbine of efficiency 90%, therefore the DC capacity becomes:

$$P_{dc,STC} = \frac{P_{ac}(kW)}{\eta} = \frac{0.785}{0.90} = 0.872kW$$

For this designed house load, wind turbine capacity considered 1.1 times the load. So the adjusted wind turbine size for the equivalent DC load becomes:

$$P_{dc,STC(Adjusted)} = 1.1xP_{dc,STC} = 1.1x0.872 = 0.9592kW$$

Average wind speed of 8m/s was considered to calculate the rotor diameter for the required capacity of wind turbine. For the 0.9592kW capacity wind turbine, the rotor diameter becomes 3.95m and daily energy generated by this wind turbine was plotted in Figure 11.

Total energy generated from this hybrid system is 28.12kWh and compared with the DC load it was calculated that the Hybrid system support directly 8.45kWh of load as shown the common area in Figure 11. Therefore the minimum (18.47 - 8.45) = 10.02kWh of load needs

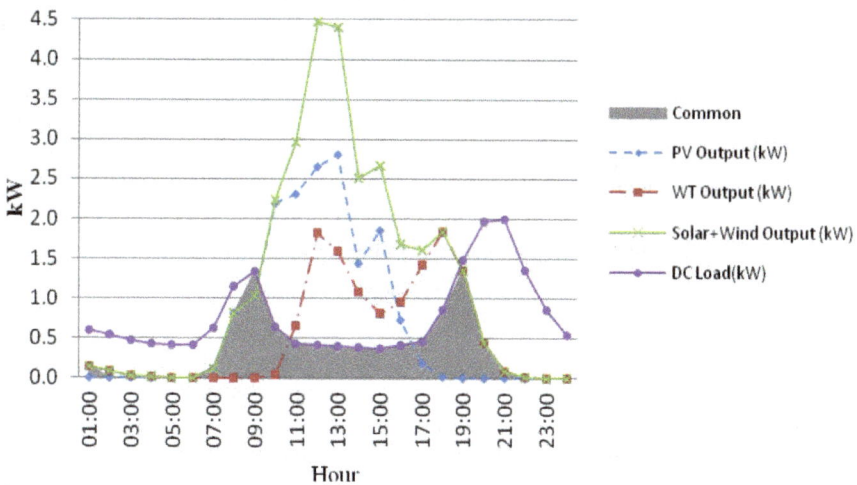

Figure 11. Hybrid system output and daily load curve shows the load on storage

to be supported by the storage system. However the hybrid system was designed to support 100% load therefore remaining generated energy (28.12 - 8.45) = 19.67kWh from hybrid system must be managed by the storage. This is the maximum load on storage.

Considering the DC system voltage as 24V, load on battery in Ah can be calculated for one day as:

$$\text{Daily minimum load in Ah @ system voltage} =$$
$$= \frac{\text{Load (Wh/day)}}{\text{System Voltage}} = \frac{10.02 \times 10^3}{24} = 417.50 \text{Ah/d}$$

$$\text{Daily maximum load in Ah @ system voltage} =$$
$$= \frac{\text{Load (Wh/day)}}{\text{System Voltage}} = \frac{19.67 \times 10^3}{24} = 819.58 \text{Ah/d}$$

Considered MDOD of Lead-Acid batteries is 80%, therefore for one day discharge the battery needs to store the energy as:

$$\text{Battery storage (minimum)} =$$
$$= \frac{\text{Load (Ah/day) x No of days}}{\text{MDOD}} = \frac{417.50 \times 1}{0.80} = 521.875 \text{Ah}$$

$$\text{Battery storage (maximum)} =$$
$$= \frac{\text{Load (Ah/day) x No of days}}{\text{MDOD}} = \frac{819.875 \times 1}{0.80} = 1024.843 \text{Ah}$$

Temperature effects are considered in the storage requirements for the hybrid systems. At 25°C for C/20 (i.e. 20 hours discharge) the discharge rate is 96%, therefore finally required battery capacity becomes:

$$\text{Required minimum Battery storage}(25^\circ\text{C,20hour-rate}) =$$
$$= \frac{\text{Battery storage}}{\text{Rated capacity}} = \frac{521.875}{0.96} = 543.62 \, Ah$$

$$\text{Required maximum Battery storage}(25^\circ\text{C,20hour-rate}) =$$
$$= \frac{\text{Battery storage}}{\text{Rated capacity}} = \frac{1024.843}{0.96} = 1067.544 \, Ah$$

Therefore in grid connected configuration to support 15.7kWh/day load the required minimum and maximum storage with solar PV, Wind turbine and hybrid system are shown in Table 5. Minimum storage indicates the required storage that needs to support the daily load. However PV, wind turbine or hybrid system generates more energy than the daily required therefore maximum storage is required to manage total generated energy by supporting load or supplying to the grid in suitable time. This design considered PV array efficiency 12.5% with solar window 8 hours and overall wind turbine efficiency 25% with wind window 10 hours.

Designed system	Required system capacity	Required Storage (at 24V DC system voltage)	
		Minimum (Ah)	Maximum (Ah)
Solar PV system	2.541kW PV with 20.328m² PV array	665.90Ah	1409.64Ah
Wind turbine system	1.92kW wind turbine with 5.58m rotor diameter	582.356Ah	1010.143Ah
Hybrid system	1.27kW PV with 10.16m² PV array and 0.9592kW wind turbine with 3.95m rotor diameter	543.62Ah	1067.544Ah

Table 5. Required storage in different configurations

4. Feasibility analysis of storage

Previous section described the required storage for the residential load in different configurations and this section describes feasibility of the storage in those configurations.

G.J. Dalton et al. in [24] compared HOMER and Hybrids as RE optimization tool and found that HOMER is better in representing hourly fluctuations in supply and demand. This chapter explains the feasibility of storage by analyzing the model output. A model was developed in HOMER version 2.68 [25] as shown in Figure 12. PV array, wind turbine, storage, inverter, grid and diesel generator were used in different size and combinations. The model identified the optimized configuration of storage, PV, wind turbine with grid or diesel generator for the residential load and investigated environmental and economical benefits due to the storage systems. The model was evaluated considering the project life time of 25 years. The performance matrices considered are NPC, COE as economical factor, Renewable Fraction (RF) and greenhouse gas (GHG) emission as environmental factor. The model compared in off-grid and grid connected configurations.

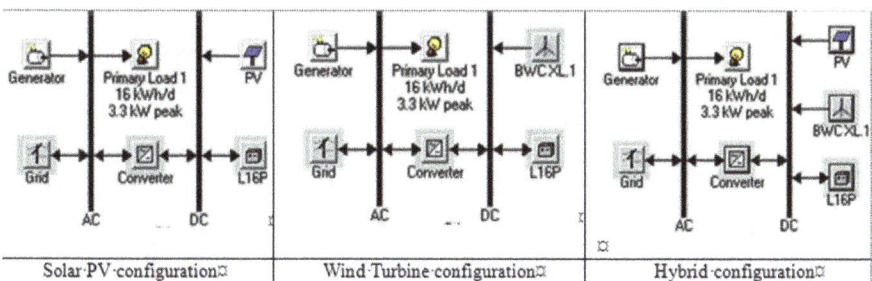

Figure 12. Model simulation in different configurations

Net Present Cost (NPC): The total net present cost of a system is the present value of all costs that it incurs over its lifetime minus the present value of all the revenue that it earns over its lifetime. NPC is the main economic output and ranked all systems accordingly. NPC can be represented [24-25] by Equation 10:

$$NPC(\$) = \frac{TAC}{CRF} \tag{10}$$

where TAC is the total annualized cost (which is the sum of the annualized costs of each system component). The capital recovery factor (CRF) is given by Equation 11:

$$CRF = \frac{i(1+i)^N}{(1+i)^N - 1} \tag{11}$$

where N denotes number of years and i means annual real interest rate (%). Model considered annual interest rate rather than the nominal interest rate. The overall annual interest rate considered as 6%.

Cost of Energy (COE): It is the average cost per kWh of useful electrical energy produced by the system. COE can be calculated by dividing the annualized cost of electricity production by the total useful electric energy production and represented [25] in Equation 12:

$$COE = \frac{C_{ann,tot} - C_{boiler} E_{thermal}}{E_{prim,AC} + E_{prim,DC} + E_{def} + E_{grid,sales}} \tag{12}$$

where $C_{ann,tot}$ is total annualized cost of the system ($/yr), C_{boiler} is boiler marginal cost ($/kWh), $E_{thermal}$ is total thermal load served (kWh/yr), $E_{prim,AC}$ is AC primary load served (kWh/yr), $E_{prim,DC}$ is DC primary load served (kWh/yr), E_{def} is deferrable load served (kWh/yr) and $E_{grid,sales}$ is total grid sales (kWh/yr).

Emission: Emission is widely accepted and understood environmental index. Greenhouse gases (CO_2, CH_4, N_2O, HFCs, PFCs, SF6) are the main concern for global warming. In addition SO_2 is another pollutant gas released by coal fired energy system. Emission is measured as yearly emission of the emitted gases in kg/year and emissions per capita in kg/kWh. Model used it as input when calculating the other O&M cost. It was represented in [25] as shown in Equation 13:

$$C_{om,other} = C_{om,fixed} + C_{cs} + C_{emission} \tag{13}$$

where $C_{om,fixed}$ is system fixed O&M cost ($/yr), C_{cs} is the penalty for capacity shortage ($/yr) and $C_{emission}$ is the penalty for emission ($/yr).

Renewable Fraction (RF): It is the total annual renewable power production divided by the total energy production. RF can be calculated [26] using Equation 14:

$$f_{pv} = \frac{E_{PV}}{E_{TOT}} \tag{14}$$

where E_{PV} and E_{TOT} are the energy generated by RE and total energy generated respectively. The overall RF (f_{ren}) can also be expressed [25] in Equation 15:

$$f_{ren} = \frac{E_{ren} + H_{ren}}{E_{tot} + H_{tot}} \tag{15}$$

where E_{ren} is renewable electric production, H_{ren} is renewable thermal production, E_{tot} is total electrical production and H_{tot} is total thermal production.

Battery dispatch: Ideally battery charging should be taken into account for future load supply. Battery dispatch strategies explained by Barley and Winn [27], named 'load following' and 'cycle charging'. Under load following strategy, a generator produces only enough power to serve the load, which does not charge the battery. Under cycle charging strategy, whenever a generator operates it runs at its maximum rated capacity, charging battery bank with any excess electricity until the battery reaches specified state of charge. Load following strategy was considered for this analysis for the better utilization of RE.

4.1. Data collection

For the simulation of the optimization model, residential load data were considered as Rockhampton resident's average load consumption. Solar radiation and wind speed data were collected from [28]. All required system components are discussed in the following sub-sections.

4.1.1. Electric load

Daily average steady state load of a 3 bed room house was estimated in section 2.1 which is 15.7kWh/d. Daily load profile of the distribution network of Capricornia region was collected from [1] and according to the electricity bill information [29] the daily average electricity consumption per house is 15.7kWh/day. Model takes a set of 24 hourly values load data or monthly average or hourly load data set of 8,760 values to represent average electric load, therefore yearly residential load (AC) becomes 5730kWh/yr. The load profile is shown in Figure 4.

4.1.2. Solar radiation data

Solar radiation data is the input to the model and hourly solar radiation data of Rockhampton was collected from [28]. Daily average extractable energy from this solar radiation is explained in section 2.2.

4.1.3. Wind speed data

Three hourly wind speed data was collected from [28] which was interpolated to generate hourly data and used as input data in the model. The extractable energy from the wind speed is explained in section 2.3.

4.1.4. Storage

For this analysis Trojan L16P Battery (6V, 360Ah) at system voltage of 24V DC is used in the model. The efficiency of this battery is 85%, min State of Charge (SoC) 30%.

4.2. System components cost

Table 6 lists the required system components with related costs in Australian currency. PV array, Wind turbine, Battery charger, Inverter, deep cycle battery, diesel generator and grid electricity costs are included for the analysis. PV array including inverter price is available, and found that 1.52kW PV array with inverter costs $3599 [30], also it is found that 1.56kW PV with inverter costs is $4991[31]. However model considered battery charger is included with PV array therefore the PV array cost is listed accordingly in Table 6 and inverter costs considered separately.

Description	Value/Information
PV array	
Capital cost	$3100.00/kW
Replacement cost	$3000.00/kW
Life Time	25 years
Operation & maintenance cost	$50.00/year
Wind Turbine (BWC XL.1 1 kW DC)	
Capacity	1kW DC
Hub Height	40m
Capital cost	$4000.00
Replacement cost	$3000.00
Life time	25 years
Operation & maintenance cost	$120/yr
Grid electricity	
Electricity price (Off peak time)	$0.30/kWh
Electricity price (Peak time)	$.42/kWh
Electricity price (Super Peak time)	$0.75/kWh
Emission factor	
CO_2	632.0 g/kWh
CO	0.7 g/kWh
Unburned hydrocarbons	0.08 g/kWh
Particulate matter	0.052 g/kWh
SO_2	2.74 g/kWh
NOx	1.34 g/kWh
Inverter	
Capital cost	$400.00/kW
Replacement cost	$325.00/kW
Life time	15 years
Operation & maintenance cost	$25.00/year
Storage (Battery)	
Capital cost	$170.00/6V 360Ah
Replacement cost	$130.00/6V 360Ah
System Voltage	24 volts

Description	Value/Information
Generator	
Capital cost	$2200.00/kW
Replacement cost	$2000.00/kW
Operation & maintenance cost	$0.05/hr
Life time	15000hrs
Fuel cost	$1.53/ltr

Table 6. Technical Data and Study assumptions

SMA Sunny Boy Grid Tie Inverter (7000Watt SB7000US) price is $2823 [32], however Sunny Boy 1700W inverter price is $699 [33]. 1kW BWC XL.1 wind turbine with 24V DC charge controller price is $3560 [34] and 10kW Bergey BWC Excel with battery charging or grid tied option wind turbine cost is $29,250 [35]. Grid electricity cost in Rockhampton is found from Ergon Energy's electricity bill [36] and for Tariff-11, it is $0.285/kWh (including GST & service). However Government's decision to impose carbon tax at the rate of $23/ton of GHG emission which will increase this electricity bill as well as the cost of conventional energy sources, therefore off-peak electricity cost is considered as $0.30/kWh for analysis. Trojan T-105 6V, 225AH (20HR) Flooded Lead Acid Battery price is $124.79 [37]. Fuel cost for generator is considered at the current price available in Rockhampton, Australia.

The significance of storage was analyzed from the optimized model to evaluate environmental and economical advantages of storage in off-grid and grid-connected configurations in fourteen different cases. All these cases were analyzed considering same load 15.7kWh/d or 5730kWh/yr.

Category-1: Off-grid Configuration

Case-1: Diesel Generator only
Case-2: PV with Diesel Generator
Case-3: PV with Storage and Diesel Generator
Case-4: Wind turbine with Diesel Generator
Case-5: Wind turbine with Storage and Diesel Generator
Case-6: Hybrid (PV & Wind turbine) with Diesel Generator
Case-7: Hybrid (PV & Wind turbine) with Storage and Diesel Generator

Category-2: Grid-connected Configuration

Case-1: Grid only
Case-2: PV with Grid and Diesel generator
Case-3: PV with Storage, Grid and Diesel generator
Case-4: Wind turbine with Grid and Diesel generator
Case-5: Wind turbine with Storage, Grid and Diesel generator
Case-6: Hybrid (PV & Wind turbine) with Grid and Diesel generator
Case-7: Hybrid (PV & Wind turbine) with Storage, Grid and Diesel generator

5. Results and discussion

Simulation was conducted to get optimized configuration of RE resources. Simulation results and findings are discussed below.

5.1. Category - 1(Off grid configuration)

Case 1. Diesel Generator only

In this configuration 10kW Diesel generator was used to support total load of 5730kWh/yr which consumed enough fuel (8440L/yr) and emitted significant amount of GHG & pollutant gas to the air. Generator required frequent maintenance and fuel cost was also high therefore NPC was high and COE was $5.342/kWh. This configuration was the costliest and environmentally most vulnerable.

Case 2. PV with Diesel Generator configuration

In this off-grid configuration, model used 12kW PV with 5kW Inverter and 10kW Diesel generator as required resources. Results showed that, although PV generates electricity more than the total load demand but could not meet the load demand during night. Total 12,781kWh/yr electricity was generated from PV and diesel generator. PV alone generates 8908kWh/yr i.e. RF became 69.7% but most of the energy from PV array was wasted. Diesel generator directly supplied 3873kWh/yr to the load which was 67.6% of load demand, although compared to the total production; generator contribution was only 30.3%. The remaining load demand, (5730 -3873) = 1857kWh/yr was supported by PV array through inverter. Therefore a significant amount of electricity from PV array was wasted. Wasted electricity was (8908 - 1857/0.94) = 6932.46kWh/yr which is 54.2% of total electricity production but compared to the total PV electricity production, the wasted electricity was 77.82%. To reduce this great amount of loss, this system should have some way to store the energy and could reduce the use of diesel generator.

Case 3. PV with Storage and Diesel Generator configuration

In this off-grid configuration model 11kW PV, 48 number of Trojan L16P Battery (@ 6V, 360Ah) at 24V system voltage with 5kW Inverter was used. The optimized configuration shaded out diesel generator, therefore 100% load supported by PV and storage. Results showed that, PV generates electricity more than the load demand and battery stored the excess electricity to maintain the load demand.

PV generates 8166kWh/yr of electricity from which a good amount of energy was stored in the battery and used at other time. Total AC load supported directly by PV array during day time and by battery during morning & night. Inverter converts 6096kWh/yr of DC electricity to AC. Battery stored 4281kWh/yr of energy and supplied 3692kWh/yr to support the load. Battery stored 52.42% of PV generated energy and supported 64.43% of load while PV directly supports 35.56% of load. However still 1480kWh/yr of excess energy generated by the PV array and was wasted that could be sold to the grid. This model configuration supports 100% load by PV array and storage which makes it environment friendly off-grid configuration.

Case 4. Wind turbine with Diesel Generator configuration

This off-grid configuration used 10kW BWC XL.1 wind generator with 5kW inverter and 10kW diesel generator as required resources to support 5730kWh/yr of load. Result showed that, wind turbine generates much more electricity than the total load demand but could not meet the load demand for 24 hours period.

Total 41,023kWh/yr of electricity was generated from wind turbine and diesel generator, where 38,781kWh/yr from wind turbine i.e. 94.5% of total production from RE but most of it was wasted as wind turbine supports 3488kWh/yr of load, which is 60.87% of load demand. Diesel generator contributes 2242kWh/yr or 39.13% of load demand, although compared to the total production; diesel generator contribution was only 5.5%. Total 35,070kWh/yr or 85.5% of total electricity production was wasted but compared to the total wind turbine output 90.43% was wasted. By adding storage this huge loss of electricity could be minimized and that could reduce the use of diesel generator.

Case 5. Wind turbine with Storage and Diesel Generator configuration

This off-grid configuration model used 3kW BWC XL.1 wind generator, 40 numbers of Trojan L16P Battery (@ 6V, 360Ah) at 24V system voltage and 5kW Inverter. This optimized configuration shaded out diesel generator, therefore 100% load was supported by wind turbine and storage. Result showed that, wind turbine generates electricity more than the load demand and battery stored the excess electricity to support at other time.

Wind turbine generates 11,634kWh/yr of electricity. Inverter converts 6096kWh/yr of DC electricity to AC. Battery stored 2364kWh/yr of energy and supplied 2037kWh/yr to the load. Battery stored 20.32% of wind turbine generated energy and supported 35.55% of load. However still 5211kWh/yr of excess energy generated by the wind turbine, i.e 44.79% of total generated energy was wasted that could be sold to the grid.

Case 6. Hybrid system with Diesel generator configuration

This off-grid hybrid configuration model used 3kW PV, 5kW BWC XL.1 wind generator, 5kW Inverter and 4kW diesel generator. Result showed that, although PV and wind turbine generates much more electricity than the total load demand but could not meet the load demand for 24 hours period.

Total electricity generated from RE (PV and Wind turbine) and diesel generator was 24,434kWh/yr where 2,227kWh/yr from PV, 19,390kWh/yr from wind turbine and 2,817kWh/yr from diesel generator. PV contributed 9.1%, wind turbine 79.35% and diesel generator 11.53% of total production, therefore overall RE contribution was 88.5% of total production. Diesel generator contributed 49.16% of load demand. Inverter converts 3,099kWh/yr of DC electricity to 2,913kWh/yr of AC electricity from RE generation which was 50.84% of load demand. A significant amount of electricity (18,518kWh/yr) from RE was wasted which is 75.78% of total electricity production and 85.66% compared to the total RE production. Storage could be used to reduce this huge energy loss and to minimize the use of diesel generator.

Case 7. Hybrid system with Storage and Diesel Generator configuration

This off-grid hybrid configuration model used 1kW PV, 3kW BWC XL.1 wind generator, 5kW Inverter and 32 Trojan L16P batteries at 24V system voltage to support the same load. Hybrid system (PV and wind turbine) output with storage supports 100% load and shaded out the use of diesel generator. Result showed that, storage managed the electricity from hybrid system and met the load demand 24 hours a day, but a significant amount of energy was wasted that could be sold to the grid.

Total 12,376kWh/yr of electricity was generated from hybrid system, where 742kWh/yr or 6% from PV and 11,634kWh/yr or 94% from wind turbine. Battery stored 2073kWh/yr and supported 1,788kWh/yr or 31.20% of load demand. This configuration supplied 100% load demand from RE, however 5,995kWh/yr was wasted which could be sold to the grid.

Summary of Category-1 or Off-grid configurations

The results of standalone configurations can be summarized that storage minimized the use of resources which reduced the project cost, improved load support that reduced GHG emission and reduced the loss of generated RE and showed the scope to sell excess energy to the grid. Table 7 summarizes these findings. Load support describes, percentage of load supported by RE and storage. Energy loss describes percentage of energy loss compared to total RE production.

5.2. Category-2 (Grid connected configuration)

Case 1. Grid only configuration

This is the present configuration of most residential electricity connection. Grid supplies total load demand of 5730kWh/yr. Grid electricity tariff varies with time, season and application [36, 38]. This configuration model considered 3 different price of grid electricity, depending on demand time. These are off-peak (0.30$/kWh), peak (0.42$/kWh) and super peak rate (0.75$/kWh). 6:00PM to 7:00PM considered super peak, 8:00PM to 10:00PM and 8:00AM to 9:00AM considered peak time and rest are off peak time. In this case yearly average COE becomes $0.422/kWh. As grid electricity mainly comes from conventional sources therefore a good amount of GHG and pollutant gas emits to the air.

Case 2. PV in Grid connected configuration

In this optimized model configuration diesel Generator was shaded out, however PV array still contributed a small portion of load demand. To meet load demand this model used 3kW PV, 5kW Inverter and grid supply. Total 7,182kWh/yr electricity was produced, where grid supplied 4,955kWh/yr or 69% of total production or 86.47% of total load demand. PV array produced 2,227kWh/yr or 31% of total production or 13.53% of the load demand. Total 549kWh/yr of energy was sold back to the grid and 818kWh/yr of PV generated electricity was wasted due to mismatch in timely demand which could be stored and supplied to the load.

	PV	Wind	Inverter	Storage	RE use	
				(at 24V DC system voltage)	Load support	RE energy loss
Case-2 (PV +Gen)	12kW	-	5kW	-	32.40%	77.82%
Case-3: (PV+Storage+Gen)	11kW	-	5kW	48 nos. (103.68kWh)	100%	18.12%
Case-4 (Wind turbine+Gen)	-	10kW	5kW	-	60.87%	90.43%
Case-5 (Wind turbine+Storage+Gen)	-	3kW	5kW	40 nos. (86.4kWh)	100%	44.79%
Case-6 (Hybrid +Gen)	3kW	5kW	5kW	-	50.84%	85.66%
Case-7 (Hybrid+Storage+Gen)	1kW	3kW	5kW	32 nos. (69.12kWh)	100%	48.44%

Table 7. Category-1 or off-grid configuration results

Case 3. PV with Storage in Grid connected configuration

This configuration model is very interesting compared to the earlier case that, by adding sufficient amount of storage, system improved PV contribution for same load demand. To meet load demand this model used 5kW PV, 12 numbers of Trojan L16P battery, 5kW inverter and grid supply. Total 6208kWh/yr of electricity produced where grid supplied 2496kWh/yr or 40% of total production or 43.56% of total load demand. PV array produced 3712kWh/yr or 60% of total production. Loss of energy was very insignificant. Battery stored 1918kWh/yr and supplied 1648kWh/yr to the load or 28.76% of total load. However PV array directly supported (5730-2496-1648) = 1586kWh/yr of load which was 27.68% of total load demand.

Case 4. Wind turbine in Grid connected configuration

In this configuration model, wind turbine generates enough electricity but was unable to meet the timely load demand therefore consumed sufficient amount of grid electricity. To meet the load demand this optimized model used 3kW BWC XL.1 wind turbine, 5kW inverter and grid supply. Total 14,389kWh/yr of electricity was produced where wind generator produced 11,634kWh/yr or 80.9% of total production. Grid supplied 2755kWh/yr or 19.1% of total production or 48.08% of load demand. Wind turbine supported (5730 - 2755) = 2975kWh/yr or 51.92% of load demand and 7121kWh/yr of electricity was sold back to the grid. Total 894kWh/yr of electricity was unused.

Case 5. Wind turbine with Storage in Grid connected configuration

This configuration model used 3kW BWC XL.1 wind generator, 16 numbers of Trojan L16P battery, 5kW inverter and grid supply. Total 11,784kWh/yr of electricity was produced where grid supplied only 150kWh/yr or 1.3% of total production or only 2.6% of total load demand. Wind turbine produced 11,634kWh/yr which was 98.7% of total production. Battery stored 2202kWh/yr and supplied 1895kWh/yr or 33.07% of total load demand. However wind turbine directly supported (5730-1895-150) = 3685kWh/yr or 64.31% of total load demand. Total 5068kWh/yr or 53.65% of wind production was sold back to the grid. Significant amount of electricity was sold back to the grid therefore overall GHG emission was reduced.

Case 6. Hybrid system without Storage in Grid connected configuration

In this configuration both PV and wind turbine was used. This hybrid configuration model used 1kW PV, 1kW BWC XL.1 wind generator, 1kW inverter and grid supply for the same load of 5730kWh/yr. Results showed that, the hybrid system was optimized such that minimum RE components were required but could not met the load demand for 24 hours period. Total 9,021kWh/yr of electricity was produced where grid supplied 4,401kWh/yr or 76.80% of load demand or 48.78% of total production. PV and wind hybrid system produced 4,620kWh/yr or 51.21% of total production. Hybrid system supplied 1329kWh/yr of electricity or 23.19% of load demand. However hybrid system generated enough electricity and sold 1,236kWh/yr to the grid, still 1,891kWh/yr of electricity wasted which was 40.93% of total RE production. This wasted electricity could be utilized and grid use could be minimized by adding storage.

Case 7. Hybrid system with Storage in Grid connected configuration

This hybrid configuration model used 1kW PV, 3kW BWC XL.1 wind generator, 3kW inverter, 12 Trojan L16P battery and grid supply. This configuration improved RE contribution in supporting load. Total 12,546kWh/yr of electricity was produced where grid supplied only 170kWh/yr which is 1.4% of total production or 2.96% of load demand. PV generates 742kWh/yr and wind turbine 11,634kWh/yr i.e. RE production was 98.6% of total electricity generation. This hybrid system sold back 5,527kWh/yr of electricity to the grid. Storage helped in improving RE utilization & minimized loss. Battery stored 1911kWh/yr and supported 1642kWh/yr of load or 28.65% of load demand.

Summary of Category-2 or Grid-connected configurations

The results of grid connected configurations can be summarized that storage improved load support which reduced GHG emission. Storage optimized the RE sources by minimizing grid use and reduced loss of energy. Table 8 summarizes these findings. Load support describes, percentage of load supported by RE and storage. Energy loss describes percentage of energy loss compared to total RE production. Grid sales describes, energy sold to grid compared to total RE production.

	PV	Wind	Inverter	Storage (at 24V DC system voltage)	RE use		
					Load support	Grid sales	RE energy loss
Case-2 (PV +Grid)	3kW	-	5kW	-	38.86%	24.65%	36.73%
Case-3 (PV+Storage+Grid)	5kW	-	5kW	12 nos. (25.92kWh)	56.44%	0.027%	0.0%
Case-4 (Wind turbine + Grid)	-	3kW	5kW	-	51.92%	61.19%	7.68%
Case-5 (Wind turbine + Storage +Grid)	-	3kW	5kW	16 nos. (34.56kWh)	97.38%	43.56%	0.0%
Case-6 (Hybrid +Grid)	1kW	1kW	1kW	-	23.19%	26.75%	40.93%
Case-7 (Hybrid+Storage +Grid)	1kW	3kW	3kW	12 nos. (25.92kWh)	97.03%	44.66%	2.53%

Table 8. Category-2 or grid connected configuration results

5.3. Findings

The optimization was done in two configuration categories and seven cases in each category. Four different factors were compared in each case. These factors were GHG & Pollutant gas emission, RF, COE and NPC. The comparative findings of these factors are explained below.

GHG & Pollutant gas emission

Figure 13 shows GHG and pollutant gas emissions in different case configurations. It was found that by adding storage in stand-alone system, emission of GHG and other pollutant gas was eliminated. In Grid connected configuration, it was also evident that storage minimized emission by improving RE utilization. By selling excess energy back to the grid storage with wind and hybrid system further helped in reducing GHG emission from grid which is shown in negative values in Figure 13.

Renewable Fraction (RF)

RF is the measuring index of how much electricity produced from RE, out of total production. In stand-alone system it was found that storage eliminates the use of diesel generator therefore RF became 100%. In Grid connected configuration Storage again improves the RE utilization and RF became as high as 98.7% as shown in Figure 14.

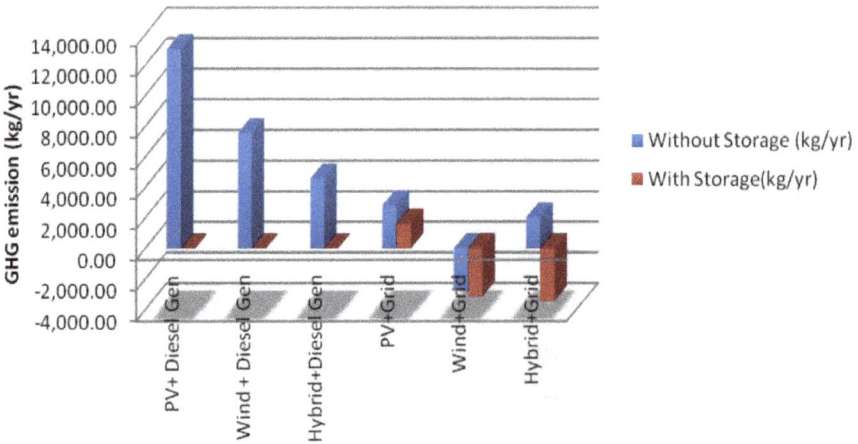

Figure 13. GHG and pollutant gas emission in different cases

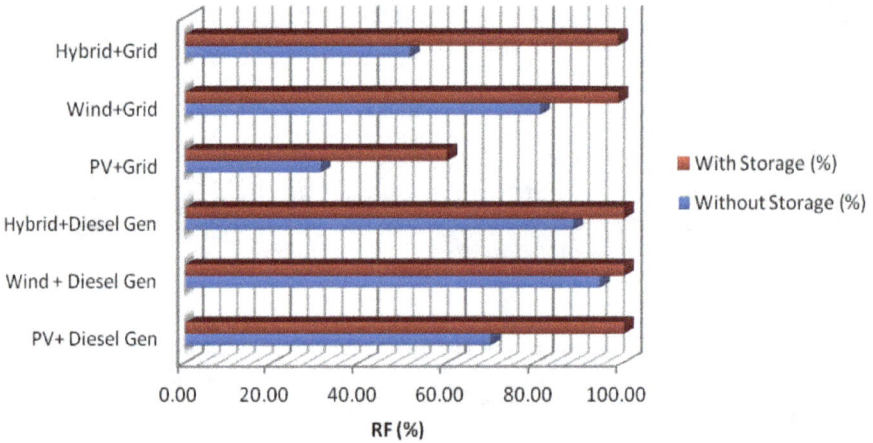

Figure 14. Renewable fraction (RF) in different cases

Cost of Energy (COE)

COE is the cost of per unit energy in $/kWh. Stand-alone configuration involving diesel generator was costly therefore COE was very high; however adding storage reduced COE to a reasonable level. In Grid connected configuration in all combination of RE sources, storage reduced the COE and in hybrid system storage reduced COE close to the grid only energy cost as shown in Figure 15.

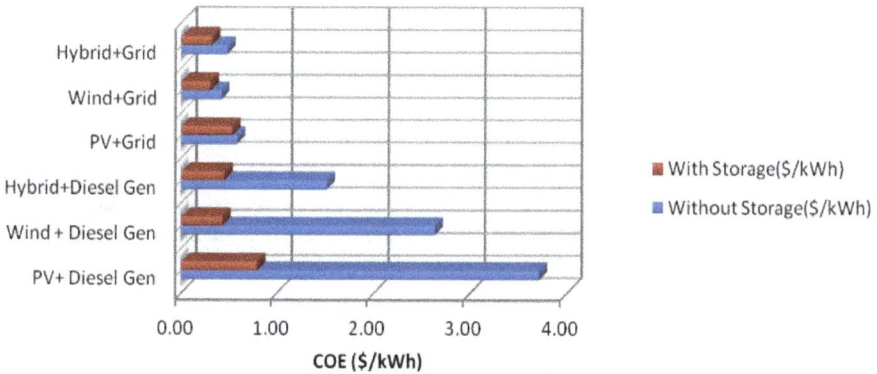

Figure 15. Cost of energy (COE) in different cases

Net Present Cost (NPC)

NPC represents present cost of the system. In standalone configuration NPC was very high however storage helped in reducing NPC to an acceptable level by improving the utilization of RE. In Grid connected configuration, storage helped in reducing NPC in every combination of RE used as shown in Figure 16.

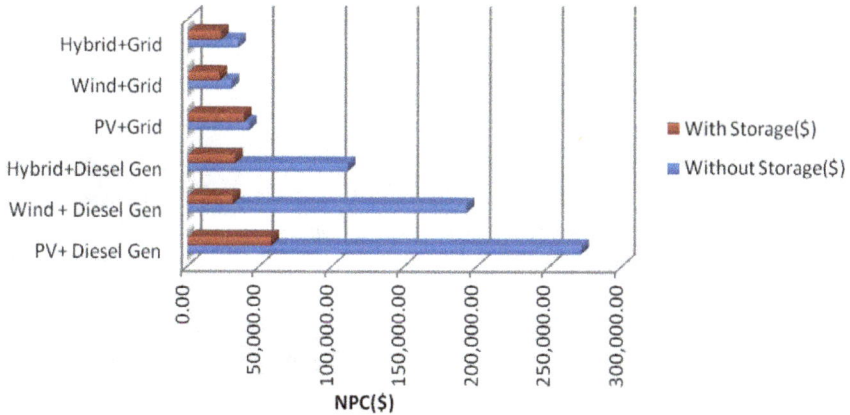

Figure 16. Net present cost (NPC) in different cases

5.4. Payback period

In this model payback was calculated by comparing one system with another. Payback is the number of years in which the cumulative cash flow switches from negative to positive by comparing storage integrated model with without storage model in grid connected configuration. Cash flow in grid connected PV with storage system compared with grid connected PV base system and it was found that payback period is 4.15 year. Similarly grid connected wind generator with storage compared with without storage system and found that payback period is 2.67 years. In case of grid connected Hybrid (PV & wind turbine) system with storage compared with same without storage system and found that payback period is 2.05 years. Storage helped in RE utilization that minimizes the use of grid electricity and increased energy sell back to the grid. Therefore it was confirmed that the investment cost of storage integration returns in very short period of time as shown in Figure 17. In Australia solar bonus scheme awards the price of electricity fed into the grid from RE at a rate of $0.44/kWh [39-40] which is much higher than the utility rate. This ensures that the payback period will be much shorter in Australia.

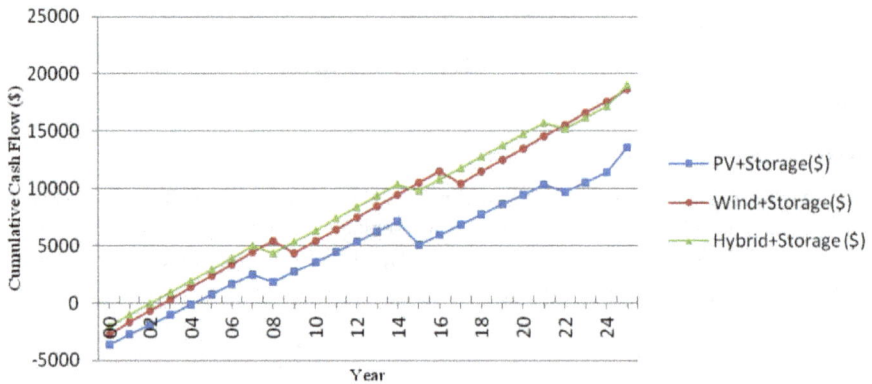

Figure 17. Payback period of storage in three different cases

6. Conclusion

Storage integrated RE system was analyzed for a residential load in Rockhampton, Australia. Estimation of required storage for the RE system was calculated. Estimation steps were developed and estimation of required storage was done in grid connected PV, wind and hybrid systems. It was found that to support daily load of 15.7kWh/day in grid connected PV system minimum 665.90Ah of storage, in grid connected wind turbine minimum 582.356Ah of storage and in hybrid system minimum 543.62Ah of storage required at 24V DC system voltage.

Model was developed for feasibility analysis of storage with RE. Model was analyzed in standalone and grid connected configurations. Analysis was conducted to observe the storage influences over the GHG emission, RF, COE and NPC indexes. It was found by analyzing the output data from the optimized model that storage has great influence on improving RE utilization.

It was evident from the analysis that storage helped significantly in reducing GHG & other pollutant gas emission, reduced COE, improved RF and reduced NPC. Comparing without and with storage system model, it was found that in grid connected PV system, storage reduced 43.35% of GHG and pollutant gas emission and in all standalone systems it was 100%. In grid connected wind and hybrid system, storage reduced GHG emission more than 100% by selling extra energy to the grid. In grid connected configuration storage improved RF where with PV, wind turbine and hybrid system RF was 59.8%, 98.7%, 98.6% respectively. Storage reduced COE and in grid connected configuration it was as low as presently available grid electricity cost. In hybrid and wind system COE was $0.316/kWh and $0.302/kWh respectively. Similarly Storage reduced NPC by 8.3%, 27.13% and 33.95% in grid connected PV, wind and hybrid configurations respectively. Moreover payback time of storage is very short therefore storage integrated RE system is more feasible for implementation.

Author details

Mohammad Taufiqul Arif and Amanullah M. T. Oo
Power Engineering Research Group, Faculty of Sciences, Engineering & Health,
Central Queensland University, Bruce Highway, Rockhampton, QLD 4702, Australia

A. B. M. Shawkat Ali
School of Computing Sciences, Faculty of Arts, Business, Informatics & Education,
Central Queensland University, Bruce Highway, Rockhampton, QLD 4702, Australia

7. References

[1] ErgonEnergy, *(2010), Capricornia region electric load, Rockhampton load data collected directly from Ergon Energy. Dated: 05/07/2010.*

[2] BoM, *Bureau of Meteorology, Australian Government. Available [Online] at:* http://reg.bom.gov.au/.

[3] Lambert, T., *How HOMER Calculates the PV Array Power Output, Software available [Online] at: http://homerenergy.com/.* Homer Help file, 2007.

[4] Ahmed M.A. Haidar, P.N.J., Mohd Shawal, *Optimal configuration assessment of renewable energy in Malaysia.* Renewable Energy 36 (2011) 881-888, 2011.

[5] Tony Burton, D.S., Nick Jenkins, Ervin Bossanyi, *Wind Energy Handbook, John Wiley & Sons Ltd. 2001.*

[6] Aaron Knoll, K.K., *Residential and Commercial scale distributed wind energy in North Dakota, USA. Wind Energy, 2009.*

[7] Masters, G.M., *Renewable and Efficient Electric Power Systems.* John Wiley & Sons, Inc., 2004.

[8] L. Cristina, A.a.J.Z.M., *Evaluation of global wind power, Journal of Geophysical Research, Vol-110, D12110, doi:1029/2004JD005462. 2005. 110.*

[9] Grigsbay, L.L., *The Electric Power Engineering Handbook.* CRC Press 2001.

[10] Paul Denholm, E.E., Brendan Kirby, and Michael Milligan, *The Role of Energy Storage with Renewable Electricity Generation*. Technical Report, NREL/TP-6A2-47187, January 2010.

[11] Wessells, C., Stanford University News, *Nanoparticle electrode for batteries could make large-scale power storage on the energy grid feasible*, Available [Online] at: http://news.stanford.edu/news/2011/november/longlife-power-storage-112311.html Access date 24 November 2011.

[12] Ter-Gazarian, A., *Energy Storage for Power Systems.* . Peter Peregrinus Ltd, 1994.

[13] Jozef, *Paska et. al, Technical and Economic Aspects of Electricity Storage Systems Co-operating with Renewable Energy Sources.* . 10th Conference EPQU 2009.

[14] Mohammad T Arif, A.M.T.O., A B M Shawkat Ali, Md. Fakhrul Islam, *Significance of Storage and feasibility analysis of Renewable energy with storage system*. Proceedings of the IASTED International Conference on Power and Energy Systems (Asia PES 2010), 2010: p. 90-95.

[15] Dan T. Ton, C.J.H., Georgianne H. Peek, and John D. Boyes, *Solar Energy Grid Integration Systems –Energy Storage (SEGIS-ES)*. SANDIA REPORT, SAND2008-4247, Unlimited Release, July 2008, 2008.

[16] Goel, G.B.S.a.L., *A study on optimal sizing of stand-alone photovoltaic stations*. IEEE Transactions on Energy Conversion, Vol. 13, No. 4, December 1998, 1998.

[17] Nabiha BRAHMI, S.S., Maher CHAABENE, *Sizing of a mini autonomous hybrid electric grid*. International Renewable Energy Congress, 2009.

[18] IEEE, *IEEE Recommended Practice for Sizing Lead-Acid Batteries for Stand-Alone Photovoltaic (PV) Systems, Available [Online] at:* http://ieeexplore.ieee.org/stamp/stamp.jsp?tp=&arnumber=4280849.   IEEE   Standard 1013™-2007.

[19] IEEE, *IEEE Std 1561-2007, IEEE Guide for Optimizing the Performance and Life of Lead-Acid Batteries in Remote Hybrid Power Systems*. Standard, 2007.

[20] IEEE, *IEEE Standard for Interconnecting Distributed Resources With Electric Power Systems, in IEEE Std 1547-2003. p. 0_1-16*. Standard, 2003.

[21] Renewable Energy, T.I.p.o.T., *Estimating PV System Size and Cost*. SECO Fact Sheet no. 24.

[22] PV, L., *LG Polycrystalline PV Module, Available [Online] at:* http://futuresustainability.rtrk.com.au/?scid=80507&kw=4858156&pub_cr_id=171647038 77 (access date: 20/03/2012).

[23] Matters, E., *SMA Sunny Boy 3800W Grid-connected Inverter, Available [Online] at: http://www.energymatters.com.au/sma-sunny-boy-3800watt-grid-connect-inverter-p-412.html.*

[24] G.J., *Dalton, Lockington, D. A. & Baldock, T. E. (2008) Feasibility analysis of stand-alone renewable energy supply options for a large hotel. Renewable Energy, vol 33, issue 7, P-1475-1490.*

[25] HOMER, *Analysis of micro powersystem options, Available [Online] at:* https://analysis.nrel.gov/homer/.

[26] Celik, A.N., *Techno-economic analysis of autonomous PV-Wind hybrid energy systems using different sizing methods. Energy Conversion and Management, 2003. 44(12): P. 1951-1968.*

[27] Barley, DC, Winn BC. *Optimal dispatch strategy in remote hybrid power systems, Solar Energy 1996;58:1 65-79.*

[28] BoM, *Bureau of Meteorology (2011), Australian Government. Available [Online] at: http://reg.bom.gov.au/ Data collected on 03/06/2011.*

[29] ErgonEnergy, *Understanding your Electricity Bill, Information about average consumption, Available [Online] at: http://www.ergon.com.au/your-business/accounts--and--billing/understanding-your-bill access date: 10/04/2012.*

[30] PV-Price, *Sun Solar System, Available [Online] at: www.sunsolarsystem.com.au (access date: 15/03/2012).*

[31] PV-Price, *Goodhew Electrical and Solar, Available [Online] at:*
http://www.goodhewsolar.com.au/customPages/goodhew-electrical-%26-solar-offers-homeowners-the-most-affordable-quality-solar-systems-on-the-market.?subSiteId=1 (access date: 15/03/2012).

[32] Inverter-Cost, *SMA Sunny Boy Grid tie Inverter 7000W SB7000US price, Available [Online] at:*
http://www.google.com/products/catalog?hl=en&q=sunny+boy+grid+tie+inverter+price&gs_sm=3&gs_upl=2378l7840l1l8534l11l10l0l1l1l0l223l2055l0.3.7l11l0&bav=on.2,or.r_gc.r_pw.,cf.osb&biw=1680&bih=831&um=1&ie=UTF-8&tbm=shop&cid=10871923140935237408&sa=X&ei=z0g6T-T0LayziQfnkZGQCg&ved=0CGkQ8wIwAQ (access date: 16/03/2012).

[33] Inverter-Cost, *SMA Sunny Boy 1700 Price, Available [Online] at:*
http://www.solarmatrix.com.au/special-offers/sunny-boy-1700?ver=gg&gclid=CMO0gIOzna4CFYVMpgod7T1OHg (access date: 16/03/2012).

[34] WindPower, B., *Wind Turbine Retail Price list, Available [Online[at: http://production-images.webapeel.com/bergey/assets/2012/3/6/98837/PriceList-March.pdf Access date: 09/04/2012.*

[35] Winturbine_Price, *Ecodirect, Clean Energy Solution, Windturbine Price, Available [Online] at: http://www.ecodirect.com/Bergey-Windpower-BWC-10kW-p/bergey-windpower-bwc-10kw-ex.htm Access date: 29/02/2012. Online Information.*

[36] ErgonEnergy, *Electricity tariffs and prices, available [Online] at:*
http://www.ergon.com.au/your-business/accounts--and--billing/electricity-prices.

[37] ALTE-Store, *Battery Price, Trojan T-105 6V, 225AH (20HR) Flooded Lead Acid Battery, Available [Online] at: http://www.altestore.com/store/Deep-Cycle-Batteries/Batteries-Flooded-Lead-Acid/Trojan-T-105-6V-225AH-20HR-Flooded-Lead-Acid-Battery/p1771/.*

[38] RedEnergy, *Pricing definition for Electricity customers, NSW, Available [Online] at: http://www.redenergy.com.au/docs/NSW-Pricing-DEFINITIONS-0311.pdf. 2011.*

[39] Future_Sustainability, *Rebates for Solar Power, Available [Online] at:*
http://futuresustainability.rtrk.com.au/?scid=80507&kw=4858156&pub_cr_id=17164703877.

[40] Depertment of Employment, E.D.I., Queensland, *Solar Bonus Scheme, Queensland Government, Office of Clean Energy, Available [Online] at:*
http://www.cleanenergy.qld.gov.au/demand-side/solar-bonus-
scheme.htm?utm_source=WWW2BUSINESS&utm_medium=301&utm_campaign=
redirection.

Chapter 2

Compressed Air Energy Storage

Haisheng Chen, Xinjing Zhang, Jinchao Liu and Chunqing Tan

Additional information is available at the end of the chapter

1. Introduction

Electrical Energy Storage (EES) refers to a process of converting electrical energy from a power network into a form that can be stored for converting back to electrical energy when needed [1-3]. Such a process enables electricity to be produced at times of either low demand, low generation cost or from intermittent energy sources and to be used at times of high demand, high generation cost or when no other generation is available[1-9].The history of EES dates back to the turn of 20th century, when power stations often shut down for overnight, with lead-acid accumulators supplying the residual loads on the then direct current (DC) networks [2-4]. Utility companies eventually recognised the importance of the flexibility that energy storage provides in networks and the first central station energy storage, a Pumped Hydroelectric Storage (PHS), was in use in 1929[2][10-15]. Up to 2011, a total of more than 128 GW of EES has been installed all over the world [9-12]. EES systems is currently enjoying somewhat of a renaissance, for a variety of reasons including changes in the worldwide utility regulatory environment, an ever-increasing reliance on electricity in industry, commerce and the home, power quality/quality-of-supply issues, the growth of renewable energy as a major new source of electricity supply, and all combined with ever more stringent environmental requirements[3-4][6]. These factors, combined with the rapidly accelerating rate of technological development in many of the emerging electrical energy storage systems, with anticipated unit cost reductions, now make their practical applications look very attractive on future timescales of only years. The anticipated storage level will boost to 10~15% of delivered inventory for USA and European countries, and even higher for Japan in the near future[4][10].

There are numerous EES technologies including Pumped Hydroelectric Storage (PHS)[11-12][17], Compressed Air Energy Storage system (CAES)[18-22], Battery[23-27], Flow Battery[3-4][6][13], Fuel Cell[24][28], Solar Fuel[4][29], Superconducting Magnetic Energy Storage system (SMES)[30-32], Flywheel[13][16][33-34] and Capacitor and Supercapacitor[4][16]. However, only two kinds of EES technologies are credible for energy storage in large scale (above 100MW in single unit) i.e. PHS and CAES. PHS is the most widely implemented large-scale form of EES. Its principle is to store hydraulic potential energy by pumping water from a lower reservoir to an elevated reservoir. PHS is a mature technology with large volume, long storage period, high efficiency and relatively low capital cost per unit energy. However, it has a major drawback of the scarcity of available sites for two large reservoirs and one or two dams. A long lead time (typically ~10 years) and a large amount of cost (typically hundreds to thousands million US dollars) for construction and environmental issues (e.g. removing trees and vegetation from the large amounts of land prior to the reservoir being flooded) are the other three major constrains in the deployment of PHS. These drawbacks or constrains of PHS make CAES an attracting alternative for large scale energy storage. CAES is the only other commercially available technology (besides the PHS) able to provide the very-large system energy storage deliverability (above 100MW in single unit) to use for commodity storage or other large-scale storage.

The chapter aims to review research and application state-of-arts of CAES including principle, function and deployments. The chapter is structured in the following manner. Section 2 will give the principle of CAES. Technical characteristics of the CAES will be described in Section 3 in terms of power rating and discharge time, storage duration, energy efficiency, energy density, cycle life and life time, capital cost etc. Functions and deployments will be given in Sections 4 and 5. And research and development of new CAES technologies will be discussed in Section 6. Finally, concluding remarks will be made in Section 7.

2. Principle

The concept of CAES can be dated back to 1949 when Stal Laval filed the first patent of CAES which used an underground cavern to store the compressed air[9]. Its principle is on the basis of conventional gas turbine generation. As shown in Figure 1, CAES decouples the compression and expansion cycle of a conventional gas turbine into two separated processes and stores the energy in the form of the elastic potential energy of compressed air. In low demand period, energy is stored by compressing air in an air tight space (typically 4.0~8.0 MPa) such as underground storage cavern. To extract the stored energy, compressed air is drawn from the storage vessel, mixed with fuel and combusted, and then expanded through a turbine. And the turbine is connected to a generator to produce electricity. The waste heat of the exhaust can be captured through a recuperator before being released to the atmosphere (figure 2).

As shown in Figure 2, a CAES system is made of above-ground and below-ground components that combine man-made technology and natural geological formations to accept, store, and dispatch energy. There are six major components in a basic CAES installation including five above-ground and one under-ground components:

1. The motor/generator that employs clutches to provide for alternate engagement to the compressor or turbine trains.
2. The air compressor that may require two or more stages, intercoolers and after-coolers, to achieve economy of compression and reduce the moisture content of the compressed air.

3. The turbine train, containing both high- and low pressure turbines.
4. Equipment controls for operating the combustion turbine, compressor, and auxiliaries and to regulate and control changeover from generation mode to storage mode.
5. Auxiliary equipment consisting of fuel storage and handling, and mechanical and electrical systems for various heat exchangers required to support the operation of the facility.
6. The under-ground component is mainly the cavity used for the storage of the compressed air.

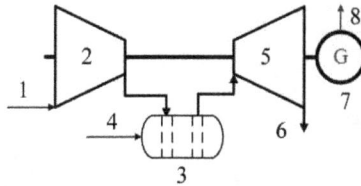

1.Air, 2.Compressor, 3.Combustor, 4.Fuel,
5.Turbine, 6.Exhaust, 7.Generator, 8.Electricity

(a) Schematic diagram of GT system

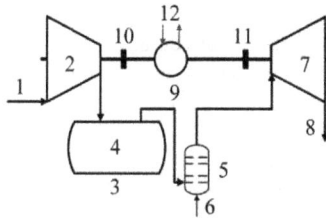

1.Air, 2.Compressor, 3.Reservoir, 4.Compressed air,
5.Combustor, 6.Fuel, 7.Turbine, 8.Exhaust,
9.Motor/Generator, 10 and 11.Clutch, 12.Electricity

(b) Schematic diagram of GT system

Figure 1. Schematic diagram of gas turbine and CAES system

The storage cavity can potentially be developed in three different categories of geologic formations: underground rock caverns created by excavating comparatively hard and impervious rock formations; salt caverns created by solution- or dry-mining of salt formations; and porous media reservoirs made by water-bearing aquifers or depleted gas or oil fields (for example, sandstone, fissured lime). Aquifers in particular can be very attractive as storage media because the compressed air will displace water, setting up a constant pressure storage system while the pressure in the alternative systems will vary when adding or releasing air.

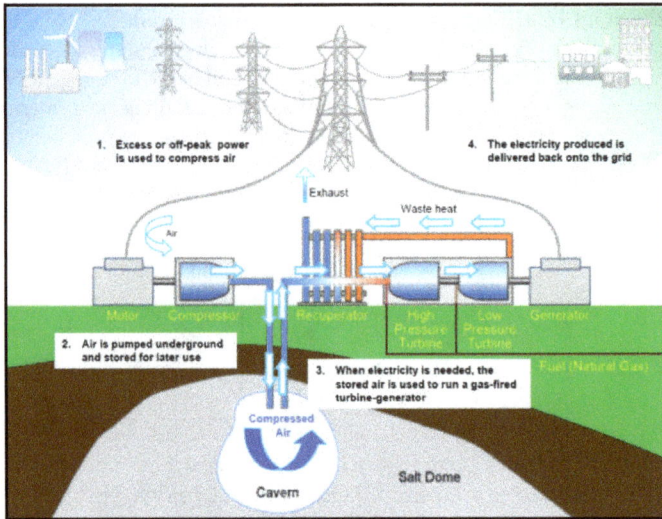

Figure 2. Components of CAES[35]

3. Technical characteristics

Figure 3 shows the comparison of technical characteristics between CAES and other EES technologies. One can see that CAES has a long storage period, low capital costs but relatively low efficiency. The typical ratings for a CAES system are in the range 50 to 300 MW and currently manufacturers can create CAES machinery for facilities ranging from 5 to 350 MW. The rating is much higher than for other storage technologies other than pumped hydro. The storage period is also longer than other storage methods since the losses are very small; actually a CAES system can be used to store energy for more than a year. The typical value of storage efficiency of CAES is in the range of 60-80%. Capital costs for CAES facilities vary depending on the type of underground storage but are typically in the range from $400 to $800 per kW. The typical specific energy density is 3-6 Wh/litre or 0.5-2 W/litre and the typical life time is 20-40 years.

Similar to PHS, the major barrier to implementation of CAES is also the reliance on favourable geography such as caverns hence is only economically feasible for power plants that have nearby rock mines, salt caverns, aquifers or depleted gas fields. In addition, in comparison with PHS and other currently available energy storage systems, CAES is not an independent system and requires to be associated the gas turbine plant. It cannot be used in other types of power plants such as coal-fired, nuclear, wind turbine or solar photovoltaic plants. More importantly, the combustion of fossil fuel leads to emission of contaminates such as nitrogen oxides and carbon oxide which render the CAES less attractive[19][36,37]. Many improved CAES are proposed or under investigation, for example Small Scale CAES with fabricated small vessels and **A**dvanced **A**diabatic CAES (AACAES) with TES[19][21], which will be discussed in Section 6.

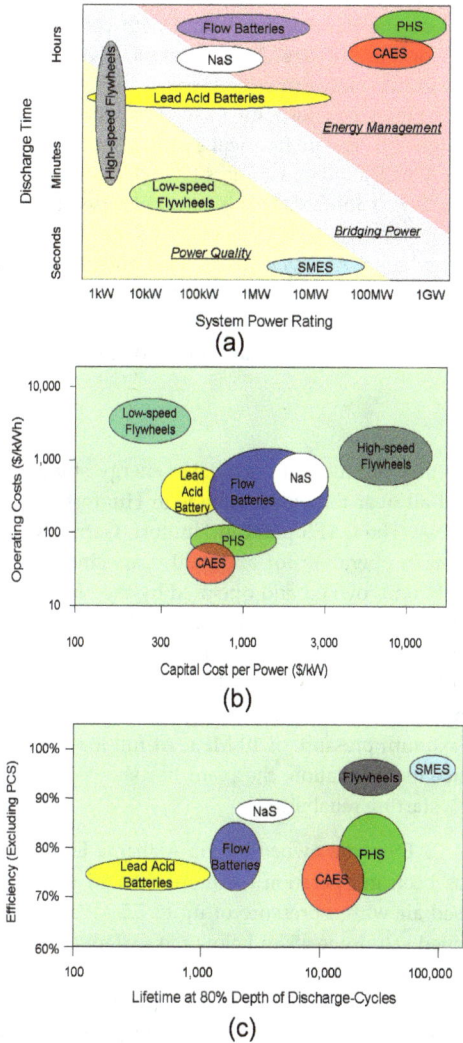

Figure 3. Technical characteristics of CAES

4. Function

CAES systems are designed to cycle on a daily basis and to operate efficiently during partial load conditions. This design approach allows CAES units to swing quickly from generation to compression modes. CAES plant can start without extra power input and take minutes to work at full power. As a result, CAES has following functions

1. Peak shaving: Utility systems that benefit from the CAES include those with load varying significantly during the daily cycle and with costs varying significantly with the generation level or time of day. It is economically important that storing and moving low-cost power into higher price markets, reducing peak power prices.
2. Load leveling: CAES plants can respond to load changes to provide load following because they are designed to sustain frequent start-up/shut-down cycles.
3. Energy Management: CAES allows customers to peak shave by shifting energy demand from one time of the day to another. This is primarily used to reduce their time-of-use (demand) charges.
4. Renewable energy: Linking CAES systems to intermittent renewable resources, it can increase the capacity credit and improve environmental characteristics.
5. Standby power: CAES could also replace conventional battery system as a standby power which decreases the construction and operation time and cost.

5. Deployment

Although CAES is a mature, commercially available energy storage technology, there are only two CAES operated all over the world. One is in Huntorf in Germany, another is in McIntosh, Alabama in USA. The CAES plant in Huntorf, Germany is the oldest operating CAES system. It has been in operation for about 30 years since 1978. The Huntorf CAES system is a 290 MW, 50Hz unit, owned and operated by the Nordwestdeutche Krafiwerke, AG. The size of the cavern, which is located in a solution mined salt dome about 600m underground, is approximately 310,000 m^3. It runs on a daily cycle with eight hours of charging required to fill the cavern. Operating flexibility, however, is greatly limited by the small cavern size. Compression is achieved through the use of electrically driven 60 MW compressors up to a maximum pressure of 10 MPa. At full load the plant can generate 290 MW for two hours. Since its installation, the plant has showed high operation ability e.g. 90% availability and 99% starting reliability.

The second commercial CAES plant, owned by the Alabama Energy Cooperative (AEC) in McIntosh, Alabama, has been in operation for more than 15 years since 1991. The CAES system stores compressed air with a pressure of up to 7.5 MPa in an underground cavern located in a solution mined salt dome 450m below the surface. The storage capacity is over 500,000 m^3 with a generating capacity of 110 MW. Natural gas heats the air released from the cavern, which is then expanded through a turbine to generate electricity. It can provide 26 hours of generation. The McIntosh CAES system utilizes a recuperator to reuse heat energy from the gas turbine, which reduces fuel consumption by 25% compared with the Huntorf CAES plant.

There are several planed or under development CAES projects:

1. The third commercial CAES is a 2700 MW plant that is planned for construction in the United States at Norton, Ohio developed by Haddington Ventures Inc.. This 9-unit plant will compress air to ~10 MPa in an existing limestone mine dome 670m under ground. The volume of the storage cavern is about 120,000,000 m^3.

2. Project Markham, Texas: This 540 MW project developed jointly by Ridege Energy Services and EI Paso Energy will consist of four 135 MW CAES units with separate low pressure and high pressure motor driven compression trains. A salt dome is used as the storage vessel.

3. Iowa stored energy project: This project under development by Iowa Association of Municipal Utilities, promises to be exciting and innovative. The compressed air will be stored in an underground aquifer, and wind energy will be used to compress air, in addition to available off-peak power. The plant configuration is for 200MW of CAES generating capacity, with 100MW of wind energy. CAES will expand the role of wind energy in the region generation mix, and will operate to follow loads and provide capacity when other generation is unavailable or non-economic. The underground aquifer near Fort Dodge has the ideal dome structure allowing large volumes of air storage at 3.6 MPa pressure.

4. Japan Chubu project: Chubu Electric of Japan is surveying its service territory for appropriate CAES sites. Chubu is Japan's third largest electric utility with 14 thermal and two nuclear power plants that generate 21,380 MWh of electicity annually. Japanese utilities recognize the value of storing off-peak power in a nation where peak electricity costs can reach $0.53/kWh.

5. Eskom project: Eskom of South Africa has expressed interest in exploring the economic benefits of CAES in one of its integrated energy plans[10].

6. Research and development

As mentioned in Section 3, there are two major barriers to implementation of CAES: the reliance on favourable caverns and the reliance on fossil fuel. To alleviate the barriers, many improved CAES systems are proposed or under research and development, typical examples are improved conventional CAES, Advanced Adiabatic CAES (AACAES) with TES[19][21] and Small Scale CAES with fabricated small vessels.

6.1. Improved conventional CAES system

Figure 4 shows the principle of the improved conventional CAES system, which is similar to Figure 3. In figure 4, there are intercoolers and aftercooler in the compression process; reheater is installed between turbine stages; and regenerator is used to preheat the compressed air by the exhausted gas. McIntosh plant can reduce fuel consumption by 25% using the improved cycle shown in figure 4.

Another improved conventional CAES system combined with a gas turbine is shown in figure 5[38-41]. When the electricity is in low-demand, the compressed air is produced and stored in underground cavity or above ground reservoir. During the high-demand period, the CAES is charging the grid simultaneously with the GT power system. The compressed air is heated by the GT exhaustion and the heated compression air expands in the high pressure (HP) turbine and then ejects to the GT turbine combustor to join GT working fluid. The CAES system shown in figure 4 can recover almost 70% of compression energy.

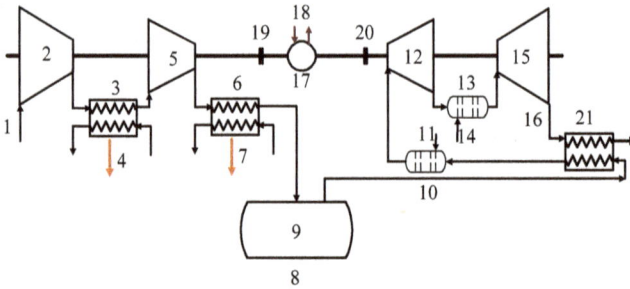

1.Air, 2 and 5.Compressor, 3 and 6.Heat Exchanger, 4 and 7. Heat, 8.Reservoir, 9.Compressed Air,
10 and13.Combustor, 11 and 14.Fuel, 12 and 15.Turbine, 16.Exhaust, 17.Motor/Generator,
18.Electricity, 19 and 20.Clutch, 21.Recuperator

Figure 4. Schematic diagrams of improved conventional CAES system

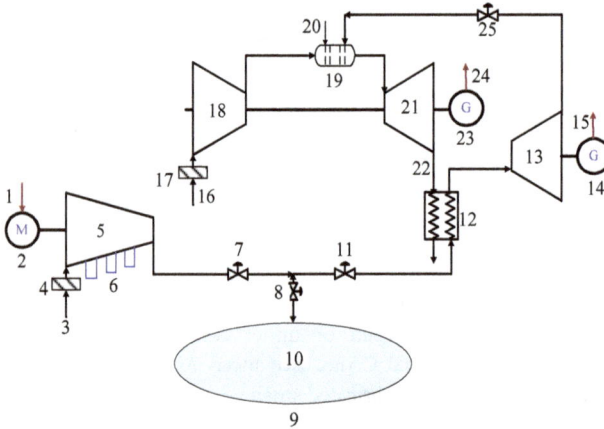

1 15 and 24.Electricity, 2.Motor, 3 and 16.Air, 4 and 17.FlltEr, 5 and 18.Compressor, 6.Intercooler, 7 8
11 And 25.Valve, 9.Underground Cavern, 10.Compressed Air, 12.Recuperator, 13 and 21 Turbine, 14
and 23 Generator, 19.Combustor, 20.Fuel, 22.Exhaust

Figure 5. CAES combined with GT system

6.2. Advanced Adiabatic CAES system

The so called Advanced Adiabatic CAES (AA-CAES) stores the potential and thermal energy of compressed air separately, and recover them during expansion (as shown in figure 6). Although the cost is about 20~30% higher than the conventional power plant, this system eliminates the combustor and is a fossil free system. IAA-CAES may be commercially viable due to the improvements of thermal energy storage (TES), compressor and turbine technologies. A project "AA-CAES" (Advanced Adiabatic – Compressed Air Energy Storage: EC DGXII contract ENK6 CT-2002-00611) committed to developing this technology to meet the current requirements of energy storage.

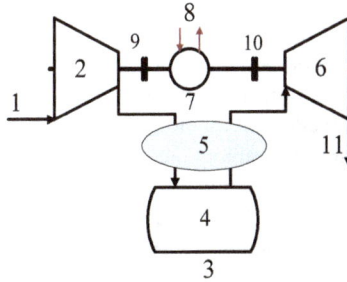

1.Air, 2.Compressor, 3.Storage Reservoir,
4.Compressed Air, 5.Thermal Energy Storage,
6.Turbine, 7.Motor/Generator, 8.Electricity, 9 and
10.Clutch, 11.Exhaust

Figure 6. Schematic diagram of AA-CAES system

6.3. Small-scale CAES System

Small-scale CAES system (<~10MW) with man-made vessels is a more adaptable solution, without need of caverns, especially for distributed generation that could be widely applicable to future power networks. Figure 7 shows a small-scale CAES used for standby power system[42]. It can replace battery with technical simplicity, low degradation of components, high reliability, low maintenance and lower life cycle cost characteristics. For a 2kW power application, CAES can work 20 years, while vented lead acid batteries (VLAB) 12 years; the installation and commissioning durations are 8 hours, respectively, while 16 and 64 hours for VLAB; with 300bar, 24,000L compressed air in cylinders, the CAES can work as a standby power for one year by charging four times. In general, there is no heat recovery/storage component in the small-scale CAES system, therefore its efficiency is lower than that of VLAB system.

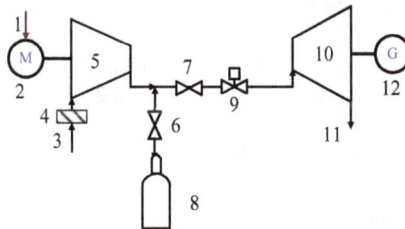

1.Electricity, 2.Motor, 3.Air, 4.Filter,
5.Compressor, 6 7 and 9.Valve, 8.Air
Reservoir, 10.Expander, 11.Exhaust,
12.Generator,

Figure 7. Schematic diagram of the CAES system as a standby power supply

7. Concluding remarks

Research and application state-of-arts of compressed air energy storage system are discussed in this chapter including principle, function, deployment and R&D status. CAES is the only other commercially available technology (besides the PHS) able to provide the very-large system energy storage deliverability (above 100MW in single unit). It has a long storage period, low capital costs but relatively low efficiency in comparison with other energy storage technologies. CAES can be used for peak shaving, load leveling, energy management, renewable energy and standby power. However, there are two major barriers to implementation of CAES: the reliance on favourable caverns and the reliance on fossil fuel. To alleviate the barriers, many improved CAES systems are under research and development such as improved conventional CAES, AACAES and Small Scale CAES.

Author details

Haisheng Chen*, Xinjing Zhang, Jinchao Liu and Chunqing Tan
Institute of Engineering Thermophysics, Chinese Academy of Sciences, Beijing, 100190, China

Acknowledgement

The authors thank the Natural Science Foundation of China under grant No. 50906079 and Beijing Natural Science Foundation under grant No.3122033 for financial supports.

8. References

[1] Mclarnon F. R., Cairns E. J. (1989) Energy storage, Annul Review of Energy, vol 14, 241-271

[2] Baker J.N. and Collinson A. (1999) Electrical energy storage at the turn of the Millennium, Power Engineering Journal, No.6, 107-112

[3] Dti Report (2004) Status of electrical energy storage systems, DG/DTI/00050/00/00, URN NUMBER 04/1878

[4] Australian Greenhouse Office (2005) Advanced electricity storage technologies programme, ISBN:1 921120 37 1

[5] Walawalkar R., Apt J., Mancini R. (2007) Economics of electric energy storage for energy arbitrage and regulation, Energy Policy, vol. 35, 2558-2568

[6] Dti Report (2004) Review of electrical energy storage technologies and systems and of their potential for the UK, DG/DTI/00055/00/00, URN NUMBER 04/1876

[7] Weinstock I. B. (2002) Recent advances in the US department of Energy's energy storage technology research and development programs for hybrid electric and electric vehicles, Journal of Power Sources, vol. 110, 471-474

[8] Koot M., Kessels J.T.B.A., Jager B., Heemels W.P.M.H., Bosch P.P. J. and Steinbuch M. (2005) Energy management strategies for vehicular electric power systems, IEEE Transactions on Vehiclular Technology, vol. 54, 771-782

---

* Corresponding Author

[9]   Chen, H., Cong, T. N., Yang, W., Tan, C., Li, Y. and Ding, Y. (2009) Progress in electrical energy storage system: A critical review. Progress in Natural Science, 19:291-312.

[10] Ahearne J. (2004) Storage of electric energy, Report on research and development of energy technologies. IUPAP working group on energy, 76-86

[11] http://en.wikipeida.org/wiki/Hydroelectric_energy_storage, 20 March 2007.

[12] Linden S. (2003) "The Commercial World of Energy Storage: A Review of Operating Facilities (under construction or planned)", presentation at the 1st Annual Conference of the Energy Storage Council, Houston, Texas, 3 March 2003.

[13] Linden S. (2006) Bulk energy storage potential in the USA, current developments and future prospects, Energy, vol.31, 3446-3457

[14] Makansi J. and Abboud J. (2002) Energy storage, the missing link in the electricity value chain, An ESC White Paper, Energy storage Council

[15] Akhil A., Swaminathan S, Sen R.K. (1997) Cost analysis of energy storage systems for electric utility applications, Sandia Report, SAND97-0443 UC-1350, Sandia National Laboratories

[16] Kondoh J., Ishii I., Yamaguchi H., Murata A. (2000) Electrical energy storage systems for energy networks, Energy Conversion& Management, vol 41, 1863-1874

[17] Bueno C. and Carta J.A. (2006), Wind powered pumped hydro storage systems, a means of increasing the penetration of renewable energy in the Canary Islands, Renewable and Sustainable Energy Reviews, Vol. 10, 312-340

[18] Najjar Y. S. H., Zaamout M. S. (1998) Performance analysis of compressed air energy storage (CAES) plant for dry regions, Energy Conversion and Management, vol 39, 1503-1511

[19] Sears J. R. (2004) TEX: The next generation of energy storage technology, Telecommunications Energy Conference, INTELEC 2004. 26th Annual International Volume, Issue, 19-23 Sept. 2004, 218 – 222

[20] Najjar Y.S.H and Jubeh N.M. (2006) Comparison of performance of compressed-air energy-storage plant with compressed-air storage with humidification, Proceeding of IMechE, Part A: Journal of Power and Energy, vol. 220, 581-588

[21] Bullough C., Gatzen C., Jakiel C., Koller M., Nowi A. and Zunft S. (2004) Advanced adiabatic compressed air energy storage for the integration of wind energy. Proceedings of the Eruopean Wind Energy Conference. London UK.

[22] Wang S., Chen G., Fang M. and Wang Q. (2006) A new compressed air energy storage refrigeration system. Energy Conversion and Management, 47, 3408-3416.

[23] Cook G.M., Spindler W.C. and Grefe G. (1991) Overview of battery power regulation and storage. IEEE Transctions on Energy Conversion, vol.6, 204-211

[24] Chalk S.G., Miller J.F. (2006), Key challenges and recent progress in batteries, fuel cells and hydrogen storage for clean energy systems, Journal of Power Sources, Vol. 159, 73-80

[25] Kashem M.A., Ledwich G. (2007) Energy requirement for distributed energy resources with battery energy storage for voltage support in three-phase distribution lines, Electric Power Systems Research, vol. 77, 10-23

[26] Kluiters E.C., Schmal D., Ter Veen W. R., Posthumus K. (1999) Testing of a sodium/nickel chloride (ZEBRA) battery for electric propulsion of ships and vehicles, Journal of Power Sources, vol 80, 261-264

[27] Karpinski A.P., Makovetski B., Russell S. J., Serenyi J. R., Williams D. C. (1999) Silver-zinc: status of technology and applications, vol 80, 53-60

[28] Weinmann O. (1999) Hydrogen-the flexible storage for electrical energy, Power Engineerign Journal, Special Feature: Electrical energy storage, 164-170

[29] Steinfeld A. and Meier A. (2004) Solar thermochemical process technology, in Encyclopedia of Energy, Elsevier Inc., Vol. 5, pp. 623-637, 2004.

[30] Kolkert W. J. andJamet F. (1999) Electric energy gun technology: status of the French-German-Netherlands programme, IEEE Transactions on Magnetics, Vol. 35, 25-30

[31] Koshizuka N., Ishikawa F., Nasu H. (2003) Progress of superconducting bearing technologies for flywheel energy storage systems, Physica C, vol. 386, 444-450

[32] Xue X., Cheng K. and Sutanto D. (2006) A study of the status and future of superconducting magnetic energy storage in power systems. Superconductor Science and Technology, 19, R31-R39.

[33] Suzuki Y., Koyanagi A., Kobayashi M. (2005) Novel applications of the flywheel energy storage system, Energy, vol 30, 2128-2143

[34] http://www.beaconpower.com/products/EnergyStorageSystems/ flywheels.htm 20 March 2007

[35] Jewitt J. (2005) Impact of CAES on Wind in Tx,OK and NM, Presentation in DOE energy storage systems research annual peer review , San Francisco, USA, Oct. 20, 2005

[36] P. Denholm, G. L. Kulcinski. (2004) Life cycle energy requirements and greehouse gas emissions from large scale energy storage systems. Energy Conversion and Management, vol 45, 2153-2172

[37] P. Denholm, T. Holloway. (2005) Improved accounting of emissions from utility energy storage system operation, Environmental Science & Technology, vol 39, 9016-9022

[38] Nakhamkin, M., Wolk, R. H., Linden, S. v. d. and Patel, M. New Compressed Air Energy Storage Concept Improves the Profitability of Existing Simple Cycle, Combined Cycle, Wind Energy, and Landfill Gas Power Plants. In: ASME, pp. 103-110.

[39] Nakhamkin, M. and Chiruvolu, M. (2007) Available Compressed Air Energy Storage (CAES) Plant Concepts. In: Power-Gen International, Minnestota.

[40] Nakhamkin, M., Chiruvolu, M., Patel, M. and Byrd, S. (2009) Second Generation of CAES Technology-Performance, Operations, Economics, Renewable Load Management, Green Energy. In: POWER-GEN International, Las Vegas Convention Center, Las Vegas, NV.

[41] Akita, E., Gomi, S., Cloyd, S., Nakhamkin, M. and Chiruvolu, M. (2007) The Air Injection Power Augmentation Technology Provides Additional Significant Operational Benefits. In: ASME Turbo Expo 2007: Power for Land, Sea and Air ASME, Montreal, Canada.

[42] Beukes, J., Jacobs, T., Derby, J., Conlon, R. and Henshaw, I. (2008) Suitability of compressed air energy storage technology for electricity utility standby power applications. In: Telecommunications Energy Conference, 2008. INTELEC 2008. IEEE 30th International, pp. 1-4.

Chapter 3

The Future of Energy Storage Systems

Luca Petricca, Per Ohlckers and Xuyuan Chen

Additional information is available at the end of the chapter

1. Introduction

During the past several years, we have witnessed a radical evolution of electronic devices. One of the major trends of this evolution has been increased portability. Laptops and smart-phone are the most common examples but also cameras or new technologies such as tablets are equal important. The request of efficient energy storage system becomes even more important if we extend it to different applications such as electrical/hybrid vehicles that require hundreds of times larger power when compared with smaller device. Unfortunately the technological improvements of batteries are slower than electronics, creating a constantly growing gap that need to be filled. For this reasons it is very important to develop an efficient energy storage system that goes beyond normal batteries. In the first part of this chapter we will give a general overview of some existing solution such as electrolytic batteries, fuel cells and microturbines. In the second part we will introduce an evolution of simple capacitors known as Supercapacitors or Ultracapacitors. This technology is very promising and it might be able to substitute, or at least improve in a considerable way current energy storage systems.

1.1. Nanotechnologies for energy related issues

Nanotechnologies (NTs) can play an important role to help to overcome to energy-related challenges and opportunities. However, what specific kinds of nanotechnologies and how can they provide such advantages? Sepeur [56] defines nanotechnologies as *"'the systematic manipulation production or alteration of structure systems materials or components in the range of atomic and molecular dimension with/into nanoscale dimensions between 1nm and 100nm"'*. In particular two subfields of NTs are interesting for energy problems: Nanofabrication and nanomaterials. By combining these two techniques we are for example able to create structures with a large surface area per unit mass, and by selective etching and deposition of different material layers we are able to fabricate very complex mechanical structures. Furthermore the use of new materials in the process allow us to create films and layers with a specific characteristic (such as conductivity, stress distribution, mechanical resistance etc). By combining nanomaterials and nanofabrication it has for example been possible to build solar cells much more efficient compared to the standard type, to build new classes of materials such

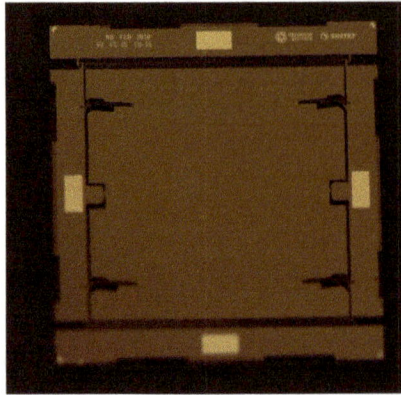

Figure 1. Example of MEMS Energy harvester device [6]

as carbon nanotubes or graphene that are revolutionizing the electronic world. Researchers have already shown that with NT it is possible to create thin film batteries Kuwata et al. [28], Ogawa et al. [44] printable on top of substrates, or creating a smart fibers that can store energy (so called *E-textiles*) Gu et al. [18], Jost et al. [25]. Furthermore, in the literature there are presented many demonstrators of energy harvesting devices that are able to "'harvest'" energy from many different physic sources (such as mechanical vibrations, temperature gradients, electro-magnetic radiations, etc) and transform it into electrical energy.

2. Current state of the art

2.1. Batteries

Nowadays electric batteries represent the most common energy storage methods for portable devices. They store the energy in a chemical way and they are able to reconvert it into electrical form. They consist of two electrodes (anode and cathode) and one electrolyte which can be either solid or liquid; In the redox reaction that powers the battery, reduction occurs at the cathode, while oxidation occurs at the anode [43]. This energy storage form has changed substantially throughout the years, even though the basic principles have been known since the invention by the Italian physicist Alessandro Volta in year 1800. The first type was consisting in a stack of zinc and copper disk separated by an acid electrolyte. Thanks also to the boost of mobile phones during the last years they have evolved to the Nickel-Cadmium (Ni-Ca) and Nickel Metal Hydrate (Ni-Mh) which dominated the market until the developments of Lithium batteries. This latter class rapidly gained market thanks to the higher specific energy (150-500 Wh/Kg versus 50-150Wh/Kg of NiMh, NiCa, See Fig.2) and are nowadays one of the most common batteries available in the market. They can further be divided into another two subclasses which are Lithium Ion (Li-Ion) batteries and Lithium Polymers (Li-Po) batteries (which basically are an evolution of the Li-Ion). The high volume of the market (around 50 billion dollar market in 2006 [39]) is expected to grow even more in the coming years, forecasted to reach around 85.76 billion dollars by 2016 with a Compounded Annual Growth rate around 7% over the next 5 years [34]. This is generating a very high volume of revenues and part of it is re-invested in battery research

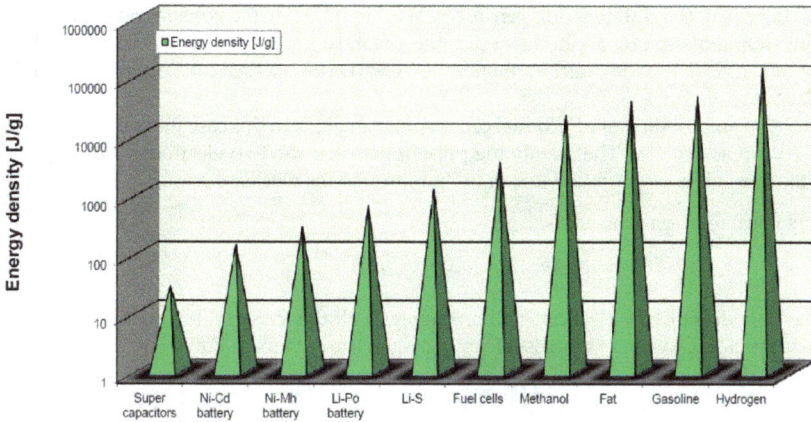

Figure 2. Energy density of different sources (adapted fromPetricca et al. [50])

which is focused to improve energy density, life time and cycling stability, without using dangerous materials that can create health hazards. Unfortunately it is difficult to find all these proprieties optimized in one material combination. One way for enhancement of the battery capacity is using nanotechnologies for increasing and structuring the surface area of the electrodes (for example by depositing nanomaterials or by growing nanostructures such as nanotubes) [43]. But despite all these efforts, the technological improvements of batteries are still much, much slower when compared with the evolutionary progress of electronics. For these reasons many researchers are trying to include smart circuitry inside the batteries for optimizing the discharge curve by optimizing the load. They call it intelligent batteries [35] and they exploit some battery-related characteristic such as charge recovery effect, for improving their lifetime.

2.2. Fuel cells

Fuel cells are one of the most developed alternatives to batteries and they are already available on the market, with many vehicles currently working on Fuel cells based engines [14]. They are electrochemical devices able to convert the chemical energy stored in the fuels into electrical energy. They mainly consist of two electrodes and a membrane which form a reaction chamber and have external stored reactants. The working principle is similar to batteries, however in this case the species at the electrodes are continuously replenished and they can be refilled, ensuring a continuous electricity supply over a long period. There are many types of fuel cells available on the market, and they mainly differ from the species used as fuel, but all of them use one element as fuel and a second element as oxidizer (commonly air) [37].

Theoretically with fuel cells we would be able to generate any power or current by changing the physical dimension of the cell and the flow rate of the fuel. However, the voltage across the single cell electrode is fixed and it is not possible to change it. In general this voltage is very low (less than 1V for realistic operating condition [37]) and thus multiple cell stacks connected in series are needed to achieve larger potentials. Mixed series and parallel connections between different cells can also be used for increase the voltage and the maximum current supplied. Among all the fuel cell types the most promising are the Proton Exchange Membrane (PEM) and the direct Methanol fuel cells (DMFC) which can be considered a PEM special case.

A graphic representation of a PEM fuel cell is shown in Fig.3; In this case the cell use Hydrogen and Oxygen as species. The membrane present between the two electrodes allows passing only the ions while electrons are forced to "'go'" trough the electric circuit.

The chemical reaction at anode is:

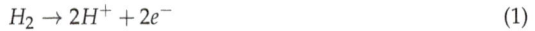

$$H_2 \rightarrow 2H^+ + 2e^- \tag{1}$$

On the cathode side, the electrons will recombine with the Ions and they will react with the cathode species (Oxygen in this case) forming water through the following reaction:

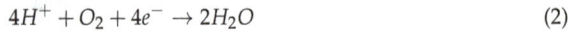

$$4H^+ + O_2 + 4e^- \rightarrow 2H_2O \tag{2}$$

Since water is the only waste product of the reaction, this type of fuel cell is very environmental friendly. Unfortunately there are many problems associated with PEM that are preventing the mass market diffusion of this technology. Hydrogen does not naturally occur in nature and must be produced in factories or laboratories. Furthermore *"'hydrogen has a very high mass energy density (143000 J/g) (See Fig.2) but a very low volumetric energy density (10790 J/L), which makes it difficult to store."'* [50]

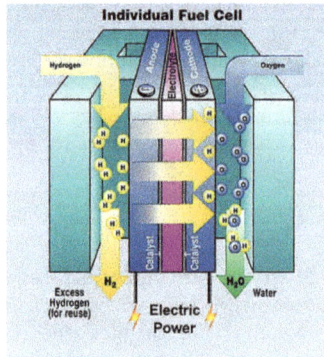

Figure 3. Fuel cell representation [unknown author]

Other problems associated with PEM are the impurity present in the hydrogen fuel, *"'such as CO, H2S, NH3, organic sulfur carbon, and carbon hydrogen compounds, and in air, such as NOx, SOx, and small organics"'*, which are brought in fuel and air feed streams into the electrodes of a PEMFC stack, causing performance degradation or membrane damages [5]. In particular they demonstrated that even small amounts of these impurity materials can poison the anode, cathode or the membrane of the cell, causing a sharp performance drop. To overcome some of these problems (especially the one associated with the hydrogen production and storage), researcher from University of Southern California's Loker Hydrocarbon Institute developed a new type of fuel cells which use a direct oxidation of Methanol instead of Hydrogen. Unlike Hydrogen, Methanol offer also the advantage of being liquid at room temperature making storage and transportation much easier. This new class of fuel cells are called Direct Methanol Fuel Cell (DMFC) and the working principle is similar to the PEM fuel cell, but with more complicated reactions at the anode (3) and at the cathode (4) [19]:

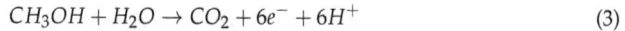

$$CH_3OH + H_2O \rightarrow CO_2 + 6e^- + 6H^+ \tag{3}$$

$$12H^+ + 6O_2 + 12e^- \rightarrow 6H_2O \tag{4}$$

Methanol has energy volume and mass densities of 4380 Wh/l and 5600 Wh/kg, which are about 11 times higher than current Li-ion batteries ($\approx$ 500Wh/l). *"This means that the FP/FC unit has superior energy density even with an overall conversion efficiency as low as 7%. In the case where no water recycling is employed to minimize system complexity, water has to be carried with methanol for the reforming. With a stoichiometric steam to carbon ratio of 1:1, this reduces the net energy density to 3000 Wh/l (4550 Wh/kg)"'* [57]. Unfortunately, unlike Hydrogen PEM, where we can assume that all of the polarization losses are located at the cathode, in DMFC the losses at the anode and cathode are comparable. Furthermore DMFC *"'utilizes cathode Pt sites for the direct reaction between methanol and oxygen, which generates a mixed potential that reduces cell voltage"'* [19]. Despite these problems, many companies are already present on the market offering DMFC [17, 53] or disposable methanol fuel cartridges [64].

2.3. Micro engines and micro turbines

It has been more than 150 years since the first Internal Combustion Engine (ICE) was developed and nowadays it is the most common power source for vehicles and large engine-generators. However in the last years thanks to the improvements of microtechnologies, many researchers started to design and develop micro internal combustion engines that may be used in the future as power source for small electrical devices. They consist of three main parts [45],

1. Combustion Chamber

2. Ignition

3. Moving Parts

forming a few cubic millimeters system able to generate power in mW to Watts range. However, Micro engines are not only a smaller version of the large size counterparts [58]. Some of the technical issue present at this scale are resumed in Fernandez-Pello [13], Sher et al. [58], Suzuki et al. [62], Walther & Ahn [65]. In particular the main challenge is to obtain a genuine combustion in a limited volume of the combustor [45]. Furthermore, at this scale the relative heat losses increase and may cause quenching of the reaction (fuel inside the combustion chamber that rapidly cools down, prevents it from burning) with consequent degradation of performances [13, 45, 65].Moreover, the engine speed and the gap width between the piston and the cylinder walls are two key parameters that can create issues in standard ICE (cylinder-piston engine)as reported by Sher et al. [58]. They simulated the miniaturization limits of a standard ICE with a rotation speed of 48000 rpm, a gap width of 10 um and a compression rate larger than 18 and they found a minimum size limit between 0.3 and 0.4 cc. This limit has already been passed by [62] which has developed a microfabricated standard ICE of 5mmŒ3mmŒ1mm in dimension (0.015 cc) supplied by a mixture of Hydrogen and Oxygen, able to generate a mechanical power of 29.1mW. However, the tests of this silicon engine were performed at 3 rpm and it has shown a compression ratio around 1.4. A similar class of micro engines is based on rotary engine (Wankel design) instead of cylinder-piston design. Example of these MEMS engines can be found in [7, 29]. Interesting is the prototype developed by [7] which consisted of a 13 mm rotor diameter coupled with a dynamo meter, able to generate up to 4W of electric power (other versions of 90mW and 50W are under development). Another interesting alternative is using micro turbines. We are already used to find in the market large gas turbines for electric energy production that may generate up to several hundreds of Megawatts [51]. However, in the last years, with the advent of MNTs, many researchers started to investigate the possibility to create a small micro turbine that are able to generate few Watts, enough to supply most of today's electronic devices. The basic concept is similar to large scale turbine which consists on an upstream rotating compressor, a combustion chamber and a downstream turbine. Furthermore, since it will be used for generating electric power, there will also be an alternator coupled with the turbine. The fabrication material is mainly Silicon, thanks to the well established and controlled processing technologies available for this material. Furthermore, when compared with common nickel alloy, single crystal silicon has an *"higher specific strength, it is quite oxidation-resistant and has thermal conductivity approaching that of copper, so it is resistant to thermal shock"'* [11]. Several authors [1, 33, 49] successfully designed micro turbines by using silicon (Si) as material. However, silicon has some limitation on high temperatures and for these reasons some other authors reported micro turbine fabricated whit other materials. In particular Peirs et al. [48] reported an example of a 36g micro turbine made of stainless steel. This 10mm diameter turbine was able to generate a maximum mechanical power of 28W with an efficiency of 18%. The turbine was then coupled to a small brushless dc motor, which was used as a three-phase generator. The total system was around 53mm long and 66g in weight, capable to generate 16W of electric power, corresponding to a total efficiency of 10.5%. A similar efficiency is also expected by another microturbine developed by Jacobson et al. [22] in which the preliminary test done so far are very encouraging. We should notice that even if the total efficiency is relative low, hydrogen and hydrocarbon fuels have a much higher energy density when compared with batteries (see fig.2), so the result is a substantial increase of the net energy density of the system.

In conclusion, is certainly possible that ICE and MEMS gas turbines may one day be very useful as compact power sources for portable electronics, equipment, and small vehicles [11].

3. Supercapacitors

As we briefly state above, batteries suffer from various limitation, such as limited life cycles, high manufacturing cost and relative low power density. Furthermore, in case of large batteries, they require also several hours for being fully charged. On the other hand, standard capacitors offer high power density, almost unlimited life cycles, and fast charge. However their energy density is currently too low for been used as primary energy storage system. Supercapacitors may combine the advantages of both battery and capacitor for creating a system that has high power density, virtual unlimited life cycles keeping at the same time acceptable energy density. In these devices, the internal leakage current (in the form of dipoles relax and/or charges re-combination) will determine how long the energy can be stored, while

ADVANTAGES	1. High Power Density
	2. Virtually unlimited life cicles
	3.Long shell life
	4. Fast charge
	5. High efficiency
	6. Wide temperature operation range
	7. Safety
DISADVANTAGES	1. Lower energy density
	2. Higher cost
	3. High self-discharging rate
	4. Low cell voltage
	5.Linear voltage drop

Table 1. Super capacitors: Advantages and Drawbacks

the maximum power will depend on the internal resistance (ESR) [55]. In TABLE 1 advantages and drawbacks of supercapacitors respect to batteries are listed.

Figure4 is the Ragone plot for all three technologies. As we can see, supercapacitors will be able to fill the gap between standard capacitors and batteries. It should be noticed that despite supercapacitors there will certainly have a lower energy density of batteries, which can be easily recharged from any power network in seconds or fraction of seconds.

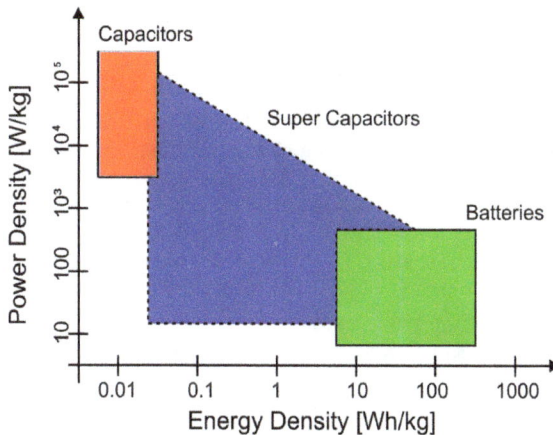

Figure 4. Ragone plot for Capacitors, Super Capacitors and Batteries; adapted from Everett [12], Halper & Ellenbogen [20], Kotz & Carlen [27]

For better understanding the working principle of supercapacitors, it is convenient to start with some basic capacitor theory. The simplest capacitor (plane capacitor) consists of a two electrodes separated by a dielectric. Calling d the distance between the two electrodes, S their overlapped area and ϵ the dielectric constant of the dielectric, the total capacitance C can be defined as:

$$C = \epsilon S / d \qquad (5)$$

which corresponds a stored energy E:

$$E = \frac{CV^2}{2} \tag{6}$$

where V is the voltage between the electrodes. From equations 5 and 6 it is clear that the energy density is proportional to the overlapped area and inversely proportional to the distance between the two electrodes. Thanks to nanotechnology it is possible to drastic reduce the effective distance d and create structures with large surface, resulting of an increasing of total capacitance. Moreover, it is possible to combine special types of dielectrics and ionic conducting liquid (electrolyte) in order to store not only electrostatic energy but also electrochemical energy. Supercapacitors can be classified into double-layer capacitor (EDLC) and electrochemical pseudo capacitor (EPC). Based on the storage mechanism we can divide supercapacitors into three categories [20]:

1. Electric Double Layer Capacitors (EDLC)
2. Electrochemical pseudo capacitors (EPC)
3. Hybrid Supercapacitors

In the following sections we will give an overview of each class.

3.1. Electric double layer capacitors

These supercapacitors are called also non-Faradaic supercapacitors since they do not involve any charge transfer between electrode and electrolyte. The energy storage mechanism is thus similar to standard capacitor where the area is much larger and the distance is in the atomic range of charges [70]. EDLC consist of two electrodes, one membrane between the two electrodes which separates the electrodes and electrolyte that can be either aqueous or non-aqueous depending on EDLC [23]. The material of the electrode is very important for the final supercapacitors performances. For the supercapacitors of today's innovation, the most common material of the electrodes is activated carbons because it is cheap, has large surface area and is easy to process [27, 70]. This material is organized in small hexagonal rings organized into graphene sheets [24]. The result is a large surface area due to the porous structure composed by micropores (< 2nm wide), mesopores (2 - 50 nm), and macropores (>50 nm) [20]. The basic structure of a carbon activated EDLC is shown in Fig.5.

For the analytic model of these capacitors equation 5 can still be considered true, where ϵ is the electrolyte dielectric constant, S is the specific surface area of the electrode accessible to the electrolyte ions, and d is the Debye length [70], however "determination of the effective dielectric constant ϵ_{eff} of the electrolyte and thickness of the double-layer formed at the interface is complex and not well understood" [3]. Indeed we would expect that doubling the area of the active carbons would double also the capacitance. However experimental data are in contrast with the theory, since empiric measurements were showing a smaller capacitance than expected. Many scientist explained this phenomena by electrolyte ions that are too large to diffuse into smaller micropores and thus unable to support electrical double layer [20, 24, 69]. For this reason many authors have affirmed that mesopores are high desirable in EDLC electrodes since they can optimize their performances [8, 69]. However, recently [40] showed the important role of small pores in the EDCL and they affirmed that ions can

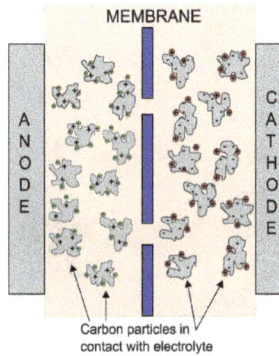

Figure 5. Graphic representation (not in scale) of active carbon EDLC, adapted from [66, 70]

penetrate dissolved in the nanopores. Furthermore it also concludes that the optimal pore size of the electrode mainly depends from the current load due to the distortion of cations and/or intercalation-like behavior [40]. This new discovery lead many scientists to re-consider roles of the micropores. The analytical models can be modified by splitting the capacitive behavior in two different parts depending on the pore size [40]. Despite of all these studies the EDLC is still not completely understood yet [59]. Active carbons super capacitors also change their capacitance respect to the electrolyte materials, aqueous electrolytes allow higher capacitances (ranging from $100 \, F/g$ to $300 \, F/g$) than organic electrolytes (less than $150 \, F/g$). [70].

For the carbon electrodes, one valid alternative to activated carbon is carbon nanotube (CNT), which consists of carbon atoms organized in cylindrical nanostructures and can be considered as rolled-up graphene sheets (which consist in carbons atoms organized in 2-D cells see fig.6). The roll up orientation is expressed by two indexes (n and m) and is very important in CNTs since different directions result different proprieties. The two indexes n and m are used for calculate the roll up direction as shown in figure6.

Both SingleWalled (SWNTs) and Multiple Walled (MWNTs) were investigated for EDLC electrodes. Thanks to their high conductivity and their open shape both SWNTs and MWNTs are particularly suitable for high power density capacitors. Indeed their quickly accessible surface area and their easily tunable pore size enable electrolyte ions to diffuse into the mesopores (fig.7), therefore, reduced internal resistance (ESR) and increased maximum power can be achieved [9, 20, 70].

Unfortunately the specific surface area of CNT ($< 500 \, m^2/g$) is much smaller than that of activated carbons ($1,000 \text{Ű} 3,000 \, m^2/g$) [32, 70], resulting in lower energy density for the capacitor (in average between $1Wh/kg$ and $10Wh/kg$)[52]. This, together with their limited availability and high cost, currently limits their usage [52].

Beside Active Carbons and Carbon Nanotubes, in literature are presented many other materials that can be used for the EDCL electrodes. Among this we should cite carbon aerogel [54], (similar to gels but where the internal liquid is replaced with gas), xerogels [15] and carbon fibers [30]. Carbon aerogel electrode material gave promising capacitive properties, despite the difficulties in preparation [70].

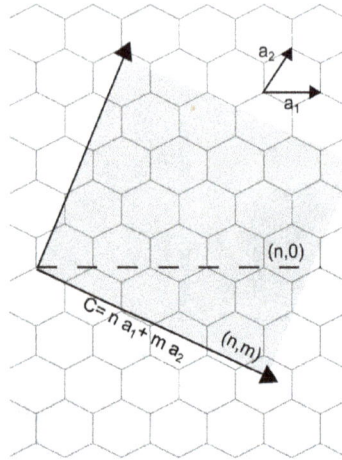

Figure 6. Carbon nanotube: orientation indexes: adapted from GNU license images

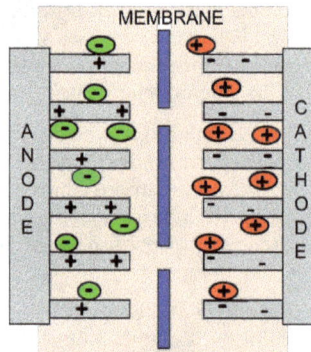

Figure 7. Graphic representation (not in scale) of carbon nanotube EDLC, adapted from [24]

3.2. Electrochemical pseudo supercapacitors

Unlike EDLC, Electrochemical pseudo supercapacitors use charge transfer between electrode and electrolyte for storing energy. This Faradic process is achieved mainly by [20, 66, 70]:

- Rapid and reversible Red-Ox reactions between the electrodes and the electrolyte
- Surface adsorption of ions from the electrolyte
- Doping and undoping of active conducting polymer material of the electrode

The first two processes belong to surface mechanism, so the capacitance will strongly depend on the surface materials of the electrodes, while the third one is more bulk-based process and thus the capacitance will weakly depend on the surface materials of the electrodes [3]. Since these processes are more battery-like rather than capacitor-like, thus, the capacitors are named as the electrochemical pseudo supercapacitors (EPC). When compared with EDLC, EPC have smaller power density since Faradic processes are normally slower than that of non Faradic reactions. However EPCs can reach much higher capacitances [3, 59] and thus they store much higher energy density.

Currently research efforts are focused on investigating two types of materials for achieving large pseudo-capacitance: metal oxides and conductive polymers. Among all the metal oxide present in nature, Ruthenium oxide (RuO_2) has been largely investigated [26, 31, 47, 63], thanks to its intrinsic reversibility for various surface redox couples and high conductivity [70]. In particular, the research has focused to explore the chemical reactions of RuO_2 in acid electrolytes. The results have explained the pseudo capacitance of EPC as adsorption of protons at Ruthenium Oxide surface, combined with a quick and reversible electron transfer, as in 7 [59, 66]:

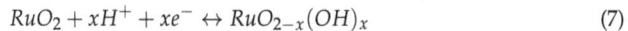

$$RuO_2 + xH^+ + xe^- \leftrightarrow RuO_{2-x}(OH)_x \tag{7}$$

where $0 < x < 2$. With this type of electrode, the specific capacitance over $700F/g$ [31]. However, due to rarity and high cost, the commercial applications of RuO_2 supercapacitors have been limited. For this reason many researcher started to investigate others oxides that may provide the same performances with a lower costs. Manganese oxide (MgO_x) has been an interesting candidate because of its low cost, nontoxic and large theoretical maximum capacitance about $1300F/g$.. However the poor electronic and ionic conductivity, low surface area and difficulty to achieve long term cycling stability are some of the issue that need to be addressed before to make Manganese oxide usable in practice [31, 66]. Other metal oxides including (but not limited) NiO, $Ni(OH)_2$, Co_2O_3, IrO_2, FeO, TiO_2, SnO_2, V_2, O_5 and MoO can also be applied [23].

Utilization of conducting polymers as the electrodes (polymer EPC) is also investigated. Since the polymer EPCs are based on a bulk process, a higher specific capacitance respect to carbon based capacitors can be achieved, and thus an expected larger energy density [36]. Furthermore conducting polymers are more conductive than the inorganic materials, thus, the reduced ESR and consequently the increased power capability respect to standard battery can be obtained [60]. Conductive polymers, as their name suggest are organic polymers that conduct electricity, are able to combine both advantages of metals (such as high conductivity) polymers (such as low cost, flexibility, low weight). They are made by doping conjugated polymers [21]. The synthesis of conjugated polymers can be done by chemical oxidation of the monomer or electrochemical oxidation of the monomer [21]. Polymer EPCs have three basic configurations depending on the type of the polymer used [60]:

- Type I: Both electrodes use the same p-doped polymer (symmetric configuration).

- Type II: Electrodes use two different p-doped polymers with a different range of electro-activity (asymmetric configuration).

- Type III: Both electrodes use the same polymer ,with the p-doped for the positive electrode and the n-doped for the negative electrode (symmetric configuration)

Among the three configurations, TYPE III is the most promising one for commercialization [36, 60]. Despite it is possible to synthesize many types of conductive polymers, three in particular are commonly used for polymer EPCs as listed below (the relative structures are shown in fig.8) [60, 66] :

- Polyaniline (PANI)
- Polypyrrole (PPy)
- Thiophene-based polymers (PTh)

PANI has shown a high specific capacitance (around 800 F/g [66]) and a good life cycle stability ($\approx -20\%$ in capacitance after 9000 cycles) [60]. Polymer EPCs based on PPy and PTh are also successfully fabricated [36, 60, 66] with a specific capacitances between 200 and 300 F/g.

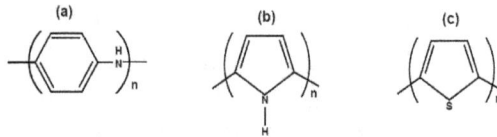

Figure 8. Polymers structures: polyaniline PANI(a), polypyrrole PPy(b), polythiophene PTh (c); adapted from [60]

In general, polymer EPCs suffer of poor cycle stability. The rapid performance degradation is due to the mechanical stress derived from the volumetric changes during the doping/dedoping process. (Swelling and shrinking)[16, 20, 27, 66]. For this reason further research is needed in order to improve performance of polymer EPCs.

3.3. Hybrid supercapacitors

Hybrids supercapacitors are a class of devices that attempt to combine the advantages of both EDLC and EPC, which are able to exploit at the same time both Faradic and non- Faradic processes for storing energy. The idea is using the propriety of EDCL for obtaining high power and the propriety of EPC for increasing the energy density [59]. According to Halper & Ellenbogen [20], the hybrid supercapacitors can be classified, based on the electrode material, in:

1. Composite
2. Asymmetric
3. Battery type

In composite supercapacitors, each electrode is formed by a combination of a carbon material as the frame and a metal oxide material or a conductive polymer deposited on top of the carbon material (e.g. [67, 68]). BBy doing this the carbon material create a large surface area for a large capacitance. Furthermore the polymer material further increases the capacitance with a result of high energy density and cycling stability comparable with EDCL [20]. For example, in the case of Carbon/PPy based electrode, the high cycling stability is due to the carbon frame below the polymer that mitigates the polymer stress for increasing life cycle. Very recently

Wang et al. [67] has reported one interesting example of this class of supercapacitors. The electrodes were made by graphene/RuO_2 and graphene/$Ni(OH)_2$. The device has shown an energy density $\approx 48Wh/kg$ at a power density of $\approx 0.23kW/kg$. Furthermore a high cycling stability has also been achieved. After 5000 cycles of charging and discharging at a current density of 10 A/g it has shown 8% reduction of the original capacitance. Another example of using carbon electrodes coated by conductive polymers was reported by An et al. [2]. They used polypyrrole-carbon aerogel for the capacitor electrodes.

For the Asymmetric supercapacitors, carbon material for one electrode and metal oxide or polymer material as second electrode are applied. One example of this technique is reported by Staiti & Lufrano [61], which uses manganese oxide and activated carbon.

The battery type supercapacitors are the most interesting candidates for the relative high energy density. Similar to the asymmetric, in this type of supercapacitors one electrode is made of carbon material, and the second one is made of a typical battery electrode material (such as Lithium). In particular one lithium- intercalated compounds (Nanostructured $Li_4Ti_5O_12$ also known as LTO)has been extensively studied [4, 10, 38, 41, 42, 46], which can enable the cycling stability during Li intercalation/ deintercalation processes [38]. When coupled with carbon electrode, the charge storage can be realized by the mechanisms of a Li-ion battery at the negative electrode and a supercapacitor at the positive electrode [10]. The energy density for battery like electrode is very high compared to capacitors. Naoi et al. [41] reported an energy density for as high as 55 Wh/Kg. However the power density for the battery type supercapacitor is in general lower than the other classes and, faradic principle leads to an increase in the energy density at the cost of life cycle. This is one major drawback of hybrid devices (when compared with EDLCs), and *"'it is important to avoid transforming a good supercapacitor into a mediocre battery"'* [59]

4. Conclusions

Tracking a clear path for the future energy storage systems is not easy. Fuel cells in the near future can became mature enough to be used as primary energy source for large vehicles. However, despite small fuel cells (in particular the direct methanol fuel cells) can also be used for supply laptop or other small electronic device, it is believed that battery will continue to lead this market for several years. Microturbines and internal combustion engines are also very promising technology but further research is needed for improving current prototypes. Supercapacitors have a great potential to play an important role in future energy storage systems, in particular to all the application that require high peek powers. Nowadays then can be used as secondary energy storage system, for example, in vehicle applications they can be used in parallel with fuel cells or batteries for overcoming the power peaks during acceleration [4]. Supercapacitors will be able to be full recharged in a very little time and this provides the key advantage that the customers and the market are waiting for.

Author details

Luca Petricca, Per Ohlckers and Xuyuan Chen
Vestfold University College, Norway

5. References

[1] Amit Mehra, I. A. W. & Schmidt, M. A. [1999]. Combustion tests in the 6-wafer static structure of a micro gas turbine engine, *TRANSDUCERS '99*.

[2] An, H., Wang, Y., Wang, X., Zheng, L., Wang, X., Yi, L., Bai, L. & Zhang, X. [2010]. Polypyrrole/carbon aerogel composite materials for supercapacitor, *Journal of Power Sources* 195(19): 6964–6969.
URL: *http://www.sciencedirect.com/science/article/pii/S0378775310007093*

[3] Andrew, B. [2000]. Ultracapacitors: why, how, and where is the technology, *Journal of Power Sources* 91(1): 37–50.
URL: *http://www.sciencedirect.com/science/article/pii/S0378775300004857*

[4] Balducci, A., Bardi, U., Caporali, S., Mastragostino, M. & Soavi, F. [2004]. Ionic liquids for hybrid supercapacitors, *Electrochemistry Communications* 6(6): 566–570.
URL: *http://www.sciencedirect.com/science/article/pii/S1388248104000803*

[5] Cheng, X., Shi, Z., Glass, N., Zhang, L., Zhang, J., Song, D., Liu, Z.-S., Wang, H. & Shen, J. [2007]. A review of pem hydrogen fuel cell contamination: Impacts, mechanisms, and mitigation, *Journal of Power Sources* 165(2): 739–756.
URL: *http://www.sciencedirect.com/science/article/pii/S0378775306025304*

[6] D. S. Nguyen, E. Halvorsen, G. U. J. & Vogl, A. [2010]. Fabrication and characterization of a wideband mems energy harvester utilizing nonlinear springs, *J. Micromech. Microeng* 20: 125009.

[7] D.C. Walther, A. P. [2003]. Mems rotary engine power system: Project overview and recent research results, *SEM Annual Conference & Exposition on Experimental and Applied Mechanics*.

[8] Ding Sheng Yuan, Jianghua Zeng, J. C. Y. L. [2009]. Highly ordered mesoporous carbon synthesized via in situ template for supercapacitors, *Int. J. Electrochem. Sci.* 4: 562–570.

[9] Du, C. & Pan, N. [2006]. High power density supercapacitor electrodes of carbon nanotube films by electrophoretic deposition, *Nanotechnology* 17(21): 5314–.
URL: *http://stacks.iop.org/0957-4484/17/i=21/a=005*

[10] Du Pasquier, A., Plitz, I., Menocal, S. & Amatucci, G. [2003]. A comparative study of li-ion battery, supercapacitor and nonaqueous asymmetric hybrid devices for automotive applications, *Journal of Power Sources* 115(1): 171–178.
URL: *http://www.sciencedirect.com/science/article/pii/S0378775302007188*

[11] Epstein, A. H. [2004]. Millimeter-scale, micro-electro-mechanical systems gas turbine engines, *J. Eng. Gas Turbines Power* 126(2): 205–226.
URL: *http://dx.doi.org/10.1115/1.1739245*

[12] Everett, M. A. [2008]. Ultracapacitors: A mainstream energy storage and power delivery solution, KiloFarad International.

[13] Fernandez-Pello, A. C. [2002]. Micropower generation using combustion: Issues and approaches, *Proceedings of the Combustion Institute* 29(1): 883–899.
URL: *http://www.sciencedirect.com/science/article/pii/S1540748902801134*

[14] Folkesson, A., Andersson, C., Alvfors, M. & Overgaard, L. [2003]. Real life testing of a hybrid pem fuel cell bus, *Journal of Power Sources* 118(1–2): 349–357.
URL: *http://www.sciencedirect.com/science/article/pii/S0378775303000867*

[15] Frackowiak, E. & Beguin, F. [2001]. Carbon materials for the electrochemical storage of energy in capacitors, *Carbon* 39(6): 937–950.
URL: *http://www.sciencedirect.com/science/article/pii/S0008622300001834*

[16] Frackowiak, E., Khomenko, V., Jurewicz, K., Lota, K. & Béguin, F. [2006]. Supercapacitors based on conducting polymers/nanotubes composites, *Journal of Power Sources* 153(2): 413–418.
   URL: *http://www.sciencedirect.com/science/article/pii/S0378775305007391*

[17] FuelCellStore [2012].
   URL: *http://www.fuelcellstore.com/en/pc/viewPrd.asp?idproduct=182&idcategory=40*

[18] Gu, J. F., Gorgutsa, S. & Skorobogatiy, M. [2010]. Soft capacitor fibers using conductive polymers for electronic textiles, *Smart Materials and Structures* 19(11): 115006–.
   URL: *http://stacks.iop.org/0964-1726/19/i=11/a=115006*

[19] Gurau, B. & Smotkin, E. S. [2002]. Methanol crossover in direct methanol fuel cells: a link between power and energy density, *Journal of Power Sources* 112: 339–352.

[20] Halper, M. S. & Ellenbogen, J. C. [2006]. Supercapacitors : A brief overview, *Group* 1(March): 1–41.

[21] Harun, M.H., S. E. K. A. Y. N. . M. E. [2007]. Conjugated conducting polymers: A brief overview, *UCSI Academic Journal: Journal for the Advancement of Science & Arts* 2: 63–68.

[22] Jacobson, S., Das, S., Savoulides, N., Steyn, J., Lang, J., Li, H., Livermore, C., Schmidt, M., Teo, C., Umans, S., Epstein, A., Arnold, D. P., Park, J., Zana, I. & Allen, M. [2004]. Progress toward a microfabricated gas turbine generator for soldier portable power applications, *Proc. 24th Army Science Conf.*, Orlando, FL.

[23] Jayalakshmi, M. & Balasubramanian, K. [2008]. Simple capacitors to supercapacitors - an overview, *Int. J. Electrochem. Sci* 3: 1196 – 1217.

[24] J.M. Boyea, R.E. Camacho, S. P. T. & Ready, W. J. [2007]. Carbon nanotube-based supercapacitors: Technologies and markets, *Nanotechnology Law and Business* 4: 585–593.

[25] Jost, K., Perez, C. R., McDonough, J. K., Presser, V., Heon, M., Dion, G. & Gogotsi, Y. [2011]. Carbon coated textiles for flexible energy storage, *Energy Environ. Sci.* 4(12): 5060–5067.
   URL: *http://dx.doi.org/10.1039/C1EE02421C*

[26] Kim, I.-H. & Kim, K.-B. [2001]. Ruthenium oxide thin film electrodes for supercapacitors, *Electrochem. Solid-State Lett.* 4(5): A62–A64.
   URL: *http://dx.doi.org/10.1149/1.1359956*

[27] Kotz, R. & Carlen, M. [2000]. Principles and applications of electrochemical capacitors, *Electrochimica Acta* 45(15–16): 2483–2498.
   URL: *http://www.sciencedirect.com/science/article/pii/S0013468600003546*

[28] Kuwata, N., Kawamura, J., Toribami, K., Hattori, T. & Sata, N. [2004]. Thin-film lithium-ion battery with amorphous solid electrolyte fabricated by pulsed laser deposition, *Electrochemistry Communications* 6(4): 417–421.
   URL: *http://www.sciencedirect.com/science/article/pii/S1388248104000347*

[29] Lee, C., Jiang, K., Jin, P. & Prewett, P. [2004]. Design and fabrication of a micro wankel engine using mems technology, *Microelectronic Engineering* 73–74(0): 529–534.
   URL: *http://www.sciencedirect.com/science/article/pii/S0167931704002060*

[30] Leitner, K., Lerf, A., Winter, M., Besenhard, J., Villar-Rodil, S., Suarez-Garcia, F., Martinez-Alonso, A. & Tascan, J. [2006]. Nomex-derived activated carbon fibers as electrode materials in carbon based supercapacitors, *Journal of Power Sources* 153(2): 419–423.
   URL: *http://www.sciencedirect.com/science/article/pii/S0378775305007238*

[31] Lokhande, C., Dubal, D. & Joo, O.-S. [2011]. Metal oxide thin film based supercapacitors, *Current Applied Physics* 11(3): 255–270.
    URL: *http://www.sciencedirect.com/science/article/pii/S1567173910004773*

[32] Lu, W. & Dai, L. [2010-03-01 N1]. Carbon nanotube supercapacitors, *Carbon Nanotube*, InTech, pp. –.

[33] Luc G. Frechette, Changgu Lee, S. A. & Liu, Y.-C. [2003]. Preliminary design of a mems steam turbine power plant-on-a-chip, *3rd Int'l Workshop on Micro & Nano Tech. for Power Generation & Energy Conv. (PowerMEMS'03)*.

[34] Lucintel [2011]. Growth opportunities in global battery market 2011-2016: Market size, market share and forecast analysis. Lucintel's report.

[35] Mandal, S., Bhojwani, P., Mohanty, S. & Mahapatra, R. [2008]. Intellbatt: Towards smarter battery design, *Design Automation Conference, 2008. DAC 2008. 45th ACM/IEEE*, pp. 872 –877.

[36] Mastragostino, M., Arbizzani, C. & Soavi, F. [2002]. Conducting polymers as electrode materials in supercapacitors, *Solid State Ionics* 148(3–4): 493–498.
    URL: *http://www.sciencedirect.com/science/article/pii/S0167273802000930*

[37] Mench, M. M. [2008]. *Fuel cell engines*, John Wiley and Sons.

[38] Mladenov, M., Alexandrova, K., Petrov, N., Tsyntsarski, B., Kovacheva, D., Saliyski, N. & Raicheff, R. [n.d.]. Synthesis and electrochemical properties of activated carbons and li<sub>4</sub>ti<sub>5</sub>o<sub>12</sub> as electrode materials for supercapacitors.
    URL: *http://dx.doi.org/10.1007/s10008-011-1424-6*

[39] Muller, D. C. [2007]. Energy storage: Strong momentum in high-end batteries. Credit Suisse, Suisse.

[40] Mysyk, R., Raymundo-Pinero, E. & Baguin, F. [2009]. Saturation of subnanometer pores in an electric double-layer capacitor, *Electrochemistry Communications* 11(3): 554–556.
    URL: *http://www.sciencedirect.com/science/article/pii/S1388248108006310*

[41] Naoi, K., Ishimoto, S., Isobe, Y. & Aoyagi, S. [2010]. High-rate nano-crystalline li4ti5o12 attached on carbon nano-fibers for hybrid supercapacitors, *Journal of Power Sources* 195(18): 6250–6254.
    URL: *http://www.sciencedirect.com/science/article/pii/S0378775310000327*

[42] Ni, J., Yang, L., Wang, H. & Gao, L. [n.d.]. A high-performance hybrid supercapacitor with li<sub>4</sub>ti<sub>5</sub>o<sub>12</sub>-c nano-composite prepared by in situ and ex situ carbon modification.
    URL: *http://dx.doi.org/10.1007/s10008-012-1704-9*

[43] Odile BERTOLDI, S. B. [2009]. Report on energy, *Technical report*, European Commission-Observatory nano.

[44] Ogawa, M., Kanda, R., Yoshida, K., Uemura, T. & Harada, K. [2012]. High-capacity thin film lithium batteries with sulfide solid electrolytes, *Journal of Power Sources* 205(0): 487–490.
    URL: *http://www.sciencedirect.com/science/article/pii/S0378775312001851*

[45] Park, D.-E., Lee, D.-H., Yoon, J.-B., Kwon, S. & Yoon, E. [2002]. Design and fabrication of micromachined internal combustion engine as a power source for microsystems, *Micro Electro Mechanical Systems, 2002. The Fifteenth IEEE International Conference on*, pp. 272 –275.

[46] Pasquier, A. D., Plitz, I., Gural, J., Menocal, S. & Amatucci, G. [2003]. Characteristics and performance of 500 f asymmetric hybrid advanced supercapacitor prototypes, *Journal of*

Power Sources 113(1): 62–71.
URL: *http://www.sciencedirect.com/science/article/pii/S0378775302004913*

[47] Patake, V., Lokhande, C. & Joo, O. S. [2009]. Electrodeposited ruthenium oxide thin films for supercapacitor: Effect of surface treatments, *Applied Surface Science* 255(7): 4192–4196.
URL: *http://www.sciencedirect.com/science/article/pii/S0169433208023234*

[48] Peirs, J., Reynaerts, D. & Verplaetsen, F. [2004]. A microturbine for electric power generation, *Sensors and Actuators A: Physical* 113(1): 86–93.
URL: *http://www.sciencedirect.com/science/article/pii/S0924424704000081*

[49] Pelekies, S. O., Schuhmann, T., Gardner, W. G., Camacho, A. & Protz, J. M. [2010]. Fabrication of a mechanically aligned single-wafer mems turbine with turbocharger, *Proc. SPIE*, Vol. 7833, SPIE, Toulouse, France, pp. 783306–9.
URL: *http://dx.doi.org/10.1117/12.862729*

[50] Petricca, L., Ohlckers, P. & Grinde, C. [2011]. Micro- and nano-air vehicles: State of the art.
URL: *http://dx.doi.org/10.1155/2011/214549*

[51] Phil Ratliff, P. G. & Fischer, W. [2007]. *The New Siemens Gas Turbine SGT5-8000H for More Customer Benefit*, Siemens. URL: *http://www.energy.siemens.com/us/pool/hq/power-generation/gas-turbines/downloads/SGT5-8000H_benefits.pdf*

[52] Pint, C. L., Nicholas, N. W., Xu, S., Sun, Z., Tour, J. M., Schmidt, H. K., Gordon, R. G. & Hauge, R. H. [2011]. Three dimensional solid-state supercapacitors from aligned single-walled carbon nanotube array templates, *Carbon* 49(14): 4890–4897.
URL: *http://www.sciencedirect.com/science/article/pii/S0008622311005549*

[53] PowerStream, Lund Instrument Engineering, I. C. [2012].
URL: *http://www.powerstream.com/methanol-fuel-cell.htm*

[54] Probstle, H., Schmitt, C. & Fricke, J. [2002]. Button cell supercapacitors with monolithic carbon aerogels, *Journal of Power Sources* 105(2): 189–194.
URL: *http://www.sciencedirect.com/science/article/pii/S0378775301009387*

[55] Reynolds, C. [2009]. Evolution of supercapacitors, AVX Application Engineer.

[56] Sepeur, S. [2008]. *Nanotechnology: technical basics and applications*, Vincentz Network GmbH.

[57] Shah, K. & Besser, R. [2007]. Key issues in the microchemical systems-based methanol fuel processor: Energy density, thermal integration, and heat loss mechanisms, *Journal of Power Sources* 166: 177–193.

[58] Sher, I., Levinzon-Sher, D. & Sher, E. [2009]. Miniaturization limitations of hcci internal combustion engines, *Applied Thermal Engineering* 29(2–3): 400–411.
URL: *http://www.sciencedirect.com/science/article/pii/S1359431108001427*

[59] Simon, P. & Gogotsi, Y. [2008]. Materials for electrochemical capacitors, *Nat Mater* 7(11): 845–854.
URL: *http://dx.doi.org/10.1038/nmat2297*

[60] Snook, G. A., Kao, P. & Best, A. S. [2011]. Conducting-polymer-based supercapacitor devices and electrodes, *Journal of Power Sources* 196(1): 1–12.
URL: *http://www.sciencedirect.com/science/article/pii/S0378775310010712*

[61] Staiti, P. & Lufrano, F. [2010]. Investigation of polymer electrolyte hybrid supercapacitor based on manganese oxide-carbon electrodes, *Electrochimica Acta* 55(25): 7436–7442.
URL: *http://www.sciencedirect.com/science/article/pii/S0013468610000824*

[62] Suzuki, Y., Okada, Y., Ogawa, J., Sugiyama, S. & Toriyama, T. [2008]. Experimental study on mechanical power generation from mems internal combustion engine, *Sensors and*

Actuators A: Physical 141(2): 654–661.
URL: *http://www.sciencedirect.com/science/article/pii/S0924424707006334*

[63] Timothy J. Boyle, Louis J. Tribby, T. N. L. S. M. H. C. D. E. L. P. F. F. [2008]. Metal oxide coating of carbon supports for supercapacitor applications, *Technical Report SAND-2008-3813*, Sandia Report.

[64] VIASPACEGreenEnergyInc. [2012].
URL: *http://www.viaspace.com/ae_cartridges.php*

[65] Walther, D. C. & Ahn, J. [2011]. Advances and challenges in the development of power-generation systems at small scales, *Progress in Energy and Combustion Science* 37(5): 583–610.
URL: *http://www.sciencedirect.com/science/article/pii/S0360128510000754*

[66] Wang, G., Zhang, L. & Zhang, J. [2012]. A review of electrode materials for electrochemical supercapacitors, *Chem. Soc. Rev.* 41(2): –.
URL: *http://dx.doi.org/10.1039/C1CS15060J*

[67] Wang, H., Liang, Y., Mirfakhrai, T., Chen, Z., Casalongue, H. & Dai, H. [2011-08-01]. Advanced asymmetrical supercapacitors based on graphene hybrid materials.
URL: *http://dx.doi.org/10.1007/s12274-011-0129-6*

[68] Xiao, Q. & Zhou, X. [2003]. The study of multiwalled carbon nanotube deposited with conducting polymer for supercapacitor, *Electrochimica Acta* 48(5): 575–580.
URL: *http://www.sciencedirect.com/science/article/pii/S0013468602007272*

[69] Xing, W., Qiao, S., Ding, R., Li, F., Lu, G., Yan, Z. & Cheng, H. [2006]. Superior electric double layer capacitors using ordered mesoporous carbons, *Carbon* 44(2): 216–224.
URL: *http://www.sciencedirect.com/science/article/pii/S0008622305004653*

[70] Zhang, L. L. & Zhao, X. S. [2009]. Carbon-based materials as supercapacitor electrodes, *Chem. Soc. Rev.* 38(9): –.
URL: *http://dx.doi.org/10.1039/B813846J*

Chapter 4

Electrochemical Energy Storage

Petr Krivik and Petr Baca

Additional information is available at the end of the chapter

1. Introduction

Electrochemical energy storage covers all types of secondary batteries. Batteries convert the chemical energy contained in its active materials into electric energy by an electrochemical oxidation-reduction reverse reaction.

At present batteries are produced in many sizes for wide spectrum of applications. Supplied powers move from W to the hundreds of kW (compare battery for power supply of pace makers and battery for heavy motor vehicle or for power station).

Common commercially accessible secondary batteries according to used electrochemical system can be divided to the following basic groups:

Standard batteries (lead acid, Ni-Cd) modern batteries (Ni-MH, Li–ion, Li-pol), special batteries (Ag-Zn, Ni-H2), flow batteries (Br2-Zn, vanadium redox) and high temperature batteries (Na-S, Na–metalchloride).

2. Standard batteries

2.1. Lead acid battery

Lead acid battery when compared to another electrochemical source has many advantages. It is low price and availability of lead, good reliability, high voltage of cell (2 V), high electrochemical effectivity, cycle life is from several hundreds to thousands of cycles. Thanks to these characteristics is now the most widely used secondary electrochemical source of electric energy and represent about 60% of installed power from all types of secondary batteries. Its disadvantage is especially weight of lead and consequently lower specific energy in the range 30-50 Wh/kg.

Lead-acid batteries are suitable for medium and large energy storage applications because they offer a good combination of power parameters and a low price.

2.1.1. Battery composition and construction

Construction of lead acid (LA) battery depends on usage. It is usually composed of some series connected cells. Main parts of lead acid battery are electrodes, separators, electrolyte, vessel with lid, ventilation and some other elements.

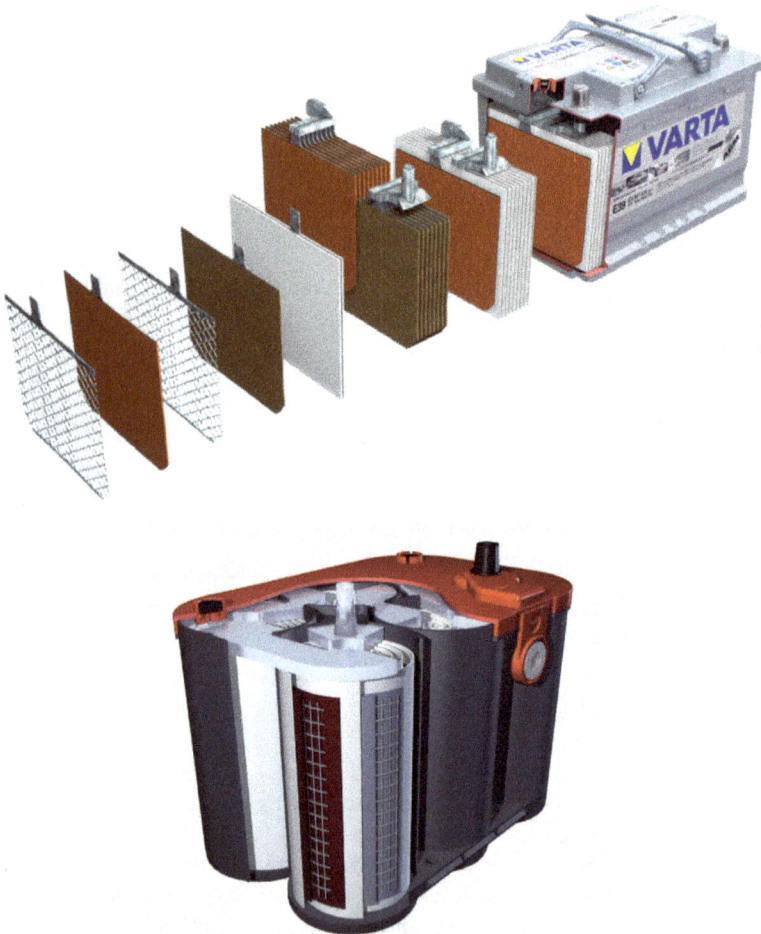

Figure 1. Scheme of prismatic and spiral wound construction of LA battery

Electrode consists of grid and of active mass. Grid as bearing structure of electrode must be mechanically proof and positive electrode grid must be corrosion proof. Corrosion converts lead alloy to lead oxides with lower mechanical strength and electric conductivity. Grids are made from lead alloys (pure lead would be too soft); it is used Pb-Ca or Pb-Sb alloys, with mixture of additives as Sn, Cd and Se, that improve corrosion resistance and make higher mechanical strength.

Active material is made from lead oxide PbO pasted onto a grid and then electrochemically converted into reddish brown lead dioxide PbO2 on positive electrode and on grey spongy lead Pb on negative electrode.

Separators electrically separate positive electrode from negative. They have four functions:

1.   to provide electrical insulation between positive and negative plate and to prevent short circuits,
2.   to act as a mechanical spacer which holds the plates in the prescribed position,
3.   to help retain the active materials in close contact with the grid,
4.   to permit both the free diffusion of electrolyte and the migration of ions.

The materials used for separators can be wood veneers, cellulose (paper), usually stiffened with a phenol-formaldehyde resin binder, and those made from synthetic materials, e.g., rubber, polyvinyl chloride (PVC), polyethylene (PE), and glass-microfibre.

Electrolyte is aqueous solution of H2SO4 with density of 1.22-1.28 g/cm^3. Mostly it is liquid, covered battery plates. Sometimes it is transformed to the form of gel, or completely absorbed in separators.

Vessel must to withstand straining caused by weight of inner parts of battery and inner pressure from gas rising during cycling. The most used material is polypropylene, but also, PVC, rubber etc. If overpressure rises inside classical battery during charging, problem is solved by valve placed mostly in lid.

There are some major types of battery construction: prismatic construction with grid or tubular plates, cylindrical construction (spiral wound or disc plates) or bipolar construction.

2.1.2. Principle of operation

Overall chemical reaction during discharge is:

$$PbO_2 + Pb + 2H_2SO_4 \rightarrow 2PbSO_4 + 2H_2O \qquad E^0 = +2.048 \text{ V} \qquad (1)$$

Reaction proceeds in opposite direction during charge.

2.1.3. Types of LA batteries

According to the usage and construction, lead acid batteries split into stationary, traction and automotive batteries.

Stationary battery ensures uninterrupted electric power supply in case of failure in distributing network. During its service life battery undergo only few cycles. Battery life is as many as 20 years.

Traction battery is used for power supply of industrial trucks, delivery vehicles, electromobiles, etc. It works in cyclic regime of deep charge–discharge. Cycle life of the battery is about 5 years (1000 of charge–discharge cycles).

Automotive battery is used for cranking automobile internal combustion engines and also for supporting devices which require electrical energy when the engine is not running. It must be able of supplying short but intense discharge current. It is charged during running of engine.

According to the maintenance operation lead acid batteries could be branched into conventional batteries (i.e., those with free electrolyte, so-called 'flooded' designs), requiring regular maintenance and valve-regulated lead-acid (VRLA) maintenance free batteries.

2.1.4. VRLA batteries

Originally, the battery worked with its plates immersed in a liquid electrolyte and the hydrogen and the oxygen produced during overcharge were released into the atmosphere. The lost gases reflect a loss of water from the electrolyte and it had to be filled in during maintenance operation. Problems with water replenishing were overcome by invention of VRLA (valve regulated lead acid) batteries.

The VRLA battery is designed to operate with help of an internal oxygen cycle, see Fig. 2. Oxygen liberated during the latter stages of charging, and during overcharging, on the positive electrode, i.e.

$$H_2O \rightarrow 2H^+ + 1/2O_2 + 2e^- , \tag{1a}$$

travels through a gas space in separator to the negative electrode where is reduced to the water:

$$Pb + 1/2O_2 + H_2SO_4 \rightarrow PbSO_4 + H_2O + Heat \tag{1b}$$

The oxygen cycle, defined by reactions (1a) and (1b), moves the potential of the negative electrode to a less negative value and, consequently, the rate of hydrogen evolution decreases. The small amount of hydrogen that could be produced during charging is released by pressure valve. The produced lead sulphate is immediately reduced to lead via the reaction (1c), because the plate is simultaneously on charge reaction:

$$PbSO_4 + 2H^+ + 2e^- \rightarrow Pb + H_2SO_4 \tag{1c}$$

The sum of reactions (1a), (1b) and (1c) is zero. Part of the electrical energy delivered to the cell is consumed by the internal oxygen recombination cycle and it is converted into heat.

There are two designs of VRLA cells which provide the internal oxygen cycle. One has the electrolyte immobilized as a gel (gel batteries), the other has the electrolyte held in an AGM separator (AGM batteries). Gas can pass through crack in the gel, or through channels in the AGM separator.

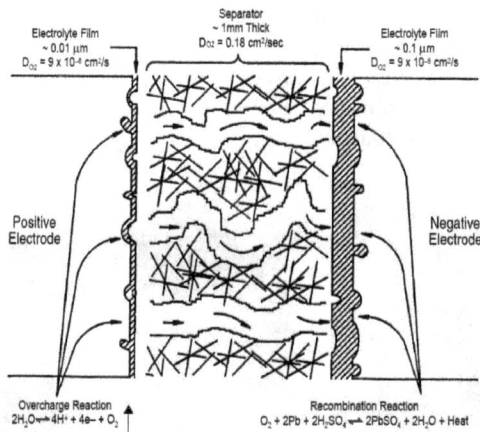

Figure 2. Internal oxygen cycle in a valve regulated lead acid cell (Nelson, 2001).

When the cell is filled with electrolyte, the oxygen cycle is impossible because oxygen diffuses through the electrolyte very slow. On the end of charge, first oxygen (from the positive), and then both oxygen (from the positive) and hydrogen (from the negative), are liberated and they are released through the pressure valve. Gassing causes loss of water and opens gas spaces due to drying out of the gel electrolyte or a liquid electrolyte volume decrease in the AGM separators). It allows the transfer of oxygen from the positive to the negative electrode. Gas release from the cell then falls rapidly (Rand et al., 2004).

2.1.5. Failure mechanisms of LA batteries

Lead acid batteries can be affected by one or more of the following failure mechanisms:

1. positive plate expansion and positive active mass fractioning,
2. water loss brought about by gassing or by a high temperature,
3. acid stratification,
4. incomplete charging causing active mass sulphation,
5. positive grid corrosion,
6. negative active mass sulphation (batteries in partial state of charge (PSoC) cycling - batteries in hybrid electric vehicles (HEV) and batteries for remote area power supply (RAPS) applications).

Repetitive discharge and charge of the LA battery causes expansion of the positive active mass because product of the discharge reaction PbSO4 occupies a greater volume than the positive active material PbO2. Charging of the cell restores most of the lead dioxide, but not within the original volume. The negative active mass does not show the same tendency to expand. Reason could be that lead is softer than lead dioxide and that is why the negative active material is more compressed during discharge as the conversion from lead to the more voluminous lead sulphate proceeds. Another reason could be that spongy lead contains bigger pores than pores in lead dioxide and therefore is more easily able to absorb a lead sulphate without expansion of a negative active mass. Progressive expansion of the positive electrode causes an increasing fraction of the positive active material. This material becomes to be electrically disconnected from the current collection process and it causes decreasing of the cell capacity. (Calabek et al., 2001).

Figure 3. Positive active mass fractioning

Gas evolving during overcharge leads to reduction of the volume of the electrolyte. Some of the active material consequently loses contact with the electrodes. Drying out increases the internal resistance of the battery which causes excessive rise of temperature during charging and this process accelerates water loss through evaporation.

During charge, sulphuric acid is produced between the electrodes and there is a tendency for acid of higher concentration, which has a greater relative density, to fall to the bottom of the lead acid cell. Acid stratification can be caused also by preferential discharge of upper parts of the cell, because of lower ohmic resistance of these parts. Concentration of electrolyte in the upper part of the cell is temporarily lower than on the bottom of the cell. It leads to discharge of the bottom parts and charge of the upper parts of the cell. The vertical concentration gradient of sulphuric acid can give rise to non uniform utilization of active mass and, consequently, shortened service life through the irreversible formation of $PbSO_4$ (Ruetschi, 2004).

When the electrodes are repeatedly not fully charged, either because of a wrong charging procedure or as a result of physical changes that keep the electrode from reaching an adequate potential (antimony poisoning of negative electrode), then a rapid decreasing in battery capacity may occur because of progressive accumulation of lead sulphate in active mass. Sulphation is creation of insulation layer of lead sulphate on the electrode surface. It leads to inhibition of electrolyte contact with active mass. Sulphation grows during the long term standing of the battery in discharge state, in case of electrolyte stratification, or incomplete charging. In the course of sulphation originally small crystals of lead sulphate grow to big ones. Big crystals of lead sulphate increase internal resistance of the cell and during charging it is hardly possible to convert them back to the active mass.

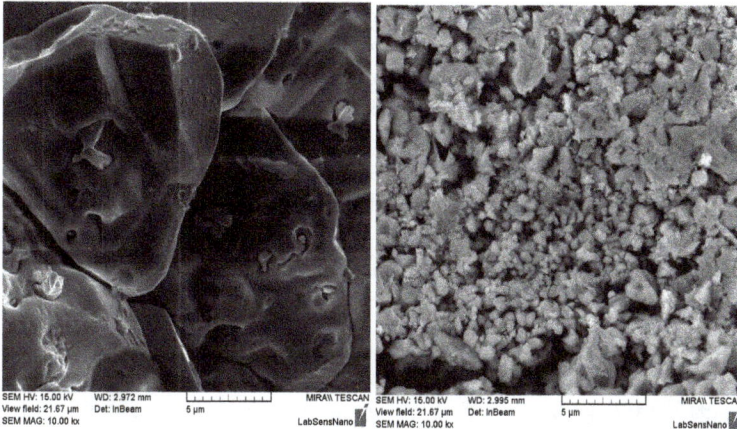

Figure 4. SEM images of negative active mass. Sulphation on the left, healthy state on the right

During charge the positive grid is subject to corrosion. Lead collector turns on lead dioxide or lead sulphate. The rate of this process depends on the grid composition and microstructure, also on plate potential, electrolyte composition and temperature of the cell. The corrosion products have usually a bigger electric resistance than positive grid. In extreme cases, corrosion could result to disintegration of the positive grid and consequently to collapse of the positive electrode.

Figure 5. Positive grid corrosion

2.2. Ni-Cd battery

They are main representative of batteries with positive Nickel electrode; other possible systems could be system Ni-Fe and Ni–Zn, Ni-H2 or Ni–MH.

2.2.1. Battery composition and construction

The nickel cadmium cell has positive electrode from nickel hydroxide and negative electrode from metallic cadmium, an electrolyte is potassium hydroxide. The nickel cadmium battery is produced in a wide range of commercially important battery systems from sealed maintenance free cells (capacities of 10 mAh - 20 Ah) to vented standby power units (capacities of 1000 Ah and more). Nickel cadmium battery has long cycle life, overcharge capability, high rates of discharge and charge, almost constant discharge voltage and possibility of operation at low temperature. But, the cost of cadmium is several times that of lead and the cost of nickel cadmium cell construction is more expensive than that of lead acid cell. And there is also problem with the manipulation of toxic cadmium. But also low maintenance and good reliability have made it an ideal for a number of applications such (emergency lighting, engine starting, portable television receivers, hedge trimmers, electric shavers, aircraft and space satellite power systems).

Depending on construction, nickel cadmium cells have energy densities in the range 40-60 Wh/kg (50-150 Wh/dm^3). Cycle life is moving from several hundreds for sealed cells to several thousands for vented cells.

Cell construction is branched to two types. First using pocket plate electrodes (in vented cells). The active material is found in pockets of finely perforated nickel plated sheet steel. Positive and negative plates are then separated by plastic pins or ladders and plate edge insulators. Second using sintered, bonded or fibre plate electrodes (in both vented and sealed cells). In sintered plate electrodes, a porous sintered nickel electrode is sintered in belt furnace in reducing atmosphere at 800 to 1000°C. Active material is distributed within the pores. In sintered plate cells, a special woven or felted nylon separator is used. It permits oxygen diffusion (oxygen cycle) in sealed cells. In the most common version, a spiral or prismatic construction of cells is used.

The electrolyte is an aqueous solution KOH (concentration of 20-28% by weight and a density of 1.18-1.27 g/cm^3 at 25°C). 1-2% of LiOH is usually added to electrolyte to minimize coagulation of the NiOOH electrode during charge/discharge cycling. For low temperature applications, the more concentrated KOH solution is used. When it is operating at high temperature it is sometimes used aqueous NaOH electrolyte.

2.2.2. Principle of operation

The overall cell reaction during discharge:

$$2NiOOH + Cd + 2H_2O \rightarrow 2Ni(OH)_2 + Cd(OH)_2 \qquad E^0 = +1.30 \text{ V} \qquad (2)$$

It is notable that amount of water in the electrolyte falls during discharge. Ni-Cd batteries are designed as positive limited utilizing oxygen cycle. The oxygen evolved at the positive electrode during charge difuses to the negative electrode and reacts with cadmium to form Cd(OH)2.

In addition, carbon dioxide in the air can react with KOH in the electrolyte to form K2CO3, and CdCO3 can be formed on the negative plates. Both of these compounds increase the internal resistance and lower the capacity of the Ni-Cd batteries.

Ni-Cd batteries suffer from the memory effect (see also chapter Ni-MH battery). Besides Ni-Cd batteries also suffer from high rate of self-discharge at high temperatures.

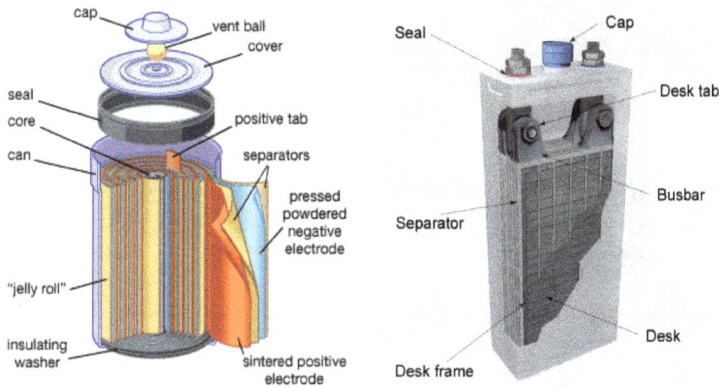

Figure 6. Scheme of spiral wound and prismatic construction of Ni-Cd battery

3. Modern batteries

3.1. Ni-MH battery

3.1.1. Battery composition and construction

The sealed nickel metal hydride cell has with hydrogen absorbed in a metal alloy as the active negative material. When compare with Ni-Cd cell it is not only increases the energy density, but also it is a more environmentally friendly power source. The nickel metal hydride cell, however, has high selfdischarge and is less tolerant to overcharge than the Ni-Cd cell.

Positive electrode is NiOOH, negative electrode contains hydrogen absorption alloys. They can absorb over a thousand times their own volume of hydrogen: Alloys usually consist of two metals. First absorbs hydrogen exothermically, a second endothermically. They serve as a catalyst for the dissociative adsorption of atomic hydrogen into the alloy lattice. Examples of used metals: Pd, V, Ti, Zr, Ni, Cr, Co, Sn, Fe, lanthanides and others. The AB_2 series ($ZrNi_2$) and the AB_5 series ($LaNi_5$) are usually used.

Design of the cylindrical and prismatic sealed Ni-MH cells are similar as with a nickel cadmium cells (see Fig. 7). Hydrophilic polypropylene separator is used in Ni-MH cell.

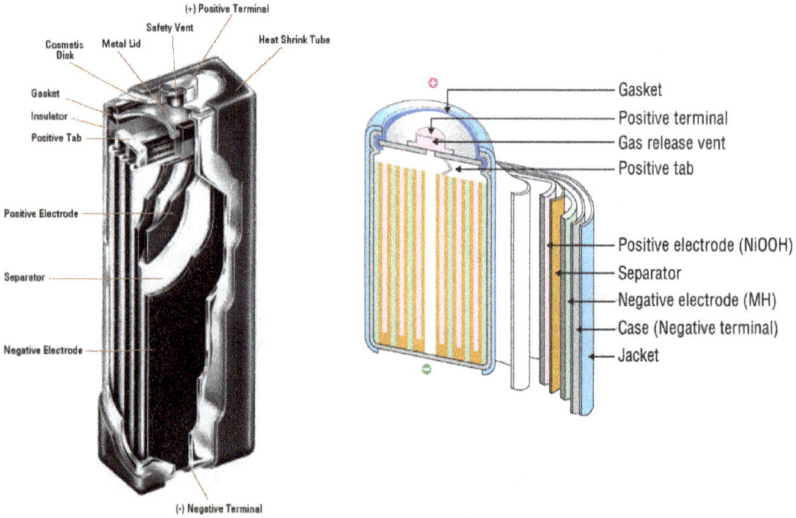

Figure 7. Scheme of prismatic and spiral wound Ni-MH battery

3.1.2. Principle of operation

The overall reaction during discharge:

$$NiOOH + MH \rightarrow Ni(OH)_2 + M \tag{3}$$

The electrolyte is concentrated potassium hydroxide, voltage is in the range 1.32-1.35 V, depending on used alloy. Water is not involved in the cell reaction.

The energy density is 25% higher than a Ni-Cd cell (80 Wh/kg), power density around 200 W/kg, cycle life over 1000 cycles. Self-discharge is high - up to 4-5% per day. It is caused especially by the hydrogen dissolved in the electrolyte that reacts with the positive electrode.

Ni-MH batteries are used in hybrid electric vehicle batteries, electric razors, toothbrushes, cameras, camcorders, mobile phones, pagers, medical instruments, and numerous other high rate long cycle life applications.

3.1.3. Memory effect

Ni-MH batteries also suffer from the memory effect. It is a reversible process which results in the temporary reduction of the capacity of a Ni-Cd and Ni-MH cell. It is caused by shallow charge-discharge cycling.

After shallow cycling there is a voltage step during discharge, i.e. as if the cell remembers the depth of the shallow cycling. The size of the voltage reduction depends on the number of preceding shallow cycles and the value of the discharge current. But the capacity of the cell is not affected if the cell is now fully discharged (to 0.9 V) and then recharged. Deep discharge then shows a normal discharge curve. It seems that some morphological change occurs in the undischarged active material during the shallow cycling. It could cause a reduction of the cell voltage during folowing discharge. The effect is probably based on an increase in the resistance of the undischarged material (γ-NiOOH formation on overcharge during the shallow cycles) (Vincent & Scrosati, 2003).

Progressive irreversible capacity loss can be confused with the reversible memory effect. The former is caused different mechanisms. For example by a reduction in the electrolyte volume due to evaporation at high temperatures or prolonged overcharge. Irreversible capacity loss can also be caused by internal short circuits.

3.2. Li-ion battery

Lithium is attractive as a battery negative electrode material because it is light weight, high reduction potential and low resistance. Development of high energy density lithium-ion battery started in the 1970s. The lithium-ion cell contains no metallic lithium and is therefore much safer on recharge than the earlier, primary lithium-metal design of cell.

3.2.1. Battery composition and construction

The principle of the lithium-ion cell is illustrated schematically in Fig. 8. The lithium ions travel between one electrode and the other during charge and discharge.

The most of commercial lithium-ion cells have positive electrodes of cobalt oxide. Other possible positive electrodes are except $LiCoO_2$ and $LiNiO_2$ based on especially manganese oxide, namely, $LiMnO_2$ and $LiMn_2O_4$.

Negative electrode is carbon, in the form of either graphite or an amorphous material with a high surface-area. Carbon is an available and cheap material of low weight and also it is able to absorb a good quantity of lithium. When paired with a metal oxide as the positive electrode it gives a cell with a relatively high voltage (from 4 V in the fully charged state to 3 V in discharged state) (Dell & Rand, 2001).

Electrolyte is composed from organic liquid (ether) and dissolved salt ($LiPF_6$, $LiBF_4$, $LiClO_4$). The positive and negative active mass is applied to both sides of thin metal foils (aluminium on positive and copper on negative). Microporous polymer sheet between the positive and negative electrode works as the separator.

Lithium-ion cells are produced in coin format, as well as in cylindrical and prismatic (see Fig. 9) shapes.

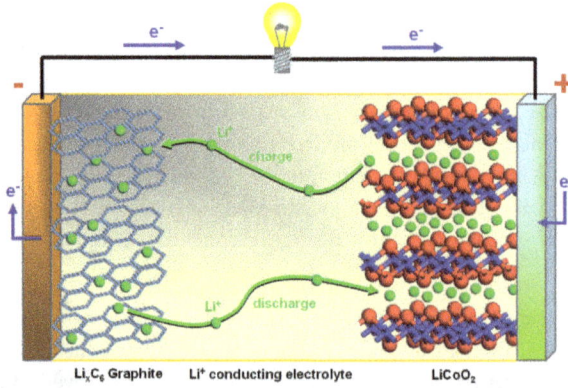

Figure 8. The principle of the lithium-ion cell

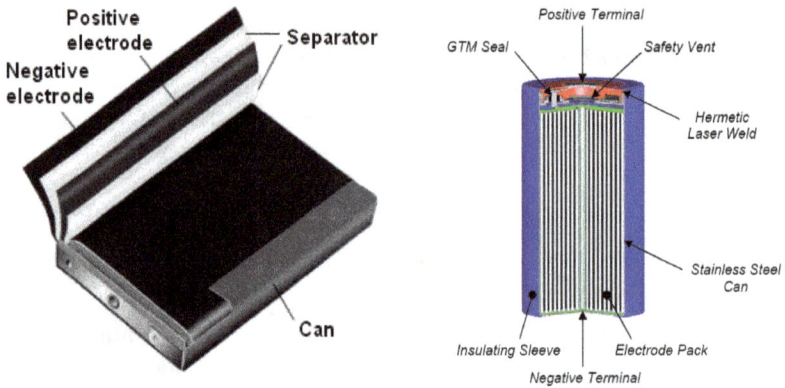

Figure 9. Prismatic and cylindrical Li-ion cell construction

3.2.2. *Principle of operation*

The positive electrode reaction is:

Positive electrode:

$$LiCoO_2 \rightarrow Li_{1-x}CoO_2 + xLi^+ + xe^- \qquad (4)$$

Negative electrode:

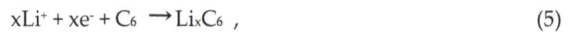

$$xLi^+ + xe^- + C_6 \rightarrow Li_xC_6 \ , \qquad (5)$$

where x moves on negative electrode from 0 to 1, on positive electrode from 0 to 0.45.

The most important advantages of lithium-ion cell are high energy density from 150 to 200 Wh/kg (from 250 to 530 Wh/l), high voltage (3.6 V), good charge-discharge characteristics, with more than 500 cycles possible, acceptably low selfdischarge (< 10% per month), absence of a memory effect, much safer than equivalent cells which use lithium metal, possibility of rapid recharging (2h).

Main disadvantage is a high price of the lithium-ion battery. There also must be controlled charging process, especially close the top of charge voltage 4.2 V. Overcharging or heating above 100°C cause the decomposition of the positive electrode with liberation of oxygen gas ($LiCoO_2$ yields Co_3O_4).

3.2.3. Li-pol battery

Polymers contained a hetero-atom (i.e. oxygen or sulfur) is able to dissolve lithium salts in very high concentrations. Some experiments were made with polyethylene oxide (PEO), which dissolves salts lithium perchlorate $LiClO_4$ and lithium trifluoromethane sulfonate $LiCF_3SO_3$ very well. But there is disadvantage - the conductivity of the solid solution of lithium ions is too low (about 10^{-5} S/m) on room temperature. But when higher temperature is reached (more than 60°C), transformation of crystalline to amorphous phase proceeds. It leads to much better electrical conductivity (10^{-1} S/m at 100°C). This value allows the polymer to serve as an electrolyte for lithium batteries. But thickness of the polymer must be low (10 to 100 μm). Polymer electrolyte is safer then liquid electrolyte, because it is not flammable (Dell & Rand, 2001).

4. Special batteries

4.1. Ag-Zn battery

4.1.1. Battery composition and construction

The zinc-silver oxide battery has one of the highest energy of aqueous cells. The theoretical energy density is 300 Wh/kg (1400 Wh/dm^3) and practical values are in the range 40-130 Wh/kg (110-320 Wh/dm^3). Cells have poor cycle life. But they can reach a very low internal resistance and also their high energy density makes them very useful for aerospace and even military purposes.

The silver positive active mass is formed by sintering of silver powder at temperatures between 400 and 700°C and it is placed on silver or silver-plated copper grids or perforated sheets.

The zinc negative electrode prepares as mixtures of zinc, zinc oxide and organic binding agents. The aim is to produce electrodes of high porosity. Other additives include surface active agents to minimize dendritic growth and mercuric ions to increase the hydrogen overvoltage of the zinc electrode (reduce gassing during charge) and so reduce corrosion.

Electrolyte is water solution of KOH (1.40 to 1.42 g/cm^3).

The separator is the most important component of zinc-silver oxide cell. It must prevent short circuit between electrodes, must prevent silver migration to the negative electrode, to control zincate migration, to preserve the integrity of the zinc electrode. The separator must have a low ion resistance with good thermal and chemical stability in KOH solution. Typical separators used in Ag-Zn battery, are of cellophane (regenerated cellulose), synthetic fiber mats of nylon, polypropylene, and nonwoven rayon fiber mats. Synthetic fiber mats are placed next to the positive electrode to protect the cellophane from oxidizing influence of that material. In most cells the separators are in form of envelopes completely enclosed the zinc electrodes (Vincent & Scrosati, 2003).

Commercial cells are generally prismatic – see Fig. 10 in shape and the case is usually plastic. Construction must be able to withstand the mechanical stress. The cells are usually sealed with safety valves. The volume of free electrolyte is very small. It is absorbed in the electrode pores and separator.

The energy density of practical zinc-silver oxide cells is some five to six times higher than that of their nickel-cadmium cells. The main disadvantage of the system is its high cost combined with a poor cycle life.

Figure 10. Ag-Zn prismatic and submarine torpedo battery

4.1.2. Principle of operation

The overall cell reaction during discharge:

$$Ag_2O_2 + 2H_2O + 2Zn \rightarrow 2Ag + 2Zn(OH)_2 \tag{6}$$

The cell discharge reaction takes place in two stages:

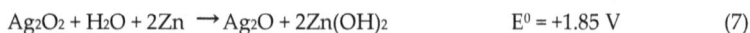

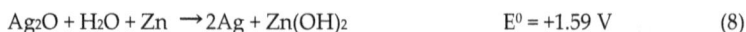

$$Ag_2O_2 + H_2O + 2Zn \rightarrow Ag_2O + 2Zn(OH)_2 \qquad\qquad E^0 = +1.85 \text{ V} \tag{7}$$

$$Ag_2O + H_2O + Zn \rightarrow 2Ag + Zn(OH)_2 \qquad\qquad E^0 = +1.59 \text{ V} \tag{8}$$

During discharge there rises metal silver inside positive electrode and that is why inner electrical resistance drops in discharged state. Maximum temperature range is from -40 to 50 °C. Self discharge of Ag-Zn battery at 25 °C is about 4% of capacity per month.

Zinc-silver oxide secondary cells with capacities of 0.5-100 Ah are manufactured for use in space satellites, military aircraft, submarines and for supplying power to portable military equipment. In space applications the batteries are used to increase the power from solar cells during period of high demand, e.g. during radio transmission or when the sun is eclipsed. At other times the batteries are charged by the solar cells.

4.2. Ni-H₂ battery

4.2.1. Battery composition and construction

The Ni-H₂ battery is an alkaline battery developed especially for use in satellites (see Fig. 11). It is a hybrid battery combining battery and fuel cell technology.

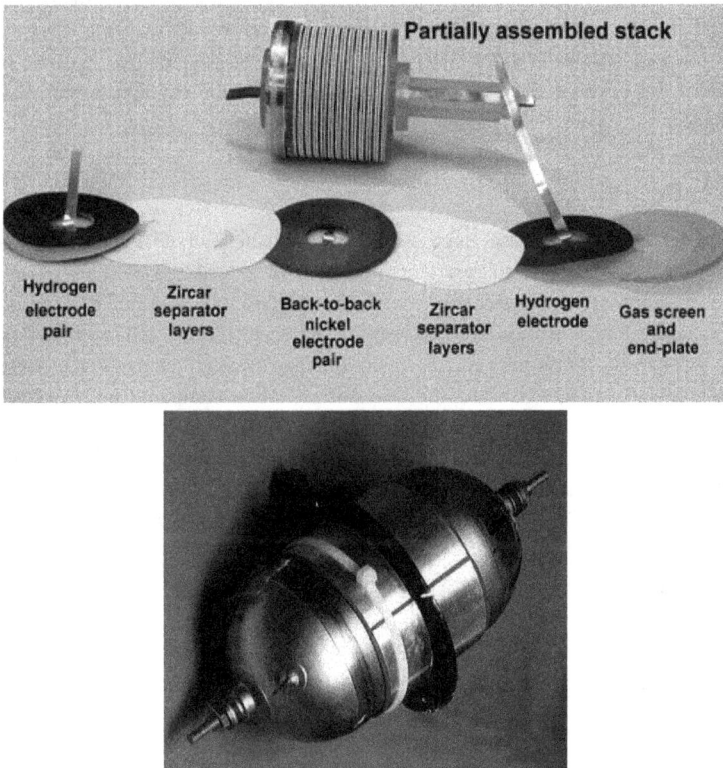

Figure 11. Scheme of a nickel-hydrogen battery (Zimmerman, 2009)

The battery has a sintered, nickel-oxide positive electrode and a negative electrode from platinum black catalyst supported with Teflon bonding dispersed on carbon paper. Two types of the separators are being used. First is formed from a porous ceramic paper, made from fibres of yttria-stabilized zirconia, second from asbestos paper (Linden & Reddy, 2002). Separators absorb the potassium hydroxide electrolyte. The battery was developed to replace Ni-Cd battery in space applications and it has some higher specific energy (50 Wh/kg) together with a very long cycle life. Standard voltage of the Ni-H$_2$ cell is 1.32 V.

4.2.2. Principle of operation

The overall reaction during discharge:

$$2NiOOH + H_2 \longrightarrow 2Ni(OH)_2 \tag{9}$$

The hydrogen gas liberated on charging is stored under pressure within the cell pressure vessel. Shape of the vessel is cylindrical with hemi-spherical end caps made from thin, Inconel alloy. Pressure of hydrogen inside the vessel grows to 4 MPa during charge whereas in the discharged state falls to 0.2 MPa. The cells may be overcharged because liberated oxygen from the positive electrode recombines rapidly at the negative electrode into the water.

5. Flow batteries

Flow batteries store and release electrical energy with help of reversible electrochemical reactions in two liquid electrolytes. An electrochemical cell has two loops physically separated by an ion or proton exchange membrane. Electrolytes flow into and out of the cell through separate loops and undergo chemical reaction inside the cell, with ion or proton exchange through the membrane and electron exchange through the external electric circuit. There are some advantages to using the flow battery when compared with a conventional secondary battery. The capacity of the system is possible to scale by increasing the amount of solution in electrolyte tanks. The battery can be fully discharged and has little loss of electrolyte during cycling. Because the electrolytes are stored separately, flow batteries have a low selfdischarge. Disadvantage is a low energy density and specific energy.

5.1. Br$_2$-Zn battery

5.1.1. Battery composition and construction

The zinc-bromine cell is composed from the bipolar electrodes. The bipolar electrode is from a lightweight, carbon-plastic composite material. Microporous plastic separator between electrodes allows the ions to pass through it. Cells are series-connected and the battery has a positive and a negative electrode loop. The electrolyte in each storage tank is circulated through the appropriate loop.

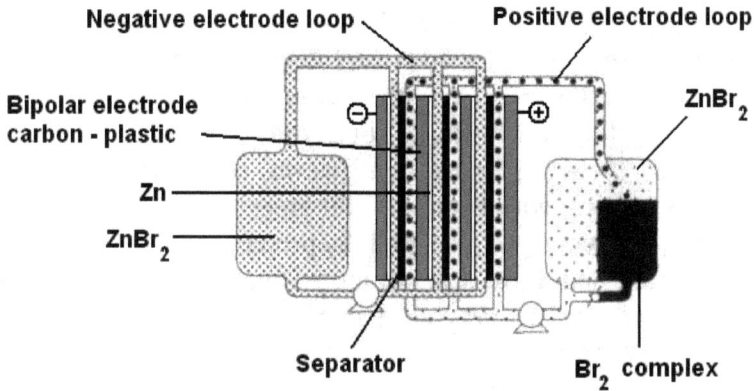

Figure 12. Scheme of zinc-bromine battery (Dell & Rand, 2001)

5.1.2. Principle of operation

The overall chemical reaction during discharge:

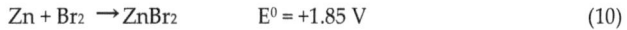

$$Zn + Br_2 \rightarrow ZnBr_2 \qquad E^0 = +1.85 \text{ V} \qquad (10)$$

During discharge product of reaction, the soluble zinc bromide is stored, along with the rest of the electrolyte, in the two loops and external tanks. During charge, bromine is liberated on the positive electrode and zinc is deposited on the negative electrode. Bromine is then complexed with an organic agent to form a dense, oily liquid polybromide complex. It is produced as droplets and these are separated from the aqueous electrolyte on the bottom of the tank in positive electrode loop. During discharge, bromine in positive electrode loop is again returned to the cell electrolyte in the form of a dispersion of the polybromide oil.

5.2. Vanadium redox battery

5.2.1. Battery composition and construction

A vanadium redox battery is another type of a flow battery in which electrolytes in two loops are separated by a proton exchange membrane (PEM). The electrolyte is prepared by dissolving of vanadium pentoxide (V_2O_5) in sulphuric acid (H_2SO_4). The electrolyte in the positive electrolyte loop contains $(VO_2)^+$ - (V^{5+}) and $(VO)^{2+}$ - (V^{4+}) ions, the electrolyte in the negative electrolyte loop, V^{3+} and V^{2+} ions. Chemical reactions proceed on the carbon electrodes.

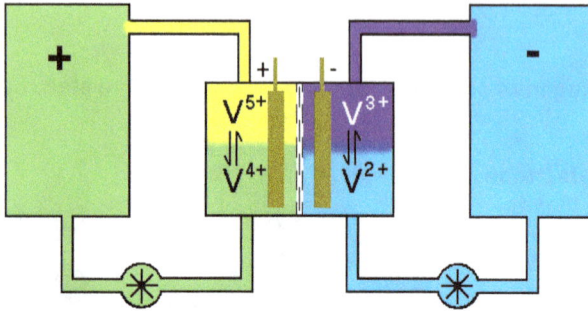

Figure 13. Scheme of vanadium redox battery

5.2.2. Principle of operation

In the vanadium redox cell, the following half-cell reactions are involved during discharge:

At the negative electrode:

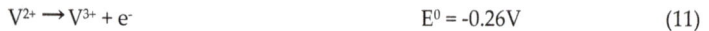

$$V^{2+} \rightarrow V^{3+} + e^- \qquad\qquad E^0 = -0.26V \qquad\qquad (11)$$

At the positive electrode:

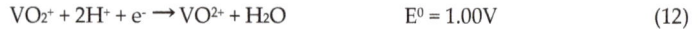

$$VO_2^+ + 2H^+ + e^- \rightarrow VO^{2+} + H_2O \qquad\qquad E^0 = 1.00V \qquad\qquad (12)$$

Under actual cell conditions, an open circuit voltage of 1.4 Volts is observed at 50% state of charge, while a fully charged cell produces over 1.6 Volts at open-circuit, fully discharged cell 1.0 Volt.

The extremely large capacities possible from vanadium redox batteries make them well suited to use in large RAPS applications, where they could to average out the production of highly unstable power sources such as wind or solar power. The extremely rapid response times make them suitable for UPS type applications, where they can be used to replace lead acid batteries. Disadvantage of vanadium redox batteries is a low energy density of about 25 Wh/kg of electrolyte, low charge efficiency (necessity using of pumps) and a high price.

6. High temperature batteries

6.1. Na-S battery

6.1.1. Battery composition and construction

Sodium, just like lithium, has many advantages as a negative-electrode material. Sodium has a high reduction potential of -2.71V and a low atomic weight (23.0). These properties allow to made a battery with a high specific energy (100-200 Wh/kg). Sodium salts are highly found in nature, they are cheap and non-toxic. Sulphur is the positive electrode material which can be used in combination with sodium to form a cell. Sulphur is also highly available in nature and very cheap.

The problem of a sodium-sulphur cell is to find a suitable electrolyte. Aqueous electrolytes cannot be used and, unlike the lithium, no suitable polymer was found. That is why a ceramic material beta-alumina (β–Al_2O_3) was used as electrolyte. It is an electronic insulator, but above 300 °C it has a high ionic conductivity for sodium ions.

In each cell, the negative electrode (molten sodium) was contained in a vertical tube (diameter from 1 to 2 cm). The positive electrode (molten sulphur) is absorbed into the pores of carbon felt (serves as the current-collector) and inserted into the annulus between the ceramic beta-alumina electrolyte tube and the cylindrical steel case (Fig. 14). Between molten sodium and beta-alumina electrolyte also could be found a safety liner with a pin-hole in its base.

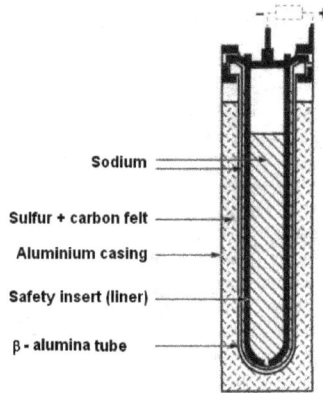

Figure 14. Schematic cross-section of Na-S cell (Dell & Rand, 2001)

6.1.2. Principle of operation

The cell discharges at 300 to 400 °C. Sodium ions pass from the sodium negative electrode, through the beta-alumina electrolyte, to the sulphur positive electrode. There they react with the sulphur to form sodium polysulphides. Standard voltage of the cell is about 2 V.

The cell discharges in two steps:

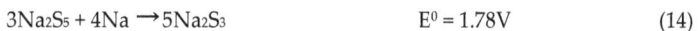

$$5S + 2Na \longrightarrow Na_2S_5 \qquad\qquad E^0 = 2.076V \qquad\qquad (13)$$

$$3Na_2S_5 + 4Na \longrightarrow 5Na_2S_3 \qquad\qquad E^0 = 1.78V \qquad\qquad (14)$$

Uncontrolled chemical reaction of molten sodium and sulphur could cause a fire and corrosion inside the cell and consequently destruction of the cell. It often happens after the fracture of the electrolyte tube. This problem is solved by inserting of safety liner to the beta-alumina tube. This allows a normal flow of sodium to the inner wall of the beta-alumina electrolyte, but prevents the flow in the case of tube fracture.

6.2. Na-metalchloride battery

6.2.1. Battery composition and construction

In the sodium-metalchloride battery the sulphur positive electrode there is replaced by nickel chloride or by a mixture of nickel chloride ($NiCl_2$) and ferrous chloride ($FeCl_2$) – see Fig. 15. The specific energy is 100-200 Wh/kg.

The negative electrode is from molten sodium, positive electrode from metalchloride and electrolyte from the ceramic beta-alumina (the same as in the sodium-sulphur battery). The second electrolyte, to make good ionic contact between the positive electrode and the electrolyte from beta-alumina, is molten sodium chloraluminate ($NaAlCl_4$).

The positive electrode is from a mixture of metal powder (Ni or Fe) and sodium chloride (NaCl). During charge, these materials are converted into the corresponding metal chloride and sodium. Iron powder is cheaper than nickel powder, but nickel cells have higher voltage and could operate over a wider temperature range (200 to 400 °C) than iron cells (200 to 300 °C) (Dell & Rand, 2001).

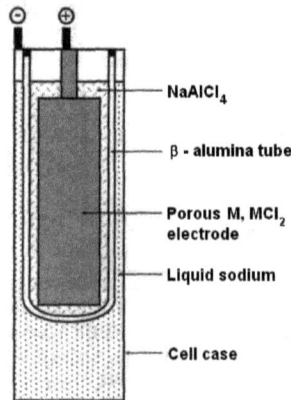

Figure 15. Schematic cross-section of Na-metalchloride cell (Rand, 1998)

6.2.2. Principle of operation

The basic cell reactions during discharge are simple, i.e.

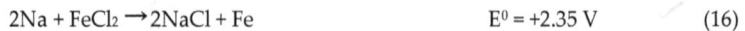

$$2Na + NiCl_2 \rightarrow 2NaCl + Ni \qquad\qquad E^0 = +2.58 \text{ V} \qquad (15)$$

$$2Na + FeCl_2 \rightarrow 2NaCl + Fe \qquad\qquad E^0 = +2.35 \text{ V} \qquad (16)$$

Advantage of the sodium metalchloride cell over the sodium sulphur cell is that there is possibility of both an overcharge and overdischarge reaction, when the second electrolyte (molten sodium chloraluminate) reacts with metal (overcharge) or with sodium (overdischarge).

Overcharge reaction for sodium nickelchloride cell:

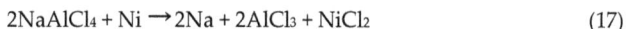

$$2NaAlCl_4 + Ni \rightarrow 2Na + 2AlCl_3 + NiCl_2 \qquad (17)$$

Overdischarge reaction for sodium nickelchloride cell:

$$3Na + NaAlCl_4 \rightarrow Al + 4NaCl \qquad (18)$$

Another advantage of the sodium metalchloride system is safety of operation. When the beta-alumina electrolyte tube cracks in this system, the molten sodium first encounters the $NaAlCl_4$ electrolyte and reacts with it according the overdischarge reaction.

7. Conclusion

This chapter is focused on electrochemical storage or batteries that constitute a large group of technologies that are potentially suitable to meet a broad market needs. The five categories of electrochemical systems (secondary batteries) were selected and discussed in detail: standard batteries (lead acid, Ni-Cd) modern batteries (Ni-MH, Li–ion, Li-pol), special batteries (Ag-Zn, Ni-H2), flow batteries (Br2-Zn, vanadium redox) and high temperature batteries (Na-S, Na–metalchloride). These batteries appear to be promising to meet the requirements for end-user applications.

However, the use of secondary batteries involves some technical problems. Since their cells slowly self-discharge, batteries are mostly suitable for electricity storage only for limited periods of time. They also age, which results in a decreasing storage capacity.

For electrochemical energy storage, the specific energy and specific power are two important parameters. Other important parameters are ability to charge and discharge a large number of times, to retain charge as long time as possible and ability to charge and discharge over a wide range of temperatures.

Author details

Petr Krivik and Petr Baca
The Faculty of Electrical Engineering and Communication,
Brno University of Technology, Czech Republic

Acknowledgement

This chapter is supported by the EU project CZ.1.05/2.1.00/01.0014 and by the internal grant FEKT-S-11-7.

8. References

Calabek, M. et al. (2001). A fundamental study of the effects of compression on the performance of lead accumulator plates, *J. Power Sources*, Vol. 95, 97 – 107, ISSN 0378-7753

Dell, R.M. & Rand, D.A.J. (2001). *Understanding Batteries*, The Royal Society of Chemistry, ISBN 0-85404-605-4, Cambridge, UK

Linden, D. & Reddy, T.B. (2002). *Handbook of Batteries, Third Edition*, McGraw-Hill, Two Penn Plaza, ISBN 0-07-135978-8, New York, USA

Nelson, R. (2001). The Basic Chemistry of Gas Recombination in Lead-Acid Batteries, Santa Fe Drive, Denver, Colorado, USA

Rand, D.A.J. et al. (1998). *Batteries for Electric Vehicles*, Research Studies Press Ltd., ISBN 0-86380-205-2, Taunton, Somerset, Great Britain

Rand, D.A.J. et al. (2004). *Valve-regulated Lead-Acid Batteries*, Elsevier B.V., ISBN 0-444-50746-9, Netherlands

Ruetschi, P. (2004). Aging mechanisms and service life of lead–acid batteries. *J. Power Sources*, Vol. 127, 33–44, ISSN 0378-7753

Vincent, C.A. & Scrosati, B. (2003). *Modern Batteries*, Antony Rowe Ltd, ISBN 0-340-66278-6, Eastbourne, Great Britain

Zimmerman, A.(2009). Nickel-Hydrogen Batteries: Principles and Practice, Available from http://www.aero.org/publications/zimmerman/chapter2.html

Chapter 5

Techno-Economic Analysis
of Different Energy Storage Technologies

Hussein Ibrahim and Adrian Ilinca

Additional information is available at the end of the chapter

1. Introduction

Overall structure of electrical power system is in the process of changing. For incremental growth, it is moving away from fossil fuels - major source of energy in the world today - to renewable energy resources that are more environmentally friendly and sustainable [1]. Factors forcing these considerations are (a) the increasing demand for electric power by both developed and developing countries, (b) many developing countries lacking the resources to build power plants and distribution networks, (c) some industrialized countries facing insufficient power generation and (d) greenhouse gas emission and climate change concerns. Renewable energy sources such as wind turbines, photovoltaic solar systems, solar-thermo power, biomass power plants, fuel cells, gas micro-turbines, hydropower turbines, combined heat and power (CHP) micro-turbines and hybrid power systems will be part of future power generation systems [2-8].

Nevertheless, exploitation of renewable energy sources (RESs), even when there is a good potential resource, may be problematic due to their variable and intermittent nature. In addition, wind fluctuations, lightning strikes, sudden change of a load, or the occurrence of a line fault can cause sudden momentary dips in system voltage [4]. Earlier studies have indicated that energy storage can compensate for the stochastic nature and sudden deficiencies of RESs for short periods without suffering loss of load events and without the need to start more generating plants [4], [9], [10]. Another issue is the integration of RESs into grids at remote points, where the grid is weak, that may generate unacceptable voltage variations due to power fluctuations. Upgrading the power transmission line to mitigate this problem is often uneconomic. Instead, the inclusion of energy storage for power smoothing and voltage regulation at the remote point of connection would allow utilization of the power and could offer an economic alternative to upgrading the transmission line.

The current status shows that several drivers are emerging and will spur growth in the demand for energy storage systems [11]. These include: the growth of stochastic generation from renewables; an increasingly strained transmission infrastructure as new lines lag behind demand; the emergence of micro-grids as part of distributed grid architecture; and the increased need for reliability and security in electricity supply [12]. However, a lot of issues regarding the optimal active integration (operational, technical and market) of these emerging energy storage technologies into the electric grid are still not developed and need to be studied, tested and standardized. The integration of energy storage systems (ESSs) and further development of energy converting units (ECUs) including renewable energies in the industrial nations must be based on the existing electric supply system infrastructure. Due to that, a multi-dimensional integration task regarding the optimal integration of energy storage systems will result.

The history of the stationary Electrical Energy Storage (EES) dates back to the turn of the 20th century, when power stations were often shut down overnight, with lead-acid accumulators supplying the residual loads on the direct current networks [13–15]. Utility companies eventually recognised the importance of the flexibility that energy storage provides in networks and the first central station for energy storage, a Pumped Hydroelectric Storage (PHS), was put to use in 1929 [13,16,17]. The subsequent development of the electricity supply industry, with the pursuit of economy of scale, at large central generating stations, with their complementary and extensive transmission and distribution networks, essentially consigned interest in storage systems up until relatively recent years. Up to 2005, more than 200 PHS systems were in use all over the world providing a total of more than 100 GW of generation capacity [16–18]. However, pressures from deregulation and environmental concerns lead to investment in major PHS facilities falling off, and interest in the practical application of EES systems is currently enjoying somewhat of a renaissance, for a variety of reasons including changes in the worldwide utility regulatory environment, an ever-increasing reliance on electricity in industry, commerce and the home, power quality/quality-of-supply issues, the growth of renewable as a major new source of electricity supply, and all combined with ever more stringent environmental requirements [14,19-20]. These factors, combined with the rapidly accelerating rate of technological development in many of the emerging EESs, with anticipated unit cost reductions, now make their practical applications look very attractive on future timescales of only a few years.

This document aims to review the state-of-the-art development of EES technologies including PHS [18,21], Compressed Air Energy Storage system (CAES) [22–26], Battery [27–31], Flow Battery [14-15,20,32], Fuel Cell [33-34], Solar Fuel [15,35], Superconducting Magnetic Energy Storage system (SMES) [36–38], Flywheel [32,39–41], Capacitor and Supercapacitor [15,39], and Thermal Energy Storage system (TES) [42–50]. Some of them are currently available and some are still under development. The applications, classification, technical characteristics, research and development (R&D) progress and deployment status of these EES technologies will be discussed in the following sections.

2. Electrical energy storage

2.1. Definition of electrical energy storage

Electrical Energy Storage (EES) refers to a process of converting electrical energy from a power network into a form that can be stored for converting back to electrical energy when needed [13–14,51]. Such a process enables electricity to be produced at times of either low demand, low generation cost or from intermittent energy sources and to be used at times of high demand, high generation cost or when no other generation means is available [13–15,19,51] (Figure 1). EES has numerous applications including portable devices, transport vehicles and stationary energy resources [13-15], [19-20], [51-54]. This document will concentrate on EES systems for stationary applications such as power generation, distribution and transition network, distributed energy resource, renewable energy and local industrial and commercial customers.

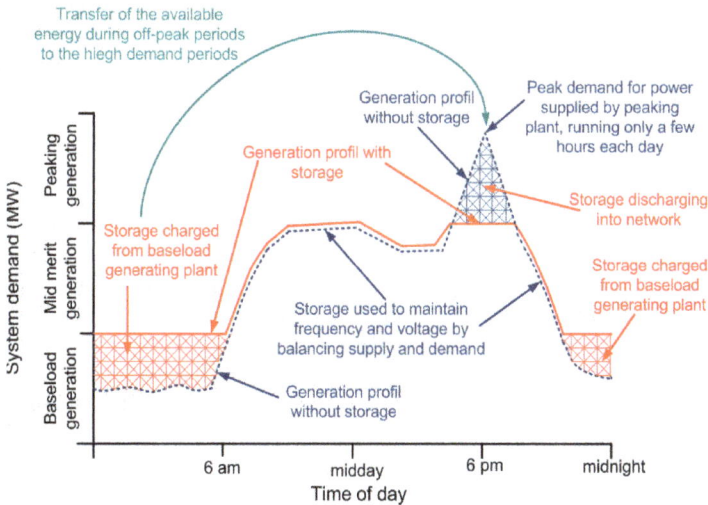

Figure 1. Fundamental idea of the energy storage [55]

2.2. Role of energy storage systems

Breakthroughs that dramatically reduce the costs of electricity storage systems could drive revolutionary changes in the design and operation of the electric power system [52]. Peak load problems could be reduced, electrical stability could be improved, and power quality disturbances could be eliminated. Indeed, the energy storage plays a flexible and multifunctional role in the grid of electric power supply, by assuring more efficient management of available power. The combination with the power generation systems by the conversion of renewable energy, the Energy Storage System (ESS) provide, in real time, the balance between production and consumption and improve the management and the reliability of the grid [56]. Furthermore, the ESS makes easier the integration of the renewable resources in the energy system, increases their penetration rate of energy and the quality of the supplied energy by better controlling frequency and voltage. Storage can be applied at the power plant, in support of the transmission system, at various points in the distribution system and on particular appliances and equipments on the customer's side of the meter [52].

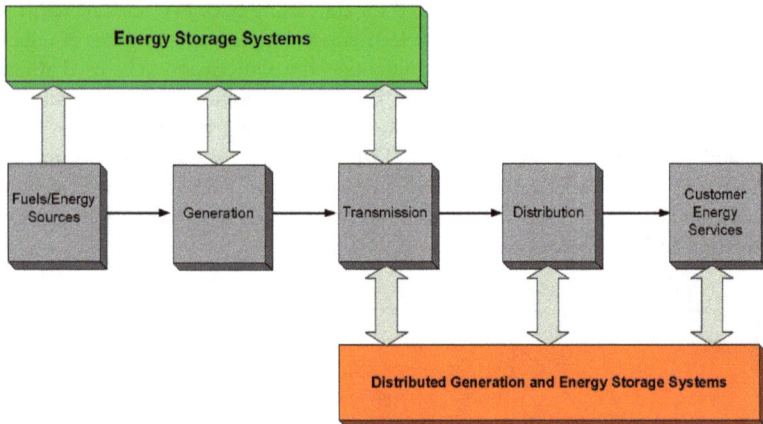

Figure 2. New electricity value chain with energy storage as the sixth dimension [11]

The ESS can be used to reduce the peak load and eliminate the extra thermal power plant operating only during the peak periods, enabling better utilization of the plant functioning permanently and outstanding reduction of emission of greenhouse gases (GHG) [57]. Energy storage systems in combination with advanced power electronics (power electronics are often the interface between energy storage systems and the electrical grid) have a great technical role and lead to many financial benefits. Some of these are summarized in the following sections. Figure 2 shows how the new electricity value chain is changing supported by the integration of energy storage systems (ESS). More details about the different applications of energy storage systems will be presented in the section 4.

3. Energy storage components

Before discussing the technologies, a brief explanation of the components within an energy storage device are discussed. Every energy storage facility is comprised of three primary components [58]:

- Storage Medium
- Power Conversion System (PCS)
- Balance of Plant (BOP)

3.1. Storage medium

The storage medium is the 'energy reservoir' that retains the potential energy within a storage device. It ranges from mechanical (Pumped Heat Electricity Storage – PHES), chemical (Battery Energy Storage - BES) and electrical (Superconductor Magnetic Energy Storage – SMES) potential energy [58].

3.2. Power Conversion System (PCS)

It is necessary to convert from Alternating Current (AC) to Direct Current (DC) and vice versa, for all storage devices except mechanical storage devices e.g. PHES and CAES (Compressed Air Energy Storage) [59]. Consequently, a PCS is required that acts as a rectifier while the energy device is charged (AC to DC) and as an inverter when the device is discharged (DC to AC). The PCS also conditions the power during conversion to ensure that no damage is done to the storage device.

The customization of the PCS for individual storage systems has been identified as one of the primary sources of improvement for energy storage facilities, as each storage device operates differently during charging, standing and discharging [59]. The PCS usually costs from 33% to 50% of the entire storage facility. Development of PCSs has been slow due to the limited growth in distributed energy resources e.g. small scale power generation technologies ranging from 3 to 10,000 kW [60].

3.3. Balance-of-Plant (BOP)

These are all the devices that [58]:

- Are used to house the equipment
- Control the environment of the storage facility
- Provide the electrical connection between the PCS and the power grid

It is the most variable cost component within an energy storage device due to the various requirements for each facility. The BOP typically includes electrical interconnections, surge protection devices, a support rack for the storage medium, the facility shelter and environmental control systems [59].

The balance-of-plant includes structural and mechanical equipment such as protective enclosure, Heating/Ventilation/Air Conditioning (HVAC), and maintenance/auxiliary devices. Other BOP features include the foundation, structure (if needed), electrical protection and safety equipment, metering equipment, data monitoring equipment, and communications and control equipment. Other cost such as the facility site, permits, project management and training may also be considered here [61].

4. Applications and technical benefits of energy storage systems

The traditional electricity value chain has been considered to consist of five links: fuel/energy source, generation, transmission, distribution and customer-side energy service as shown in Figure 3. By supplying power when and where needed, ESS is on the brink of becoming the "sixth link" by integrating the existing segments and creating a more responsive market [62]. Stored energy integration into the generation-grid system is illustrated in Figure 4 [32]. It can be seen that potential applications of EES are numerous and various and could cover the full spectrum ranging from larger scale, generation and transmission-related systems, to those primarily related to the distribution network and even 'beyond the meter', into the customer/end-user site [13]. Some important applications have been summarised in [13–15], [32], [52], [62–66]:

Challenges

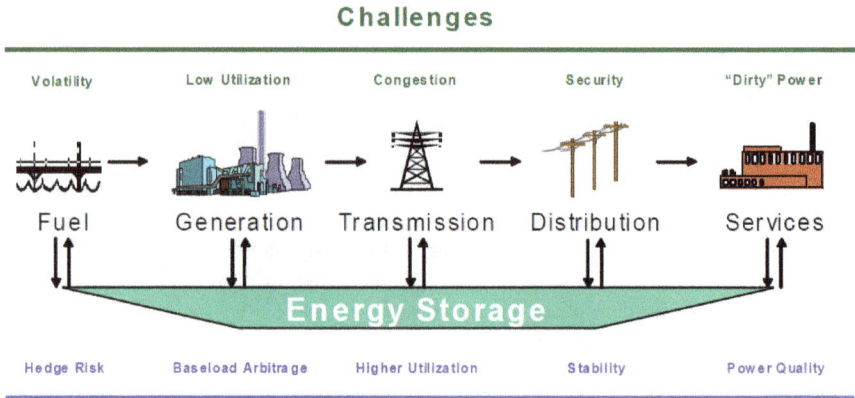

Volatility	Low Utilization	Congestion	Security	"Dirty" Power
Fuel	Generation	Transmission	Distribution	Services

Energy Storage

| Hedge Risk | Baseload Arbitrage | Higher Utilization | Stability | Power Quality |

Benefits

Figure 3. Benefits of ESS along the electricity value chain [62].

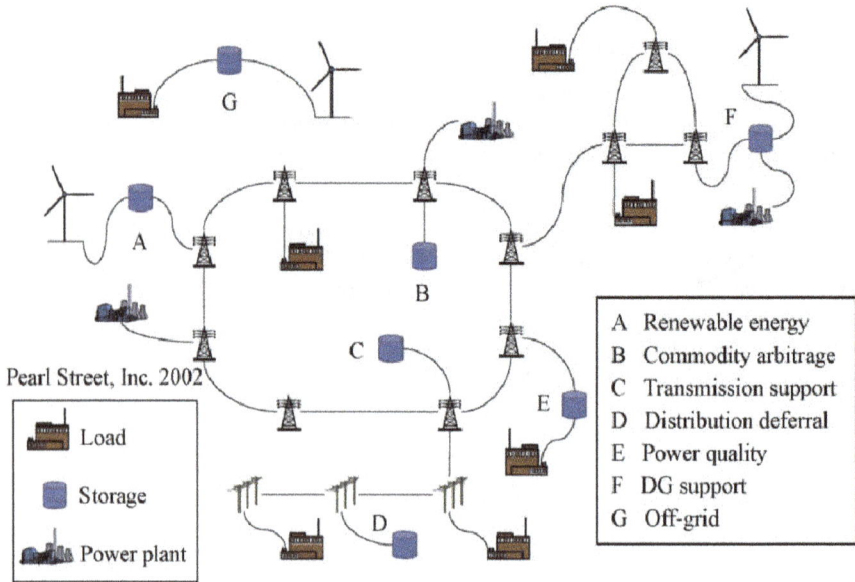

Pearl Street, Inc. 2002

A   Renewable energy
B   Commodity arbitrage
C   Transmission support
D   Distribution deferral
E   Power quality
F   DG support
G   Off-grid

Load
Storage
Power plant

Figure 4. Energy storage applications into grid [32].

4.1. Generation

- *Commodity Storage:* Storing bulk energy generated at night for use during peak demand periods during the day. This allows for arbitrating the production price of the two periods and a more uniform load factor for the generation, transmission, and distribution systems [62].
- *Contingency Service:* Contingency reserve is power capacity capable of providing power to serve customer demand should a power facility fall off-line. Spinning reserves are ready instantaneously, with non-spinning and long-term reserves ready in 10 minutes or longer. Spinning Reserve is defined as the amount of generation capacity that can be used to produce active power over a given period of time which has not yet been committed to the production of energy during this period [67].
- *Area Control:* Prevent unplanned transfer of power between one utility and another.
- *Grid Frequency Support:* Grid Frequency Support means real power provided to the electrical distribution grid to reduce any sudden large load/generation imbalance and maintain a state of frequency equilibrium for the system's 60Hz (cycles per second) during regular and irregular grid conditions. Large and rapid changes in the electrical load of a system can damage the generator and customers' electrical equipment [62].
- *Black-Start:* This refers to units with the capability to start-up on their own in order to energize the transmission system and assist other facilities to start-up and synchronize to the grid.

4.2. Transmission and distribution

- *System Stability:* The ability to maintain all system components on a transmission line in synchronous operation with each other to prevent a system collapse [62].
- *Grid Angular Stability:* Grid Angular Stability means reducing power oscillations (due to rapid events) by injection and absorption of real power.
- *Grid Voltage Support:* Grid Voltage Support means power provided to the electrical distribution grid to maintain voltages within the acceptable range between each end of all power lines. This involves a trade-off between the amount of "real" energy produced by generators and the amount of "reactive" power produced [68].
- *Asset Deferral:* Defer the need for additional transmission facilities by supplementing and existing transmission facilities—saving capital that otherwise goes underutilized for years [69].

4.3. Energy service

- *Energy Management (Load Levelling / Peak Shaving):* Load Levelling is rescheduling certain loads to cut electrical power demand, or the production of energy during off-peak periods for storage and use during peak demand periods. Whilst Peak Shaving is reducing electric usage during peak periods or moving usage from the time of peak demand to off-peak periods. This strategy allows to customers to peak shave by shifting energy demand from one time of the day to another. This is primarily used to reduce their time-of-use (demand) charges [62].

- *Unbalanced Load Compensation:* This can be done in combination with four-wire inverters and also by injecting and absorbing power individually at each phase to supply unbalanced loads.
- *Power Quality improvement:* Power Quality is basically related to the changes in magnitude and shape of voltage and current. This result in different issues including: Harmonics, Power Factor, Transients, Flicker, Sag and Swell, Spikes, etc. Distributed energy storage systems (DESS) can mitigate these problems and provide electrical service to the customer without any secondary oscillations or disruptions to the electricity "waveform" [67].
- *Power Reliability:* Can be presented as the percentage/ratio of interruption in delivery of electric power (may include exceeding the threshold and not only complete loss of power) versus total uptime. DESS can help provide reliable electric service to consumers (UPS) to 'ride-through' a power disruption. Coupled with energy management storage, this allows remote power operation [68].

4.4. Supporting the integration of intermittent renewable energy sources

The development and use of renewable energy has experienced rapid growth over the past few years. In the next 20–30 years all sustainable energy systems will have to be based on the rational use of traditional resources and greater use of renewable energy.

Decentralized electrical production from renewable energy sources yields a more assured supply for consumers with fewer environmental hazards. However, the unpredictable character of these sources requires that network provisioning and usage regulations be established for optimal system operation.

Figure 5. Integration of extrapolated (x6) wind power using energy storage on the Irish electricity grid [58]

However, renewable energy resources have two problems. First, many of the potential power generation sites are located far from load centers. Although wind energy generation facilities can be constructed in less than one year, new transmission facilities must be constructed to bring this new power source to market. Since it can take upwards of 7 years to build these transmission assets, long, lag-time periods can emerge where wind generation is "constrained-off" the system [62]. For many sites this may preclude them from delivering power to existing customers, but it opens the door to powering off-grid markets—an important and growing market.

The second problem is that the renewable resources fluctuate independently from demand. Therefore, the most of the power accessible to the grids is generated when there is low demand for it. By storing the power from renewable sources from off-peak and releasing it during on-peak, energy storage can transform this low value, unscheduled power into schedulable, high-value product (see Figure 5). Beyond energy sales, with the assured capability of dispatching power into the market, a renewable energy source could also sell capacity into the market through contingency services.

This capability will make the development of renewable resources far more cost-effective — by increasing the value of renewables it may reduce the level of subsidy down to where it is equal to the environmental value of the renewable, at which point it is no longer a subsidy but an environmental credit [62].

- *Frequency and synchronous spinning reserve support:* In grids with a significant share of wind generation, intermittency and variability in wind generation output due to sudden shifts in wind patterns can lead to significant imbalances between generation and load that in turn result in shifts in grid frequency [68]. Such imbalances are usually handled by spinning reserve at the transmission level, but energy storage can provide prompt response to such imbalances without the emissions related to most conventional solutions.
- *Transmission Curtailment Reduction:* Wind power generation is often located in remote areas that are poorly served by transmission and distribution systems. As a result, sometimes wind operators are asked to curtail their production, which results in lost energy production opportunity, or system operators are required to invest in expanding the transmission capability. An EES unit located close to the wind generation can allow the excess energy to be stored and then delivered at times when the transmission system is not congested [68].
- *Time Shifting:* Wind turbines are considered as non-dispatchable resources. EES can be used to store energy generated during periods of low demand and deliver it during periods of high demand (Figure 5). When applied to wind generation, this application is sometimes called "firming and shaping" because it changes the power profile of the wind to allow greater control over dispatch [68].
- *Forecast Hedge:* Mitigation of errors (shortfalls) in wind energy bids into the market prior to required delivery, thus reducing volatility of spot prices and mitigating risk exposure of consumers to this volatility [69].
- *Fluctuation suppression:* Wind farm generation frequency can be stabilised by suppressing fluctuations (absorbing and discharging energy during short duration variations in output) [69].

5. Financial benefits of energy storage systems

In [70] detailed analysis of energy storage benefits is done including market analysis, the following are some highlights:

1. *Cost Reduction or Revenue Increase of Bulk Energy Arbitrage:* Arbitrage involves purchase of inexpensive electricity available during low demand periods to charge the storage plant, so that the low priced energy can be used or sold at a later time when the price for electricity is high [11].

2. *Cost Avoid or Revenue Increase of Central Generation Capacity:* For areas where the supply of electric generation capacity is tight, energy storage could be used to offset the need to: a) purchase and install new generation and/or b) "rent" generation capacity in the wholesale electricity marketplace.

3. *Cost Avoid or Revenue Increase of Ancillary Services:* It is well known that energy storage can provide several types of ancillary services. In short, these are what might be called support services used to keep the regional grid operating. Two more familiar ones are spinning reserve and load following [11].

4. *Cost Avoid or Revenue Increase for Transmission Access/Congestion:* It is possible that use of energy storage could improve the performance of the Transmission and Distribution (T&D) system by giving the utilities the ability to increase energy transfer and stabilize voltage levels. Further, transmission access/congestion charges can be avoided because the energy storage is used.

5. *Reduced Demand Charges:* Reduced demand charges are possible when energy storage is used to reduce an electricity end-user's use of the electric grid during times grid is high (i.e., during peak electric demand periods) [11].

6. *Reduced Reliability-related Financial Losses:* Storage reduces financial losses associated with power outages. This benefit is very end-user-specific and applies to commercial and industrial (C&I) customers, primarily those for which power outages cause moderate to significant losses.

7. *Reduced Power Quality-related Financial Losses:* Energy storage reduces financial losses associated with power quality anomalies. Power quality anomalies of interest are those that cause loads to go off-line and/or that damage electricity-using equipment and whose negative effects can be avoided if storage is used [11].

8. *Increased Revenue from Renewable Energy Sources:* Storage could be used to time-shift electric energy generated by renewables. Energy is stored when demand and price for power are low, so the energy can be used when a) demand and price for power is high and b) output from the intermittent renewable generation is low.

The previous listed functionalities point out that those energy storages in combination with power electronics will have a huge impact in future electrical supply systems. This is why any planning and implementation strategy should be related to the real-time control and operational functionalities of the ESS in combination with Distributed Energy Resources (DER) in order to get rapid integration process.

6. Techno-economic characteristics of energy storage systems

The main characteristics of storage systems on which the selection criteria are based are the following [73]:

6.1. Storage capacity

This is the quantity of available energy in the storage system after charging. Discharge is often incomplete. For this reason, it is defined on the basis of total energy stored, which is superior to that actually retrieved (operational). The usable energy, limited by the depth of discharge, represents the limit of discharge depth (minimum-charge state). In conditions of quick charge or discharge, the efficiency deteriorates and the retrievable energy can be much lower than storage capacity (Figure 6). On the other hand, self-discharge is the attenuating factor under very slow regime.

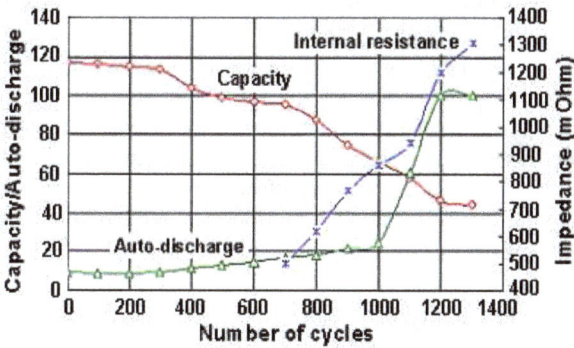

Figure 6. Variation of energy capacity, self-discharge and internal resistance of a nickel-metal-hydride battery with the number of cycles [71]

6.2. Storage System Power

This parameter determines the constitution and size of the motor-generator in the stored energy conversion chain. A storage system's power rating is assumed to be the system's nameplate power rating under normal operating conditions [73]. Furthermore, that rating is assumed to represent the storage system's *maximum* power output under *normal* operating conditions. In this document, the normal discharge rate used is commonly referred to as the system's 'design' or 'nominal' (power) rating.

6.3. Storage 'Emergency' Power Capability

Some types of storage systems can discharge at a relatively high rate (*e.g.*, 1.5 to 2 times their nominal rating) for relatively short periods of time (*e.g.*, several minutes to as much as 30 minutes). One example is storage systems involving a Na/S battery, which is capable of producing two times its rated (normal) output for relatively short durations [72].

That feature – often referred to as the equipment's 'emergency' rating – is valuable if there are circumstances that occur infrequently that involve an urgent need for relatively high power output, for relatively short durations.

Importantly, while discharging at the higher rate, storage efficiency is reduced (relative to efficiency during discharge at the nominal discharge rate), and storage equipment damage increases (compared to damage incurred at the normal discharge rate).

So, in simple terms, storage with emergency power capability could be used to provide the nominal amount of power required to serve a regularly occurring need (*e.g.*, peak demand reduction) while the same storage could provide additional power for urgent needs that occur infrequently and that last for a few to several minutes at a time [72].

6.4. Autonomy

Autonomy or discharge duration autonomy is the amount of time that storage can discharge at its rated output (power) without recharging. Discharge duration is an important criterion affecting the technical viability of a given storage system for a given application and storage plant cost [73]. This parameter depends on the depth of discharge and operational conditions of the system, constant power or not. It is a characteristic of system adequacy for certain applications. For small systems in an isolated area relying on intermittent renewable energy, autonomy is a crucial criterion. The difficulty in separating the power and energy dimensions of the system makes it difficult to choose an optimum time constant for most storage technologies [74].

6.5. Energy and power density

Power density is the amount of power that can be delivered from a storage system with a given volume or mass. Similarly, energy density is the amount of energy that can be stored in a storage device that has a given volume or mass. These criteria are important in situations for which space is valuable or limited and/or if weight is important (especially for mass density of energy in portable applications, but less so for permanent applications).

6.6. Space requirements for energy storage

Closely related to energy and power density are footprint and space requirements for energy storage. Depending on the storage technology, floor area and/or space constraints may indeed be a challenge, especially in heavily urbanized areas.

6.7. Efficiency

All energy transfer and conversion processes have losses. Energy storage is no different. Storage system round-trip efficiency (efficiency) reflects the amount of energy that comes out of storage relative to the amount put into the storage. This definition is often oversimplified because it is based on a single operation point [75]. The definition of efficiency must therefore be based on one or more realistic cycles for a specific application. Instantaneous power is a defining factor of efficiency (Figure 7). This means that, for optimum operation, the power-transfer chain must have limited losses in terms of energy transfer and self-discharge. This energy conservation measure is an essential element for daily network load-levelling applications.

Typical values for efficiency include the following: 60% to 75% for conventional electrochemical batteries; 75% to 85% for advanced electrochemical batteries; 73% to 80% for CAES; 75% to 78% for pumped hydro; 80% to 90% for flywheel storage; and 95% for capacitors and SMES [72], [76].

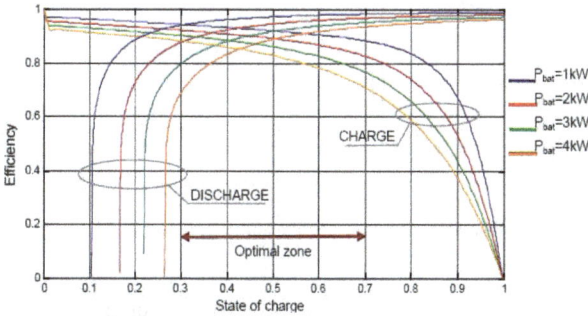

Figure 7. Power efficiency of a 48V-310Ah (15 kWh/10 h discharge) lead accumulator [77]

6.8. Storage operating cost

Storage total operating cost (as distinct from plant *capital* cost or plant financial carrying charges) consists of two key components: 1) energy-related costs and 2) operating costs not related to energy. Non-energy operating costs include at least four elements: 1) labor associated with plant operation, 2) plant maintenance, 3) equipment wear leading to loss-of-life, and 4) decommissioning and disposal cost [73].

1. *Charging Energy-Related Costs:* The energy cost for storage consists of all costs incurred to purchase energy used to charge the storage, including the cost to purchase energy needed to make up for (round trip) energy losses [73]. For a storage system with 75% efficiency, if the unit price for energy used for charging is 4¢/kWh, then the plant energy cost is 5.33¢/kWh.

2. *Labor for Plant Operation:* In some cases, labor may be required for storage plant operation. Fixed labor costs are the same magnitude irrespective of how much the storage is used. Variable labor costs are proportional to the frequency and duration of storage use [73]. In many cases, labor is required to operate larger storage facilities and/or 'blocks' of aggregated storage capacity whereas little or no labor may be needed for smaller/distributed systems that tend to be designed for autonomous operation. No explicit value is ascribed to this criterion, due in part to the wide range of labor costs that are possible given the spectrum of storage types and storage system sizes [73].

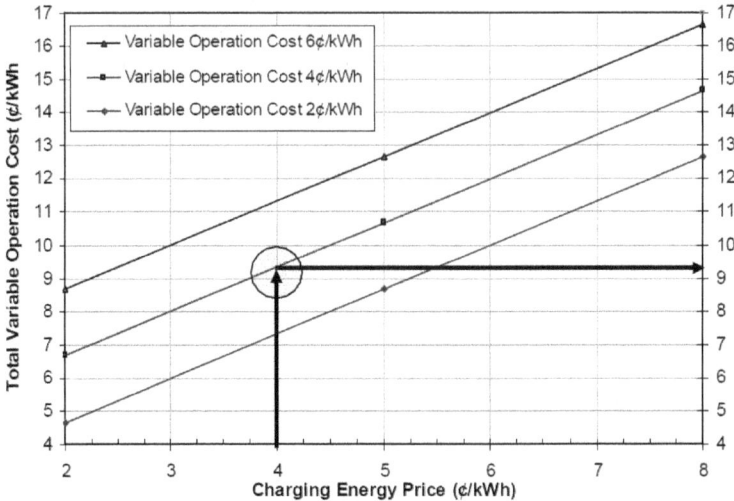

Figure 8. Storage total variable operation cost for 75% storage efficiency [73]

3. *Plant Maintenance:* Plant maintenance costs are incurred to undertake normal, scheduled, and unplanned repairs and replacements for equipment, buildings, grounds, and infrastructure. Fixed maintenance costs are the same magnitude irrespective of how much the storage is used [73]. Variable maintenance costs are proportional to the frequency and duration of storage use.

4. *Replacement Cost:* If specific equipment or subsystems within a storage system are expected to wear out during the expected life of the system, then a 'replacement cost' will be incurred. In such circumstances, a 'sinking fund' is needed to accumulate funds to pay for replacements when needed [73]. That replacement cost is treated as a variable cost (i.e., the total cost is spread out over each unit of energy output from the storage plant).

5. *Variable Operating Cost:* A storage system's total variable operating cost consists of applicable non-energy-related variable operating costs plus plant energy cost, possibly including charging energy, labor for plant operation, variable maintenance, and replacement costs. Variable operating cost is a key factor affecting the cost-effectiveness of storage [73]. It is especially important for 'high-use' value propositions involving many charge-discharge cycles.

Ideally, storage for high-use applications should have relatively high or very high efficiency and relatively low variable operating cost. Otherwise, the total cost to charge then discharge the storage is somewhat-to-very likely to be higher than the benefit. That can be a significant challenge for some storage types and value propositions.

Consider the example illustrated in Figure 8, which involves a 75% efficient storage system with a non-energy-related variable operating cost of 4¢/kWhout. If that storage system is charged with energy costing 4¢/kWhin, then the total variable operating cost – for energy output – is about 9.33¢/kWhout [73].

6.9. Durability

Lifetime or durability refers to the number of times the storage unit can release the energy level it was designed for after each recharge, expressed as the maximum number of cycles N (one cycle corresponds to one charge and one discharge) [81]. All storage systems degrade with use because they are subject to fatigue or wear by usage use (i.e., during each charge-discharge cycle). This is usually the principal cause of aging, ahead of thermal degradation. The rate of degradation depends on the type of storage technology, operating conditions, and other variables. This is especially important for electrochemical batteries [73].

For some storage technologies – especially batteries – the extent to which the system is emptied (discharged) also affects the storage media's useful life. Discharging a small portion of stored energy is a 'shallow' discharge and discharging most or all of the stored energy is a 'deep' discharge. For these technologies, a shallow discharge is less damaging to the storage medium than a deep discharge [73].

To the extent that the storage medium degrades and must be replaced during the expected useful life of the storage system, the cost for that replacement must be added to the variable operating cost of the storage system.

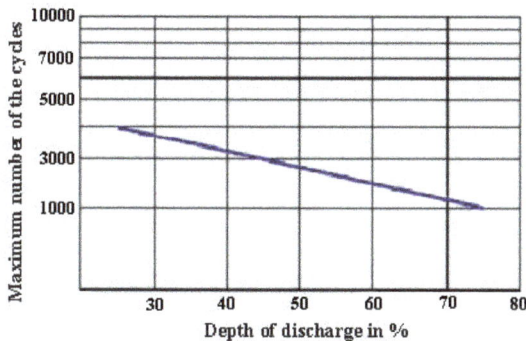

Figure 9. Evolution of cycling capacity as a function of depth of discharge for a lead-acid battery [79]

The design of a storage system that considers the endurance of the unit in terms of cycles should be a primary importance when choosing a system. However, real fatigue processes are often complex and the cycling capacity is not always well defined. In all cases, it is strongly linked to the amplitude of the cycles (Figure 9) and/or the average state of charge [78]. As well, the cycles generally vary greatly, meaning that the quantification of N is delicate and the values given represent orders of magnitude [74].

6.10. Reliability

Like power rating and discharge duration, storage system reliability requirements are circumstance-specific. Little guidance is possible. Storage-system reliability is always an important factor because it is a guarantee of on-demand service [81]. The project design engineer is responsible for designing a plant that provides enough power and that is as reliable as necessary to serve the specific application.

6.11. Response time

Storage response time is the amount of time required to go from no discharge to full discharge. At one extreme, under almost all conditions, storage has to respond quite rapidly if used to provide capacity on the margin *in lieu* of transmission and distribution (T&D) capacity. That is because the output from T&D equipment (*i.e.*, wires and transformers) changes nearly instantaneously in response to demand [73].

In contrast, consider storage used *in lieu* of generation capacity. That storage does not need to respond as quickly because generation tends to respond relatively slowly to demand changes. Specifically, some types of generation – such as engines and combustion turbines – take several seconds to many minutes before generating at full output. For other generation types, such as those fueled by coal and nuclear energy, the response time may be hours [73].

Most types of storage have a response time of several seconds or less. CAES and pumped hydroelectric storage tend to have a slower response, though they still respond quickly enough to serve several important applications.

6.12. Ramp rate

An important storage system characteristic for some applications is the ramp rate – the rate at which power output can change. Generally, storage ramp rates are rapid (*i.e.*, output can change quite rapidly); pumped hydro is the exception. Power devices with a slow response time tend also to have a slow ramp rate [73].

6.13. Charge rate

Charge rate – the rate at which storage can be charged – is an important criterion because, often, modular energy storage (MES) must be recharged so it can serve load during the next day [58]. If storage cannot recharge quickly enough, then it will not have enough energy to provide the necessary service. In most cases, storage charges at a rate that is similar to the rate at which it discharges [73]. In some cases, storage may charge more rapidly or more slowly, depending on the capacity of the power conditioning equipment and the condition and/or chemistry and/or physics of the energy storage medium.

6.14. Self-discharge and energy retention

Energy retention time is the amount of time that storage retains its charge. The concept of energy retention is important because of the tendency for some types of storage to self-discharge or to otherwise dissipate energy while the storage is not in use. In general terms, energy losses could be referred to as *standby* losses [74].

Storage that depends on chemical media is prone to self-discharge. This self-discharge is due to chemical reactions that occur while the energy is stored. Each type of chemistry is different, both in terms of the chemical reactions involved and the rate of self-discharge. Storage that uses mechanical means to store energy tends to be prone to energy dissipation. For example, energy stored using pumped hydroelectric storage may be lost to evaporation. CAES may lose energy due to air escaping from the reservoir [73].

To the extent that storage is prone to self-discharge or energy dissipation, retention time is reduced. This characteristic tends to be less important for storage that is used frequently. For storage that is used infrequently (i.e., is in standby mode for a significant amount of time between uses), this criterion may be very important [72].

6.15. Transportability

Transportability can be an especially valuable feature of storage systems for at least two reasons. First, transportable storage can be (re)located where it is needed most and/or where benefits are most significant [58]. Second, some locational benefits only last for one or two years. Given those considerations, transportability may significantly enhance the prospects that lifecycle benefits will exceed lifecycle cost.

6.16. Power conditioning

To one extent or another, most storage types require some type of power conditioning (*i.e.,* conversion) subsystem. Equipment used for power conditioning – the power conditioning unit (PCU) – modifies electricity so that the electricity has the necessary voltage and the necessary form; either alternating current (AC) or direct current (DC). The PCU, in concert with an included control system, must also synchronize storage output with the oscillations of AC power from the grid [73].

Output from storage with relatively low-voltage DC output must be converted to AC with higher voltage before being discharged into the grid and/or before being used by most load types. In most cases, conversion from DC to AC is accomplished using a device known as an *inverter* [73].

For storage requiring DC input, the electricity used for charging must be converted from the form available from the grid (*i.e.,* AC at relatively high voltage) to the form needed by the storage system (*e.g.,* DC at lower voltage). That is often accomplished via a PCU that can function as a DC 'power supply' [73].

6.17. Power quality

Although requirements for applications vary, the following storage characteristics may or may not be important. To one extent or another, they are affected by the PCU used and/or they drive the specifications for the PCU. In general, higher quality power (output) costs more.

6. *Power Factor:* Although detailed coverage of the concept of power factor is beyond the scope of this report, it is important to be aware of the importance of this criterion. At a minimum, the power output from storage should have an acceptable power factor, where acceptable is somewhat circumstance variable power factor.
7. *Voltage Stability:* In most cases, it is important for storage output voltage to remain somewhat-to-very constant. Depending on the circumstances, voltage can vary; though, it should probably remain within about 5% to 8% of the rated value.
8. *Waveform:* Assuming that storage output is AC, in most cases, the waveform should be as close as possible to that of a sine wave. In general, higher quality PCUs tend to have waveforms that are quite close to that of a sine wave whereas output from lower quality PCUs tends to have a waveform that is somewhat square.
9. *Harmonics:* Harmonic currents in distribution equipment can pose a significant challenge. Harmonic currents are components of a periodic wave whose frequency is an integral multiple of the fundamental frequency [73]. In this case, the fundamental frequency is the utility power line frequency of 60 Hz.

6.18. Modularity

One attractive feature of modular energy storage is the flexibility that system 'building blocks' provide. Modularity allows for more optimal levels and types of capacity and/or discharge duration because modular resources allow utilities to increase or decrease storage capacity, when and where needed, in response to changing conditions [72-73]. Among other attractive effects, modular capacity provides attractive means for utilities to address uncertainty and to manage risk associated with large, 'lumpy' utility T&D investments.

6.19. Storage system reactive power capability

One application (Voltage Support) and one incidental benefit (Power Factor Correction) described in this guide involve storage whose capabilities include absorbing and injecting reactive power (expressed in units of volt-Amperes reactive or VARs) [58], [72-73]. This feature is commonly referred as VAR support. In most cases, storage systems by themselves do not have reactive power capability. For a relatively modest incremental cost, however, reactive power capability can be added to most storage system types.

6.20. Feasibility and adaptation to the generating source

To be highly efficient, a storage system needs to be closely adapted to the type of application (low to mid power in isolated areas, network connection, etc.) and to the type of production (permanent, portable, renewable, etc.) (Figure 10) it is meant to support. It needs to be harmonized with the network.

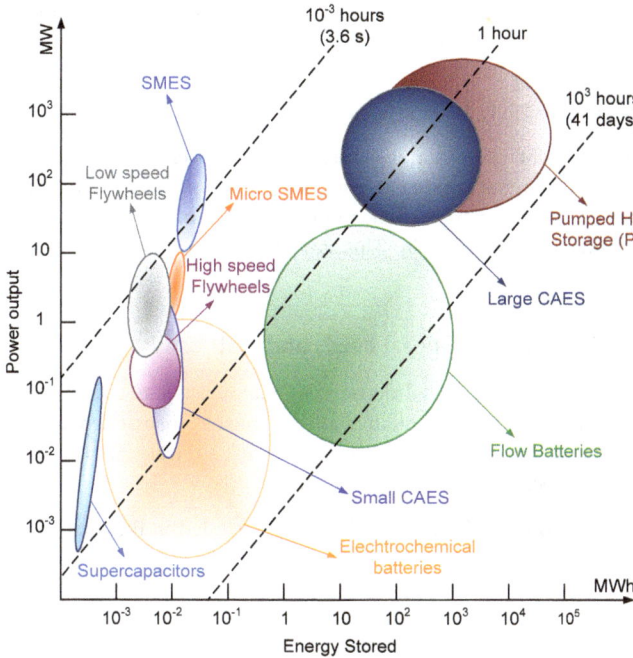

Figure 10. Fields of application of the different storage techniques according to stored energy and power output [80]

6.21. Monitoring, control and communications equipments

This equipment, on both the quality and safety of storage levels, has repercussions on the accessibility and availability of the stored energy [74]. Indeed, storage used for most applications addressed in this report must receive and respond to appropriate control signals. In some cases, storage may have to respond to a dispatch control signal. In other cases, the signal may be driven by a price or prices [73]. Storage response to a control signal may be a simple ramp up or ramp down of power output in proportion to the control signal. A more sophisticated response, requiring one or more control algorithms, may be needed.

6.22. Interconnection

If storage will be charged with energy from the grid or will inject energy into the grid, it must meet applicable interconnection requirements. At the distribution level, an important point of reference is the Institute of Electronics and Electrical Engineers (IEEE) Standard 1547 [82]. Some countries and utilities have more specific interconnection rules and requirements.

6.23. Operational constraints

Especially related to safety (explosions, waste, bursting of a flywheel, etc.) or other operational conditions (temperature, pressure, etc.), they can influence the choice of a storage technology as a function of energy needs [74].

6.24. Environmental aspect

While this parameter is not a criterium of storage-system capacity, the environmental aspect of the product (recyclable materials) is a strong sales pitch. For example, in Nordic countries (Sweden, Norway), a definite margin of the population prefers to pay more for energy than to continue polluting the country [83]. This is a dimension that must not, therefore, be overlooked.

6.25. Decommissioning and disposal needs and cost

In most cases there will be non-trivial decommissioning costs associated with almost any storage system [73]. For example, eventually batteries must be dismantled and the chemicals must be removed. Ideally, dismantled batteries and their chemicals can be recycled, as is the case for the materials in lead-acid batteries.

Ultimately, decommissioning-related costs should be included in the total cost to own and to operate storage.

6.26. Other characteristics

The ease of maintenance, simple design, operational flexibility (this is an important characteristic for the utility), fast response time for the release of stored energy, etc.

Finally, it is important to note that these characteristics apply to the overall storage system: storage units and power converters alike [74].

7. Classification of energy storage systems

There are two criteria to categorise the various ESSs: function and form. In terms of the function, ESS technologies can be categorised into those that are intended firstly for high power ratings with a relatively small energy content making them suitable for power quality or UPS [69]; and those designed for energy management, as shown in Figure 11. PHS, CAES, TES, large-scale batteries, flow batteries, fuel cells, solar fuel and TES fall into the category of energy management, whereas capacitors/super-capacitors, SMES, flywheels and batteries are in the category of power quality and reliability. This simple classification glosses over the wide range of technical parameters of energy storage devices.

Although electricity is not easy to be directly stored cheaply, it can be easily stored in other forms and converted back to electricity when needed. Storage technologies for electricity can also be classified by the form of storage into the following [69]:

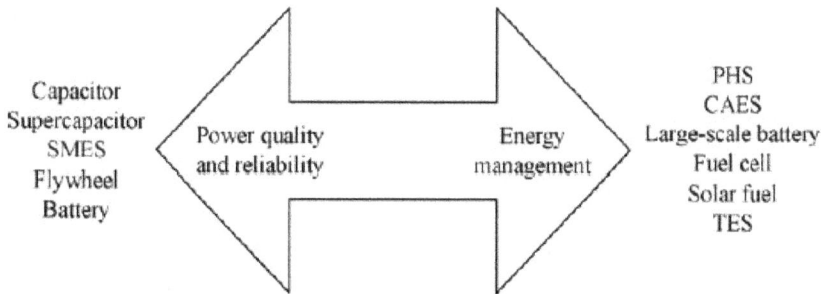

Figure 11. Energy storage classification with respect to function [69].

1. *Electrical energy storage:* (i) Electrostatic energy storage including capacitors and super-capacitors; (ii) Magnetic/current energy storage including SMES.
2. *Mechanical energy storage:* (i) Kinetic energy storage (flywheels); (ii) Potential energy storage (PHES and CAES).
3. *Chemical energy storage:* (i) electrochemical energy storage (conventional batteries such as lead-acid, nickel metal hydride, lithium ion and flow-cell batteries such as zinc bromine and vanadium redox); (ii) chemical energy storage (fuel cells, Molten-Carbonate Fuel Cells – MCFCs and Metal-Air batteries); (iii) thermochemical energy storage (solar hydrogen, solar metal, solar ammonia dissociation–recombination and solar methane dissociation–recombination).
4. *Thermal energy storage:* (i) Low temperature energy storage (Aquiferous cold energy storage, cryogenic energy storage); (ii) High temperature energy storage (sensible heat systems such as steam or hot water accumulators, graphite, hot rocks and concrete, latent heat systems such as phase change materials).

8. Description of energy storage technologies

8.1. Pumped hydro storage (PHS)

In pumping hydro storage, a body of water at a relatively high elevation represents a potential or stored energy. During peak hours the water in the upper reservoir is lead through a pipe downhill into a hydroelectric generator and stored in the lower reservoir. Along off-peak periods the water is pumped back up to recharge the upper reservoir and the power plant acts like a load in power system [72], [84].

Pumping hydro energy storage system (figure 12) consists in two large water reservoirs, electric machine (motor/generator) and reversible pump-turbine group or pump and turbine separated. This system can be started-up in few minutes and its autonomy depends on the volume of stored water.

Restrictions to pumping hydro energy storage are related with geographical constraints and weather conditions. In periods of much rain, pumping hydro capacity can be reduced.

Pumped hydroelectric systems have conversion efficiency, from the point of view of a power network, of about 65–80%, depending on equipment characteristics [72]. Considering the cycle efficiency, 4 kWh are needed to generate three. The storage capacity depends on two parameters: the height of the waterfall and the volume of water. A mass of 1 ton falling 100 m generates 0.272 kWh.

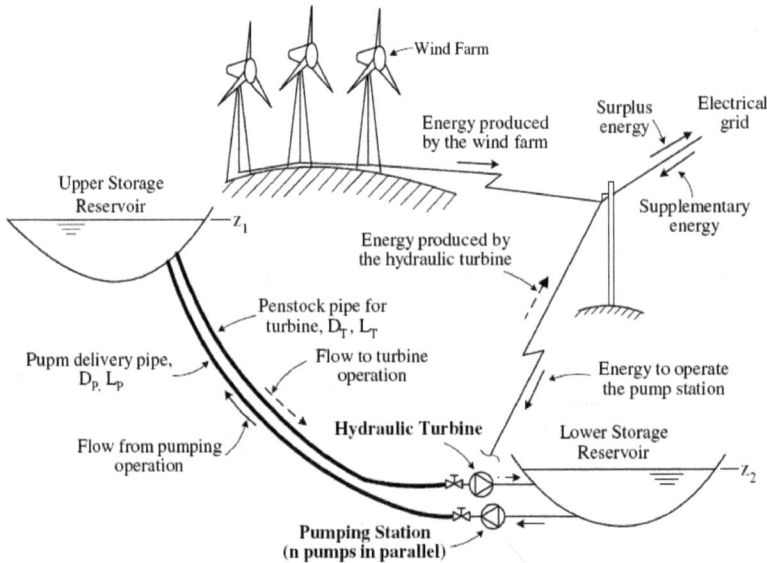

Figure 12. Wind-Pumped hydro energy storage hybrid system [69].

8.2. Batteries energy storage

Batteries store energy in electrochemical form creating electrically charged ions. When the battery charges, a direct current is converted in chemical energy, when discharges, the chemical energy is converted back into a flow of electrons in direct current form [75]. Electrochemical batteries use electrodes both as part of the electron transfer process and store the products or reactants via electrode solid-state reactions [85]. Batteries are the most popular energy storage devices. However, the term battery comprises a sort of several technologies applying different operation principals and materials. There is a wide range of technologies used in the fabrication of electrochemical accumulators (lead–acid (Figure 13), nickel–cadmium, nickel–metal hydride, nickel–iron, zinc–air, iron–air, sodium–sulphur, lithium–ion, lithium–polymer, etc.) and their main assets are their energy densities (up to 150 and 2000 Wh/kg for lithium) and technological maturity. Their main inconvenient however is their relatively low durability for large-amplitude cycling (a few 100 to a few 1000 cycles). They are often used in portable systems, but also in permanent applications (emergency network back-up, renewable-energy storage in isolated areas, etc.) [83].

The minimum discharge period of the electrochemical accumulators rarely reaches below 15 minutes. However, for some applications, power up to 100 W/kg, even a few kW/kg, can be reached within a few seconds or minutes. As opposed to capacitors, their voltage remains stable as a function of charge level. Nevertheless, between a high-power recharging operation at near-maximum charge level and its opposite, that is to say a power discharge nearing full discharge, voltage can easily vary by a ratio of two [74].

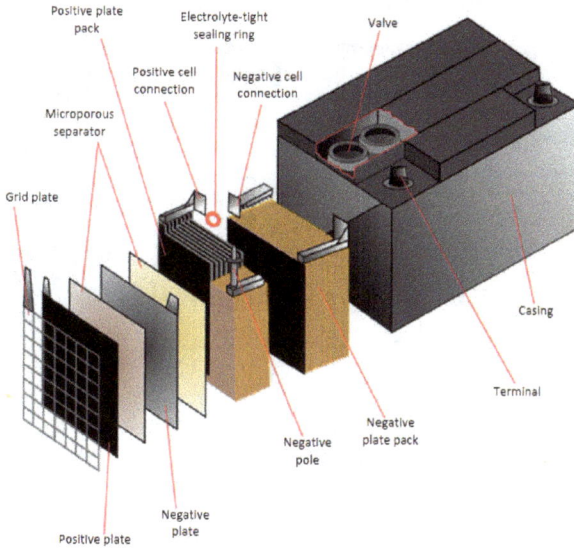

Figure 13. Structure of a lead-acid battery [86]

8.3. Flow batteries energy storage (FBES)

Flow batteries are a two-electrolyte system in which the chemical compounds used for energy storage are in liquid state, in solution with the electrolyte. They overcome the limitations of standard electrochemical accumulators (lead-acid or nickel-cadmium for example) in which the electrochemical reactions create solid compounds that are stored directly on the electrodes on which they form. This is therefore a limited-mass system, which obviously limits the capacity of standard batteries.

Various types of electrolyte have been developed using bromine as a central element: with zinc (ZnBr), sodium (NaBr) (Figure 14), vanadium (VBr) and, more recently, sodium polysulfide. The electrochemical reaction through a membrane in the cell can be reversed (charge-discharge). By using large reservoirs and coupling a large number of cells, large quantities of energy can be stored and then released by pumping electrolyte into the reservoirs.

The main advantages of the technology include the following [87]: 1) high power and energy capacity; 2) fast recharge by replacing exhaust electrolyte; 3) long life enabled by easy electrolyte replacement; 4) full discharge capability; 5) use of nontoxic materials; and 6) low-temperature operation. The main disadvantage of the system is the need for moving mechanical parts such as pumping systems that make system miniaturization difficult. Therefore, the commercial uptake to date has been limited. The best example of flow battery was developed in 2003 by Regenesys Technologies, England, with a storage capacity of 15 MW-120 MWh. It has since been upgraded to an electrochemical system based entirely on vanadium. The overall electricity storage efficiency is about 75 % [88].

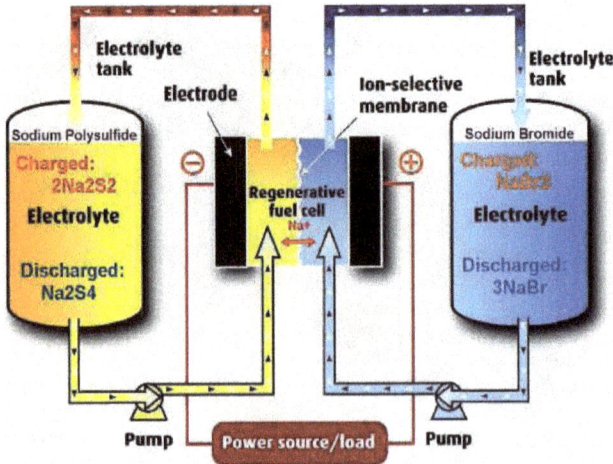

Figure 14. Illustration of a flow-battery

8.4. Flywheel energy storage (FES)

Flywheel energy accumulators are comprised of a massive or composite flywheel coupled with a motor-generator and special brackets (often magnetic), set inside a housing at very low pressure to reduce self-discharge losses (Figure 15) [9]. They have a great cycling capacity (a few 10,000 to a few 100,000 cycles) determined by fatigue design.

To store energy in an electrical power system, high-capacity flywheels are needed. Friction losses of a 200 tons flywheel are estimated at about 200 kW. Using this hypothesis and instantaneous efficiency of 85 %, the overall efficiency would drop to 78 % after 5 hours, and 45 % after one day. Long-term storage with this type of apparatus is therefore not foreseeable.

From a practical point of view, electromechanical batteries are more useful for the production of energy in isolated areas. Kinetic energy storage could also be used for the distribution of electricity in urban areas through large capacity buffer batteries, comparable to water reservoirs, aiming to maximize the efficiency of the production units. For example, large installations made up of forty 25kW-25kWh systems are capable of storing 1 MW that can be released within one hour.

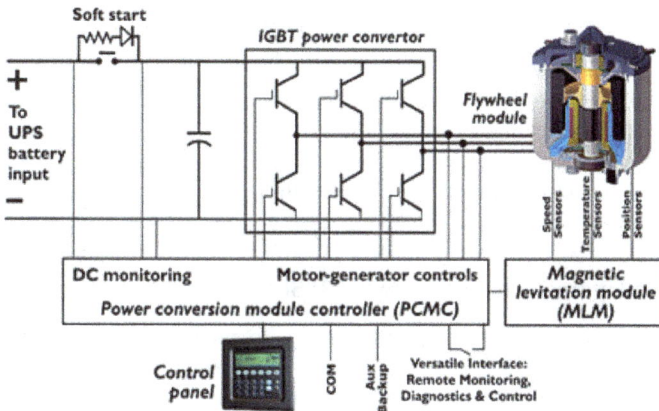

Figure 15. Flywheel energy accumulators [89]

8.5. Supercapacitors energy storage (SES)

Supercapacitors are the latest innovational devices in the field of electrical energy storage. In comparison with a battery or a traditional capacitor, the supercapacitor allows a much powerful power and energy density [15]. Supercapacitors are electrochemical double layer capacitors that store energy as electric charge between two plates, metal or conductive, separated by a dielectric, when a voltage differential is applied across the plates. As like battery systems, capacitors work in direct current.

The energy/volume obtained is superior to that of capacitors (5 Wh/kg or even 15 Wh/kg), at very high cost but with better discharge time constancy due to the slow displacement of ions in the electrolyte (power of 800–2000 W/kg). Super-capacitors generally are very durable, that is to say 8–10 years, 95% efficiency and 5% per day self-discharge, which means that the stored energy must be used quickly.

Supercapacitors find their place in many applications where energy storage is needed, like uninterruptible power supplies, or can help in smoothing strong and short-time power solicitations of weak power networks. Their main advantages are the long life cycle and the short charge/discharge time [2], [19].

8.6. Superconducting magnetic energy storage (SMES)

An emerging technology, systems store energy in the magnetic field created by the flow of direct current in a coil of cryogenically cooled, superconducting material. Due to their construction, they have a high operating cost and are therefore best suited to provide constant, deep discharges and constant activity. The fast response time (under 100 ms) of these systems makes them ideal for regulating network stability (load levelling). Power is available almost instantaneously and very high power output can be provided for a brief period of time [20-21]. These facilities currently range in size up to 3 MW units and are generally used to provide grid stability in a distribution system and power quality at manufacturing facilities requiring ultra-clean power such a chip fabrication facility.

One advantage of this storage system is its great instantaneous efficiency, near 95 % for a charge-discharge cycle [90]. Moreover, these systems are capable of discharging the near totality of the stored energy, as opposed to batteries. They are very useful for applications requiring continuous operation with a great number of complete charge-discharge cycles.

8.7. Fuel cells-Hydrogen energy storage (HES)

Fuel cells are a means of restoring spent energy to produce hydrogen through water electrolysis. The storage system proposed includes three key components: electrolysis which consumes off-peak electricity to produce hydrogen, the fuel cell which uses that hydrogen and oxygen from air to generate peak-hour electricity, and a hydrogen buffer tank to ensure adequate resources in periods of need.

Fuel cells can be used in decentralized production (particularly low-power stations – residential, emergency...), spontaneous supply related or not to the network, mid-power cogeneration (a few hundred kW), and centralized electricity production without heat upgrading. They can also represent a solution for isolated areas where the installation of power lines is too difficult or expensive (mountain locations, etc.). There are several hydrogen storage modes, such as: compressed, liquefied, metal hydride, etc. For station applications, pressurized tanks with a volume anywhere between 10^{-2} m^3 and 10,000 m^3 are the simplest solution to date. Currently available commercial cylinders can stand pressures up to 350 bars.

Combining an electrolyser and a fuel cell for electrical energy storage is a low-efficiency solution (at best 70 % for the electrolyser and 50 % for the fuel cell, and 35 % for the combination). As well, the investment costs are prohibitive and life expectancy is very limited, especially for power network applications [74].

8.8. Thermal energy storage (TES)

Thermal energy storage (TES) already exists in a wide spectrum of applications. It uses materials that can be kept at high/low temperatures in insulated containments. Heat/cold recovered can then be applied for electricity generation using heat engine cycles.

Energy input can, in principle, be provided by electrical resistance heating or refrigeration/cryogenic procedures, hence the overall round trip efficiency of TES is low (30–60%) although the heat cycle efficiency could be high (70–90%), but it is benign to the environment and may have particular advantages for renewable and commercial buildings.

TES systems can be classified into low-temperature TES and high-temperature TES depending on whether the operating temperature of the energy storage material is higher than the room temperature. More precisely, TES can be categorised into industrial cooling (below -18 °C), building cooling (at 0-12 °C), building heating (at 25-50 °C) and industrial heat storage (higher than 175 °C).

8.9. Compressed Air Energy Storage (CAES)

This method consist to use off-peak power to pressurize air into an underground reservoir (salt cavern, abandoned hard rock mine or aquifer) which is then released during peak daytime hours to power a turbine/generator for power production. CAES (Figure 16) is the only other commercially available technology (besides pumped-hydro) able to provide the very-large system energy storage deliverability (above 100 MW in single unit sizes) to use for commodity storage or other large-scale setting [74].

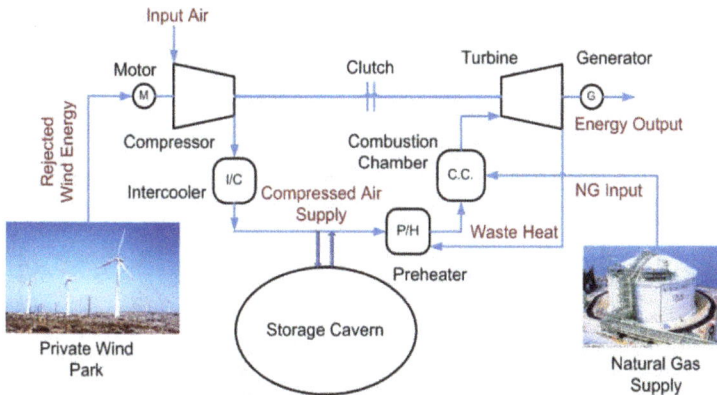

Figure 16. Illustration of compressed-air energy storage

The energy density for this type of system is in the order of 12 kWh/m³ [91], while the estimated efficiency is around 70 % [92]. Let us note that to release 1 kWh into the network, 0.7–0.8 kWh of electricity needs to be absorbed during off-peak hours to compress the air, as well as 1.22 kWh of natural gas during peak hours (retrieval). Two plants currently exist, with several more under development. The first operating unit is a 290 MW unit built in Huntorf, Germany in 1978. The second plant is a 110 MW unit built in McIntosh, Alabama in 1991. Small-scale compressed air energy storage (SSCAES), compressed air storage under high pressure in cylinders (up to 300 bars with carbon fiber structures) are still developing and seem to be a good solution for small- and medium-scale applications.

9. Assessment and comparison of the energy storage technologies

Following, some figures are presented that compare different aspects of storage technologies. These aspects cover topics such as: technical maturity, range of applications, efficiencies, lifetime, costs, mass and volume densities, etc.

9.1. Technical maturity

The technical maturity of the EES systems is shown in Figure 17. The EES technologies can be classified into three categories in terms of their maturity [69]:

1. Mature technologies: PHS and lead-acid battery are mature and have been used for over 100 years.
2. Developed technologies: CAES, NiCd, NaS, ZEBRA Li-ion, Flow Batteries, SMES, flywheel, capacitor, supercapacitor, Al-TES (Aquiferous low- temperature – Thermal energy storage) and HT-TES (High temperature – Thermal energy storage) are developed technologies. All these EES systems are technically developed and commercially available; however, the actual applications, especially for large-scale utility, are still not widespread. Their competitiveness and reliability still need more trials by the electricity industry and the market.
3. Developing technologies: Fuel cell, Meta-Air battery, Solar Fuel and CES (Cryogenic Energy Storage) are still under development. They are not commercially mature although technically possible and have been investigated by various institutions. On the other hand, these developing technologies have great potential for industrial take up in the near future. Energy costs and environmental concerns are the main drivers.

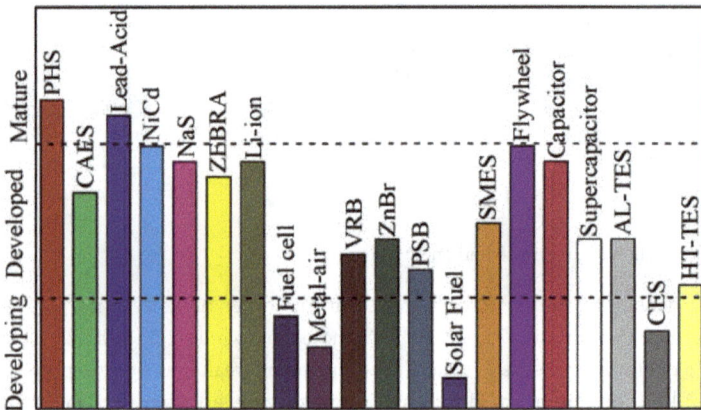

Figure 17. Technical maturity of EES systems [69]

9.2. Power rating and discharge time

The power ratings of various EESs are compared in Table 1. Broadly, the EESs fall into three types according to their applications [69], [74]:

1. *Energy management:* PHS, CAES and CES are suitable for applications in scales above 100 MW with hourly to daily output durations. They can be used for energy management for large-scale generations such as load leveling, ramping/load following, and spinning reserve. Large-scale batteries, flow batteries, fuel cells, CES and TES are suitable for medium-scale energy management with a capacity of 10–100 MW [55], [69].

2. *Power quality:* Flywheel, batteries, SMES, capacitor and supercapacitor have a fast response (~milliseconds) and therefore can be utilized for power quality such as the instantaneous voltage drop, flicker mitigation and short duration UPS. The typical power rating for this kind of application is lower than 1 MW [55].

3. *Bridging power:* Batteries, flow batteries, fuel cells and Metal-Air cells not only have a relatively fast response (<~1 s) but also have relatively long discharge time (hours), therefore they are more suitable for bridging power. The typical power rating for these types of applications is about 100 kW–10 MW [55], [74].

Systems	Power rating and discharge time		Storage duration		Capital cost		
	Power rating	Discharge time	Self discharge per day	Suitable storage duration	$/kW	$/kWh	¢/kWh-Per cycle
PHS	100–5000 MW	1–24 h+	Very small	Hours–months	600–2000	5–100	0.1–1.4
CAES	5–300 MW	1–24 h+	Small	Hours–months	400–800	2–50	2–4
Lead-acid	0–20 MW	Seconds–hours	0.1–0.3%	Minutes–days	300–600	200–400	20–100
NiCd	0–40 MW	Seconds–hours	0.2–0.6%	Minutes–days	500–1500	800–1500	20–100
NaS	50 kW–8 MW	Seconds–hours	~20%	Seconds–hours	1000–3000	300–500	8–20
ZEBRA	0–300 kW	Seconds–hours	~15%	Seconds–hours	150–300	100–200	5–10
Li-ion	0–100 kW	Minutes–hours	0.1–0.3%	Minutes–days	1200–4000	600–2500	15–100
Fuel cells	0–50 MW	Seconds–24 h+	Almost zero	Hours–months	10,000+		6000–20,000
Metal-Air	0–10 kW	Seconds–24 h+	Very small	Hours–months	100–250	10–60	
VRB	30 kW–3 MW	Seconds–10 h	Small	Hours–months	600–1500	150–1000	5–80
ZnBr	50 kW–2 MW	Seconds–10 h	Small	Hours–months	700–2500	150–1000	5–80
PSB	1–15 MW	Seconds–10 h	Small	Hours–months	700–2500	150–1000	5–80
Solar fuel	0–10 MW	1–24 h+	Almost zero	Hours–months	–	–	–
SMES	100 kW–10 MW	Milliseconds–8 s	10–15%	Minutes–hours	200–300	1000–10,000	
Flywheel	0–250 kW	Milliseconds–15 min	100%	Seconds–minutes	250–350	1000–5000	3–25
Capacitor	0–50 kW	Milliseconds–60 min	40%	Seconds–hours	200–400	500–1000	
Super-capacitor	0–300 kW	Milliseconds–60 min	20–40%	Seconds–hours	100–300	300–2000	2–20
AL-TES	0–5 MW	1–8 h	0.5%	Minutes–days		20–50	
CES	100 kW–300 MW	1–8 h	0.5–1.0%	Minutes–days	200–300	3–30	2–4
HT-TES	0–60 MW	1–24 h+	0.05–1.0%	Minutes–months		30–60	

Table 1. Comparison of technical characteristics of EES systems [69]

9.3. Storage duration

Table 1 also illustrates the self-discharge (energy dissipation) per day for EES systems. One can see that PHS, CAES, Fuel Cells, Metal-Air Cells, solar fuels and flow batteries have a very small self-discharge ratio so are suitable for a long storage period. Lead-Acid, NiCd, Li-ion, TESs and CES have a medium self-discharge ratio and are suitable for a storage period not longer than tens of days [69].

NaS, ZEBRA, SMES, capacitor and supercapacitor have a very high self-charge ratio of 10–40% per day. They can only be implemented for short cyclic periods of a maximum of several hours. The high self-discharge ratios of NaS and ZEBRA are from the high working temperature which needs to be self-heating to maintain the use of the storage energy [69].

Flywheels will discharge 100% of the stored energy if the storage period is longer than about 1 day. The proper storage period should be within tens of minutes.

9.4. Capital cost

Capital cost is one of the most important factors for the industrial take-up of the EES. They are expressed in the forms shown in Table 2, cost per kWh, per kW and per kWh per cycle. All the costs per unit energy shown in the table have been divided by the storage efficiency to obtain the cost per output (useful) energy [69]. The per cycle cost is defined as the cost per unit energy divided by the cycle life which is one of the best ways to evaluate the cost of energy storage in a frequent charge/discharge application, such as load levelling. For example, while the capital cost of lead-acid batteries is relatively low, they may not necessarily be the least expensive option for energy management (load levelling) due to their relatively short life for this type of application. The costs of operation and maintenance, disposal, replacement and other ownership expenses are not considered, because they are not available for some emerging technologies [55], [69].

Systems	Energy and power density				Life time and cycle life		Influence on environment	
	Wh/kg	W/kg	Wh/L	W/L	Life time (years)	Cycle life (cycles)	Influence	Description
PHS	0.5–1.5		0.5–1.5		40–60		Negative	Destruction of trees and green land for building the reservoirs
CAES	30–60		3–6	0.5–2.0	20–40		Negative	Emissions from combustion of natural gas
Lead-acid	30–50	75–300	50–80	10–400	5–15	500–1000	Negative	Toxic remains
NiCd	50–75	150–300	60–150		10–20	2000–2500		
NaS	150–240	150–230	150–250		10–15	2500		
ZEBRA	100–120	150–200	150–180	220–300	10–14	2500+		
Li-ion	75–200	150–315	200–500		5–15	1000–10,000+		
Fuel cell	800–10,000	500+	500–3000	500+	5–15	1000+	Negative	Remains and/or combustion of fossil fuel
Metal-Air	150–3000		500–10,000			100–300	Small	Little amount of remains
VRB	10–30		16–33		5–10	12,000+	Negative	Toxic remains
ZnBr	30–50		30–60		5–10	2000+		
PSB	–		–	–	10–15			
Solar fuel	800–100,000		500–10,000		–	–	Benign	Usage and storage of solar energy
SMES	0.5–5	500–2000		0.2–2.5	1000–4000	20+	100,000+	Negative	Strong magnetic fields
Flywheel	10–30	400–1500	20–80	1000–2000	~15	20,000+	Almost none	
Capacitor	0.05–5	~100,000	2–10	100,000+	~5	50,000+	Small	Little amount of remains
Super-			capacitor	2.5–15	500–5000			100,000+
					10–30			
20+		100,000+	Small					Little amount of remains
AL-TES	80–120		80–120		10–20		Small	
CES	150–250	10–30	120–200		20–40		Positive	Removing contaminates during air liquefaction (Charge)
HT-TES	80–200		120–500		5–15		Small	

Table 2. Comparison of technical characteristics of EES systems [69]

CAES, Metal-Air battery, PHS, TESs and CES are in the low range in terms of the capital cost per kWh. The Metal-Air batteries may appear to be the best choice based on their high energy density and low cost, but they have a very limited life cycle and are still under development. Among the developed techniques, CAES has the lowest capital cost compared to all the other systems. The capital cost of batteries and flow batteries is slightly higher than the break even cost against the PHS although the gap is gradually closing. The SMES, flywheel, capacitor and supercapacitor are suitable for high power and short duration applications, since they are cheap on the output power basis but expensive in terms of the storage energy capacity [55], [69].

The costs per cycle kWh of PHS and CAES are among the lowest among all the EES technologies, the per cycle cost of batteries and flow batteries are still much higher than PHS and CAES although a great decrease has occurred in recent years. CES is also a promising technology for low cycle cost. However, there are currently no commercial products available. Fuel cells have the highest per cycle cost and it will take a long time for them to be economically competitive. No data have been found for the solar fuels as they are in the early stage of development [55], [69].

It should also be noted that the capital cost of energy storage systems can be significantly different from the estimations given here due to, for example, breakthroughs in technologies, time of construction, location of plants, and size of the system. The information summarised here should only be regarded as being preliminary.

9.5. Cycle efficiency

The cycle efficiency of EES systems during one charge-discharge cycle is illustrated in Figure 18. The cycle efficiency is the "round-trip" efficiency defined as ratio between output energy and input energy. The self-discharge loss during the storage is not considered. One can see that the EES systems can be broadly divided into three groups:

1. *Very high efficiency:* SMES, flywheel, supercapacity and Li-ion battery have a very high cycle efficiency of > 90%.
2. *High efficiency:* PHS, CAES, batteries (except for Li-ion), flow batteries and conventional capacitor have a cycle efficiency of 60–90%. It can also be seen that storing electricity by compression and expansion of air using the CAES is usually less efficient than pumping and discharging water with PHSs, since rapid compression heats up a gas, increasing its pressure thus making further compression more energy consuming [55], [69].
3. *Low efficiency:* Hydrogen, DMFC, Metal-Air, solar fuel, TESs and CES have an efficiency lower than ~60% mainly due to large losses during the conversion from the commercial AC side to the storage system side. For example, hydrogen storage of electricity has relatively low round-trip energy efficiency (~20–50%) due to the combination of electrolyser efficiency and the efficiency of re-conversion back to electricity [55], [69].

It must be noted that there is a trade-off between the capital cost and round-trip efficiency, at least to some extent. For example, a storage technology with a low capital cost but a low round-trip efficiency may well be competitive with a high cost, high round-trip efficiency technology.

9.6. Energy and power density

The power density (W/kg or W/litre) is the rated output power divided by the volume of the storage device. The energy density is calculated as a stored energy divided by the volume. The volume of the storage device is the volume of the whole energy storage system

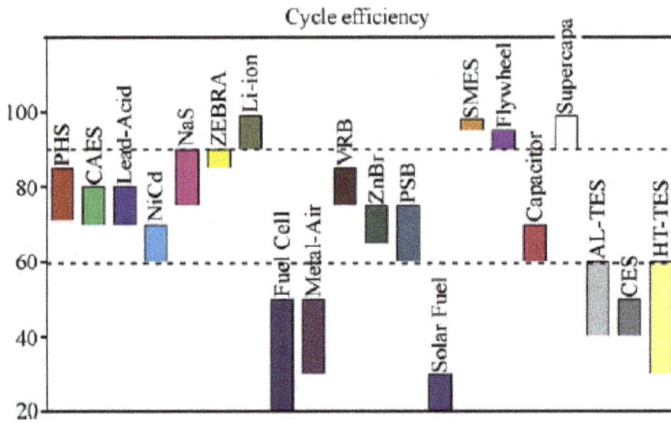

Figure 18. Cycle efficiency of EES systems. [69]

including the energy storing element, accessories and supporting structures, and the inverter system. As can be seen from Table 1, the Fuel Cells, Metal-Air battery, and Solar fuels have an extremely high energy density (typically ~1000 Wh/kg), although, as mentioned above, their cycle efficiencies are very low. Batteries, TESs, CES and CAES have medium energy density. The energy density of PHS, SMES, Capacitor/supercapacitor and flywheel are among the lowest below ~30 Wh/kg. However, the power densities of SMES, capacitor/supercapacitor and flywheel are very high which are, again, suitable for applications for power quality with large discharge currents and fast responses [69], [74]. NaS and Li-ion have a higher energy density than other conventional batteries. The energy densities of flow batteries are slightly lower than those of conventional batteries. It should be noted that there are differences in the energy density of the same type of EES made by different manufacturers [55].

9.7. Life time and cycle life

Also compared in Table 1 are life time and/or cycle life for various EESs. It can be seen that the cycle lives of EES systems whose principles are largely based on the electrical technologies are very long normally greater than 20,000. Examples include SMES, capacitor and supercapacitor. Mechanical and thermal energy storage systems, including PHS, CAES, flywheel, AL-TES, CES and HT-TES, also have long cycle lives. These technologies are based on conventional mechanical engineering, and the life time is mainly determined by the life time of the mechanical components. The cycle abilities of batteries, flow batteries, and fuel cells are not as high as other systems due to chemical deterioration with the operating time. Metal-Air battery only has a life of a few hundred cycles and obviously needs to be further developed [55], [69].

10. Conclusion

The key element of this analysis is the review of the available energy storage techniques applicable to electrical power systems.

There is obviously a cost associated to storing energy, but we have seen that, in many cases, storage is already cost effective. More and more application possibilities will emerge as further research and development is made in the field [91].

Storage is a major issue with the increase of renewable but decentralized energy sources that penetrate power networks [93]. Not only is it a technical solution for network management, ensuring real-time load leveling, but it is also a mean of better utilizing renewable resources by avoiding load shedding in times of overproduction. Coupled with local renewable energy generation, decentralized storage could also improve power network sturdiness through a network of energy farms supplying a specific demand zone.

Many solutions are available to increase system security, but they are so different in terms of specifications that they are difficult to compare. This is why we tried to bring out a group of technical and economical characteristics which could help improve performance and cost estimates for storage systems.

Based on the review, the following conclusions could be drawn [55], [69], [74]:

1. Although there are various commercially available EES systems, no single storage system meets all the requirements for an ideal EES - being mature, having a long lifetime, low costs, high density and high efficiency, and being environmentally benign. Each EES system has a suitable application range. PHS, CAES, large-scale batteries, flow batteries, fuel cells, solar fuels, TES and CES are suitable for energy management application; flywheels, batteries, capacitors and supercapacitors are more suitable for power quality and short duration UPS, whereas batteries, flow batteries, fuel cells and Metal-Air cells are promising for the bridging power.
2. PHS and Lead-Acid battery are technically mature; CAES, NiCd, NaS, ZEBRA Li-ion, flow battery, SMES, flywheel, capacitor, Supercapacitor, AL-TES and HT-TES are technically developed and commercially available; Fuel Cell, Meta-Air battery, Solar Fuel and CES are under development. The capital costs of CAES, Metal-Air battery, PHS, TES and CES are lower than other EESs. CAES has the lowest capital cost among the developed technologies. Metal-Air battery has the potential to be the cheapest among currently known EES systems.
3. The cycle efficiencies of SMES, flywheel, capacitor/supercapacitor, PHS, CAES, batteries, flow batteries are high with the cycle efficiency above 60%. Fuel Cell, DMFC, Metal-Air, solar fuel, TES and CES have a low efficiency mainly due to large losses during the conversion from commercial AC to the storage energy form.
4. The cycle lives of the EES systems based on the electrical technologies, such as SMES, capacitor and supercapacitor, are high. Mechanical and thermal based EES, including PHS, CAES, flywheel, ALTES, CES and HT-TES, also have a long cycle life. The cycle abilities of batteries, flow batteries, and fuel cells are not as high as other systems due to

chemical deterioration with the operating time. Metal-Air battery has the lowest life time at least currently.

5.   PHS, CAES, batteries, flow batteries, fuel cells and SMES are considered to have some negative effects on the environment due to one or more of the following: fossil combustion, strong magnetic field, landscape damage, and toxic remains. Solar fuels and CES are more environmentally friendly. However, a full life-cycle analysis should be done before a firm conclusion can be drawn.

Based on the contents of this study and carefully measuring the stakes, we find that:

1.   The development of storage techniques requires the improvement and optimization of power electronics, often used in the transformation of electricity into storable energy, and vice versa.
2.   The rate of penetration of renewable energy will require studies on the influence of the different storage options, especially those decentralized, on network sturdiness and overall infrastructure and energy production costs.
3.   The study of complete systems (storage, associated transformation of electicity, power electronics, control systems...) will lead to the optimization of the techniques in terms of cost, efficiency, reliability, maintenance, social and environmental impacts, etc.
4.   It is important to assess the national interest for compressed gas storage techniques.
5.   Investment in research and development on the possibility of combining several storage methods with a renewable energy source will lead to the optimization of the overall efficiency of the system and the reduction of greenhouse gases created by conventional gas-burning power plants.
6.   Assessing the interest for high-temperature thermal storage systems, which have a huge advantage in terms of power delivery, will lead to the ability of safely establish them near power consumption areas,
7.   The development of supercapacitors will lead to their integration into the different types of usage.
8.   The development of low-cost, long-life flywheel storage systems will lead to increased potential, particularly for decentralized applications.
9.   To increase the rate of penetration and use of hydrogen-electrolysor fuel-cell storage systems, a concerted R&D effort will have to be made in this field.

Author details

Hussein Ibrahim
TechnoCentre éolien, Gaspé, QC, Canada

Adrian Ilinca
Université du Québec à Rimouski, Rimouski, QC, Canada

11. References

[1]   Smith, S.C.; Sen, P.K.; Kroposki, B., Advancement of Energy Storage Devices and Applications in Electrical Power System. Power and Energy Society General Meeting - Conversion and Delivery of Electrical Energy in the 21st Century, 22-24 2008

[2]   P.C. Ghosh , B. Emonts, H. Janßen, J. Mergel, D. Stolten. "Ten years of operational experience with a hydrogen-based renewable energy supply system", Solar Energy 75, 2003, pp. 469–478

[3]   A. Bilodeau, K. Agbossou. Control analysis of renewable energy system with hydrogen storage for residential applications, Journal of Power Sources, 2005.

[4]   N.Hamsic, A.Schmelter, A.Mohd, E.Ortjohann, E.Schultze, A.Tuckey, J.Zimmermann. "Stabilising the Grid Voltage and Frequency in Isolated Power Systems Using a Flywheel Energy Storage System," The Great Wall World Renewable Energy Forum , Beijing, China, October 2006

[5]   E.Ortjohann, N.Hamsic , A.Schmelter , A.Mohd, J.Zimmermann, A.Tuckey, E.Schultze. Increasing Renewable Energy Penetration in Isolated Grids Using a Flywheel Energy Storage System, the first International Conference on Power Engineering, Energy and Electrical Drives (POWERENG,IEEE),Portugal, April 2007.

[6]   European Renewable Energy Council, "Renewable energy in Europe: building Markets and capacity," James and James science publishers, August 2004.

[7]   "FP7 Research Priorities for the Renewable Energy Sector", Bruxelles, EUREC Agency, March 2005.

[8]   Melissa M, "Flywheel Energy Storage System: The current status and future prospect," Trinity Power Corporation, February 2004, pp.8-11.

[9]   A.Ruddell, G.Schönnenbeck, R.Jones, "Flywheel Energy Storage Systems," Rutherford Appleton Lab, UK.

[10] H.Bindner, "Power Control for Wind Turbines in Weak Grids," Risø National Lab, March 1999.

[11] Alaa Mohd, Egon Ortjohann, Andreas Schmelter, Nedzad Hamsic, Danny Morton. Challenges in integrating distributed Energy storage systems into future smart grid. IEEE International Symposium on Industrial Electronics, June 30 - July 2, 2008.

[12] James A. McDowall, "Status and Outlook of the Energy Storage Market," PES 2007, Tampa, July 2007.

[13] Baker JN, Collinson A. Electrical energy storage at the turn of the millennium. Power Eng J 1999;6:107–12.

[14] Dti Report. Status of electrical energy storage systems. DG/DTI/00050/00/00, URN NUMBER 04/1878, UK Department of Trade and Industry; 2004, p. 1–24.

[15] Australian Greenhouse Office. Advanced electricity storage technologies programme. ISBN: 1 921120 37 1, Australian Greenhouse Office; 2005, p. 1–35.

[16] Ahearne J. Storage of electric energy, Report on research and development of energy technologies. International Union of Pure and Applied Physics; 2004, p. 76–86. Available online http://www.iupap.org/wg/energy/report-a.pdf

[17] http://www.en.wikipeida.org/wiki/Hydroelectric_energy_storage

[18] Linden S. The commercial world of energy storage: a review of operating facilities (under construction or planned). In: Proceeding of 1st annual conference of the energy storage council, Houston, Texas, March 3, 2003.

[19] Walawalkar R, Apt J, Mancini R. Economics of electric energy storage for energy arbitrage and regulation. Energy Policy 2007;5:2558–68.

[20] Dti Report. Review of electrical energy storage technologies and systems and of their potential for the UK. DG/DTI/00055/00/00, URN NUMBER 04/1876, UK Department of Trade and Industry; 2004, p. 1–34.

[21] Bueno C, Carta JA. Wind powered pumped hydro storage systems, a means of increasing the penetration of renewable energy in the Canary Islands. Renew Sus Energy Rev 2006;10:312–40.

[22] Najjar Y, Zaamout MS. Performance analysis of compressed air energy storage (CAES) plant for dry regions. Energy Convers Manage 1998;39:1503–11.

[23] Sears JR. TEX: The next generation of energy storage technology. Telecommunications Energy Conference, INTELEC 2004. In: 26th annual international volume, Issue, Sept. 19–23 2004; p. 218–22.

[24] Najjar Y, Jubeh N. Comparison of performance of compressed-air energy-storage plant with compressed-air storage with humidification. In: Proceeding of IMechE, Part A: Journal of Power and Energy 2006; 220:581–8.

[25] Bullough C, Gatzen C, Jakiel C, et al. Advanced adiabatic compressed air energy storage for the integration of wind energy. In: Proceedings of the European wind energy conference. London, UK, 2004.

[26] Wang S, Chen G, Fang M, et al. A new compressed air energy storage refrigeration system. Energ Convers Manage 2006;47:3408–16.

[27] Cook GM, Spindler WC, Grefe G. Overview of battery power regulation and storage. IEEE T Energy Conver 1991;6:204–11.

[28] Chalk SG, Miller JF. Key challenges and recent progress in batteries, fuel cells and hydrogen storage for clean energy systems. J Power Sources 2006;159:73–80.

[29] Kashem MA, Ledwich G. Energy requirement for distributed energy resources with battery energy storage for voltage support in threephase distribution lines. Electr Pow Syst Res 2007;77:10–23.

[30] Kluiters EC, Schmal D, Ter Veen WR, et al. Testing of a sodium/ nickel chloride (ZEBRA) battery for electric propulsion of ships and vehicles. J Power Sources 1999;80:261–4.

[31] Karpinski AP, Makovetski B, Russell SJ, et al. Silver-zinc: status of technology and applications. J Power Sources 1999;80:53–60.

[32] Linden S. Bulk energy storage potential in the USA, current developments and future prospects. Energy 2006; 31:3446–57.

[33] Chalk SG, Miller JF. Key challenges and recent progress in batteries, fuel cells and hydrogen storage for clean energy systems. J Power Sources 2006;159:73–80.

[34] Weinmann O. Hydrogen-the flexible storage for electrical energy. Power Eng J-Special Feature: Electrical energy storage 1999:164–70.

[35] Steinfeld A, Meier A. Solar thermochemical process technology. In: Encycl Energy. Elsevier Inc; 2004, 5: 623-37.

[36] Kolkert WJ, Jamet F. Electric energy gun technology: status of the French–German–Netherlands programme. IEEE T Magn 1999; 35:25–30.

[37] Koshizuka N, Ishikawa F, Nasu H. Progress of superconducting bearing technologies for flywheel energy storage systems. Physica C 2003;386:444–50.

[38] Xue X, Cheng K, Sutanto D. A study of the status and future of superconducting magnetic energy storage in power systems. Supercond Sci Tech 2006;19:39.

[39] Kondoh J, Ishii I, Yamaguchi H, et al. Electrical energy storage systems for energy networks. Energ Convers Manage 2000;41:1863–74

[40] Suzuki Y, Koyanagi A, Kobayashi M. Novel applications of the flywheel energy storage system. Energy 2005; 30:2128–43.

[41] http://www.beaconpower.com/products/EnergyStorageSystems/flywheels.htm

[42] Rosen MA. Second-law analysis of aquifer thermal energy storage systems. Energy 1999; 24:167–82.

[43] Ameri M, Hejazi SH, Montaser K. Performance and economics of the thermal energy storage systems to enhance the peaking capacity of the gas turbines. Appl Therm Eng 2005; 25:241–51.

[44] Akbari H, Sezgen O. Performance evaluation of thermal energy storage systems. Energ Buildings 1995; 22:15–24.

[45] Ordonez CA, Plummer MC. Cold thermal storage and cryogenic heat engines for energy storage applications. Energy Sources 1997; 19(4):389–96.

[46] Kessling W, Laevemann E, Peltzer M. Energy storage in open cycle liquid desiccant cooling systems. Int J Refrig 1998; 21(2):150–6.

[47] Kishimoto K, Hasegawa K, Asano T. Development of generator of liquid air storage energy system. Mitsubishi Heavy Industries, Ltd. Technical Review, 1998; 35(3):117–20.

[48] Chen HS, Ding YL. A cryogenic energy system using liquid/slush air as the energy carrier and waste heat and waste cold to maximise efficiency, specifically it does not use combustion in the expansion process. UK Patent G042226PT, 2006-02-27.

[49] Wen DS, Chen HS, Ding YL, et al. Liquid nitrogen injection into water: pressure build-up and heat transfer. Cryogenics 2006; 46: 740–8.

[50] Chen HS, Ding YL, Toby P, et al. A method of storing energy and a cryogenic energy storage system. International Patent WO/2007/ 096656, 2007-08-30.

[51] Mclarnon FR, Cairns EJ. Energy storage. Ann Rev Energy 1989;14:241–71.

[52] Dobie WC. Electrical energy storage. Power Eng J 1998; 12:177–81.

[53] Koot M, Kessels J, Jager B, et al. Energy management strategies for vehicular electric power systems. IEEE T Veh Technol 2005; 54:771–82.

[54] Weinstock IB. Recent advances in the US Department of Energy's energy storage technology research and development programs for hybrid electric and electric vehicles. J Power Sources 2002; 110:471–4.

[55] Energy Storage Association, www.electricitystorage.org

[56] Ibrahim H., A. Ilinca, R. Younès, J. Perron, T. Basbous, "Study of a Hybrid Wind-Diesel System with Compressed Air Energy Storage", IEEE Canada, EPC2007, Montreal, Canada, October 25-26, 2007.

[57] H. Ibrahim, A. Ilinca, J. Perron, "Solutions actuelles pour une meilleure gestion et intégration de la ressource éolienne". CSME/SCGM Forum 2008 at Ottawa. The Canadian Society for Mechanical Engineering, 5-8 Juin 2008

[58] David Connolly. A Review of Energy Storage Technologies For the integration of fluctuating renewable energy. Ph.D. project, University of Limerick, October 2010.

[59] Baxter, R., Energy Storage - A Nontechnical Guide, PennWell Corporation, Oklahoma, 2006.

[60] Energy Storage Systems, Sandia National Laboratories, 14th October 2007,

[61] Gonzalez, A., Ó'Gallachóir, B., McKeogh, E. & Lynch, K. Study of Electricity Storage Technologies and Their Potential to Address Wind Energy Intermittency in Ireland. Sustainable Energy Ireland, 2004.

[62] Makansi J, Abboud J. Energy storage: the missing link in the electricity value chain-An ESC White Paper. Energy storage Council 2002; p. 1-23.

[63] Moore T, Douglas J. Energy storage, big opportunities on a smaller scale. EPRI J 2006; Spring Issue, p. 16–23.

[64] Mears D. EPRI-DOE storage handbook-storage for wind resources. In: Annual peer review meeting of DOE energy storage systems research. Washington DC, USA, Nov. 10–11, 2004, p. 1–18.

[65] Ribeiro PF, Johson BK, Crow ML, et al. Energy storage systems for advanced power applications. In: Proceedings of the IEEE 2001;89:1744–56.

[66] Ratering-Schnitzler B, Harke R, Schroeder M, et al. Voltage quality and reliability from electrical energy-storage systems. J Power Sources 1997;67:173–7.

[67] Y. Rebours, D. Kirschen. "What is spinning reserve?", The University of Manchester ,Sept 2005.

[68] Market Analysis of Emerging Electric Energy Storage Systems. National Energy Technology Laboratory. DOE/NETL-2008/1330. July 31, 2008

[69] Haisheng Chen, Thang Ngoc Cong, Wei Yang, Chunqing Tan, Yongliang Li, Yulong Ding. Progress in electrical energy storage system: A critical review. Progress in Natural Science. July 2008.

[70] James M. Eyer, Joseph J. Iannucci, Garth P. Corey. "Energy Storage Benefits and Market Analysis Handbook, "Sandia National Laboratories REPORT, SAND2004-6177, December 2004.

[71] http://www.buchmann.ca

[72] Mears, D. Gotschall, H. EPRI-DOE Handbook of Energy Storage for Transmission and Distribution Applications. Electric Power Research Institute Report #1001834. December 2003.

[73] Jim Eyer, Garth Corey. Energy Storage for the Electricity Grid: Benefits and Market Potential Assessment Guide. SANDIA REPORT SAND2010-0815, February 2010.

[74] H. Ibrahim, A. Ilinca, J. Perron, Energy storage systems—Characteristics and comparisons, Renewable and Sustainable Energy Reviews, Volume 12, Issue 5, June 2008, Pages 1221–1250.

[75] Robin G, Rullan M, Multon B, Ben Ahmaed H, Glorennec PY. Solutions de stockage de l'énergie pour les systèmes de production intermittente d'électricité renouvelable. 2004.

[76] Shoenung, Dr. Susan M. Hassenzahl, William M. Long - versus Short-Term Energy Storage Technologies Analysis, A Lifecycle Cost Study. Sandia National Laboratories, Energy Storage Program, Office of Electric Transmission and Distribution, U.S. Department of Energy. Sandia National Laboratories Report #SAND2003-2783. August 2003.

[77] Gergaud O. Modélisation énergétique et optimisation économique d'un système de production éolien et photovoltaïque couplé au réseau et associé à un accumulateur. Thèse de l'ENS de Cachan;décembre 2002.

[78] Mémoire concernant la contribution possible de la production éolienne en réponse à l'accroissement de la demande québecoise d'électricité d'ici 2010. Régie de l'énergie, Québec, Canada. Dossier R-3526;Avril 2004.

[79] R. Messenger R, Ventre J. Photovoltaic systems engineering. CRC Press, 1999.

[80] Emerging Energy Storage Technologies in Europe. Rapport Frost & Sullivan; 2003.

[81] Bonneville JM. Stockage cinétique. Institut de Recherche de l'Hydro-Québec, Varennes, Québec, Canada. Rapport interne; février 1975.

[82] IEEE 1547 Standard for Interconnecting Distributed Resources with Electric Power Systems. Approved by the IEEE Standards Board in June 2003. Approved as an American National Standard in October 2003.

[83] Faure F. Suspension magnétique pour volant d'inertie. Thèse de doctorat. Institut National Polytechnique de Grenoble,France;Juin 2003.

[84] Alaa Mohd, Egon Ortjohann, Andreas Schmelter, Nedzad Hamsic, Danny Morton. Challenges in integrating distributed Energy storage systems into future smart grid. IEEE International Symposium on Industrial Electronics, June 30 - July 2, 2008.

[85] Rudell A. Storage and Fuel Cells. EPSRC SuperGen Workshop : Future Technologies for a Sustainable Electricity System. Univ. of Cambridge, 7 november2003.

[86] Lead/acid batteries, University of Cambridge, 15th October 2007, http://www.doitpoms.ac.uk/tlplib/batteries/batteries_lead_acid.php

[87] Sergio Vazquez, Srdjan M. Lukic, Eduardo Galvan, Leopoldo G. Franquelo, Juan M. Carrasco, Energy Storage Systems for Transport and Grid Applications, IEEE TRANSACTIONS ON INDUSTRIAL ELECTRONICS, VOL. 57, NO. 12, DECEMBER 2010, Pages. 3881-3895.

[88] www.cea.fr

[89] http://www.power-thru.com/flywheel_detail.html

[90] Anzano JP, Jaud P, Madet D. Stockage de l'électricité dans le système de production électrique. Techniques de l'ingénieur, traité de Génie Électrique D4030; 1989.

[91] Multon B, Ruer J. Stocker l'électricité : Oui, c'est indispensable, et c'est possible ! pourquoi, où, comment?. Publication ECRIN en contribution au débat national sur l'énergie; Avril 2003.

[92] Robyns B. Contribution du stockage de l'énergie électrique à la participation au services système des éoliennes. Séminaire SRBE – SEE – L2EP « Éolien et réseaux : enjeux », 22 mars 2005.

[93] Multon B. L'énergie électrique : analyse des ressources et de la production. Journées section électrotechnique du club EEA. Paris, France, 28-29 janvier 1999, p. 8.

Chapter 6

Analysis and Control of Flywheel Energy Storage Systems

Yong Xiao, Xiaoyu Ge and Zhe Zheng

Additional information is available at the end of the chapter

1. Introduction

Since a few years ago, electrical energy storage has been attractive as an effective use of electricity and coping with the momentary voltage drop. Above all, flywheel energy storage systems (FESS) using superconductor have advantages of long life, high energy density, and high efficiency (Subkhan & Komori, 2011), and is now considered as enabling technology for many applications, such as space satellites and hybrid electric vehicles (Samineni et al., 2006; Suvire & Mercado, 2012). Also, the contactless nature of magnetic bearings brings up low wear, absence of lubrication and mechanical maintenance, and wide range of work temperature (Bitterly, 1998; Beach & Christopher, 1998). Moreover, the closed-loop control of magnetic bearings enables active vibration suppression and on-line control of bearing stiffness (Cimuca et al., 2006; Park et al., 2008).

Active magnetic bearing is an open-loop unstable control problem. Therefore, an initial controller based on a rigid rotor model has to be introduced to levitate the rotor. In reality, the spinning rotor under the magnetic suspension may experience two kinds of whirl modes. The conical whirl mode gives rise to the gyroscopic forces to twist the rotor, thereby severely affecting stability of the rotor if not properly controlled (Okada et al, 1992; Williams et al., 1990). The translatory whirl mode constrains the rotor to synchronous motion in the radial direction so as to suppress the gyroscopic rotation, which has been extensively used in industry (Tomizuka et al, 1992; Tsao et al., 2000). The synchronization control has also been shown to be very capable in dealing with nonlinear uncertain models, and to be very effective in disturbance rejection for systems subject to synchronous motion. Until the advent of synchronization control, the prevalent use of the synchronization controller has been limited to stable mechanical systems and therefore is not readily applicable to magnetic systems which are unstable in nature and highly nonlinear (Yang & Chang, 1996).

In the past three decades the theory of optimal control has been well developed in nearly all aspects, such as stability, nonlinearity, and robustness (Summers et al., 2011; Rawlings et al., 2008; Mayne, et al., 2000). It is known that multivariable constrained control problems in state-space can be effectively handled using Linear Quadratic Gaussian (LQG). An application of the optimal control to synchronize multiple motion axes has been reported in (Zhu & Chen, 2001; Xiao & Zhu, 2006), where cross-coupling design of generalized predictive control was presented by compensating both the tracking error and the synchronous error. In this chapter, robust MPC control algorithms for the flywheel energy storage system with magnetically assisted bearings are developed. The controllers are derived through minimization of a modified cost function, in which the synchronization errors are embedded so as to reduce the synchronization errors in an optimal way.

2. Flywheel structure

Fig.1 illustrates the basic structure of a flywheel system with integrated magnetic bearings. The motor and generator with disk-type geometry are combined into a single electric machine, and the rotor is sandwiched between two stators. Each of the stators carries a set of three-phase copper winding to be fed with sinusoidal currents. Furthermore, both axial faces of the rotor contain rare-earth permanent magnets embedded beneath the surfaces. The radial magnetic bearing which consists of eight pairs of electromagnets is constructed around the circumference of hollow center. A combination of active and passive magnetic bearings allows the rotor to spin and remain in magnetic levitation.

The control of such a system normally includes two steps. First, the spinning speed and the axial displacement of the rotor are properly regulated (Zhang & Tseng, 2007). Second, a synchronization controller is introduced to suppress the gyroscopic rotation of the rotor caused by the outside disturbance and model uncertainty (Xiao et al., 2005).

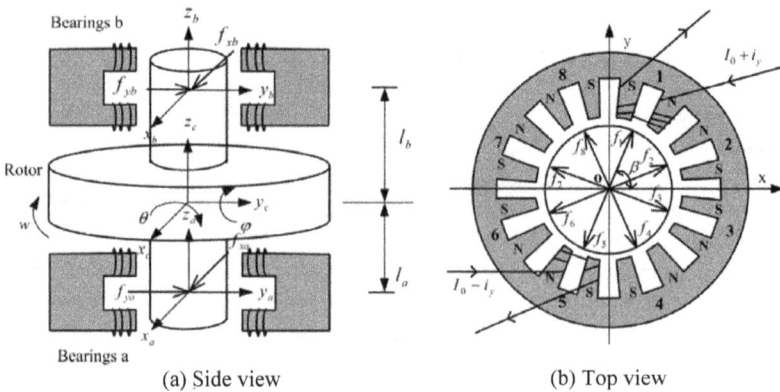

(a) Side view   (b) Top view

Figure 1. The flywheel energy storage system

3. System dynamics

Let x_c and y_c denote the displacements of the mass center of the rotor in the x and y-directions, and θ and φ the roll angles of rotation about x-axis and y-axis, respectively. Note that θ and φ are assumed to be small since the air gap is very narrow within the magnetic bearings. It is also assumed that the rotor is rigid with its inertia perfectly balanced about the z-axis so that the flexibility and eccentricity of the rotor are not considered herein; thereby, the variation effects of tensor of inertia due to the roll motion of the rotor can be negligible.

The mass center of the rotor in the radial direction can be described by

$$\begin{cases} m\ddot{x}_c = f_{xa} + f_{xb} + e_x \\ m\ddot{y}_c = f_{ya} + f_{yb} + e_y \\ J_x\ddot{\theta} = f_{yb}l_b - f_{ya}l_a - J_z w\dot{\varphi} + e_\theta \\ J_y\ddot{\varphi} = f_{xa}l_a - f_{xb}l_b + J_z w\dot{\theta} + e_\varphi \end{cases} \tag{1}$$

where $\quad x_c = \dfrac{l_b}{l_a+l_b}x_a + \dfrac{l_a}{l_a+l_b}x_b$, $y_c = \dfrac{l_b}{l_a+l_b}y_a + \dfrac{l_a}{l_a+l_b}y_b$, $\qquad \theta = \dfrac{1}{l_a+l_b}(y_b - y_a)$,

$\varphi = \dfrac{1}{l_a+l_b}(x_a - x_b)$, m is the mass of the rotor, J_x, J_y and J_z are the moments of inertia about x-axis, y-axis and z-axis respectively, w is the spinning rate about z-axis, f_{xa}, f_{xb}, f_{ya} and f_{yb} are the magnetic forces along the radial directions, e_x, e_y, e_θ and e_φ are the disturbances.

According to the Maxwell's law, the magnetic forces f_{xa}, f_{xb}, f_{ya} and f_{yb} have nonlinear relationships with the control currents and displacements of the rotor. Then, the magnetic forces at equilibriums can be linearized with Taylor's method (Zhu et al., 2009),

$$\begin{bmatrix} f_{xa} \\ f_{xb} \\ f_{ya} \\ f_{yb} \end{bmatrix} \approx K_p \begin{bmatrix} x_a \\ x_b \\ y_a \\ y_b \end{bmatrix} + K_c \begin{bmatrix} i_{xa} \\ i_{xb} \\ i_{ya} \\ i_{yb} \end{bmatrix} \tag{2}$$

where $K_p = \dfrac{8GI_0^2}{h_0^3}$, $K_c = \dfrac{8GI_0\sin\beta}{h_0^2}$, β is a constant angle corresponding to the structure of electromagnets, I_0 is the bias current, i_x and i_y are the control currents near x-axis and y-axis, respectively, h_0 is the nominal air gap at equilibrium, G is an electromagnet constant given by $G = \dfrac{1}{4}\mu_0 A_g N^2$, μ_0 is the air permeability, A_g is the cross-sectional area of air gap, and N is the number of turns of the winding circuit.

Then, the state-space model of (1) is obtained,

$$\begin{cases} \dot{X} = A_c X + B_c u + B_x d_x \\ Z = C_c X + D_z d_z \end{cases} \tag{3}$$

where $X = \left[x_a, \dot{x}_a, x_b, \dot{x}_b, y_a, \dot{y}_a, y_b, \dot{y}_b \right]^T$ is the state variable, $u = \left[i_{xa}, i_{xb}, i_{ya}, i_{yb} \right]^T$ is the forcing

vector, $Z = \left[x_a, x_b, y_a, y_b \right]^T$ is the output vector, and $T_1 = T_2 = T_3 = T_4 = \left[1, 0 \right]$ are the output

transition matrices, d_x and d_z denote model uncertainties or system disturbances with

appropriate matrices B_x and D_z,

$$A_c = \begin{bmatrix} 0 & 1 & 0 & 0 & 0 & 0 & 0 & 0 \\ A_{c21} & 0 & A_{c23} & 0 & 0 & A_{c26} & 0 & A_{c28} \\ 0 & 0 & 0 & 1 & 0 & 0 & 0 & 0 \\ A_{c41} & 0 & A_{c43} & 0 & 0 & A_{c46} & 0 & A_{c48} \\ 0 & 0 & 0 & 0 & 0 & 1 & 0 & 0 \\ 0 & A_{c62} & 0 & A_{c64} & A_{c65} & 0 & A_{c67} & 0 \\ 0 & 0 & 0 & 0 & 0 & 0 & 0 & 1 \\ 0 & A_{c82} & 0 & A_{c84} & A_{c85} & 0 & A_{c87} & 0 \end{bmatrix},$$

$$B_c = \begin{bmatrix} 0 & 0 & 0 & 0 \\ B_{c21} & B_{c22} & 0 & 0 \\ 0 & 0 & 0 & 0 \\ B_{c41} & B_{c42} & 0 & 0 \\ 0 & 0 & 0 & 0 \\ 0 & 0 & B_{c63} & B_{c64} \\ 0 & 0 & 0 & 0 \\ 0 & 0 & B_{c83} & B_{c84} \end{bmatrix}, \quad C_c = diag \left(T_1, T_2, \cdots, T_4 \right).$$

where

$$A_{c21} = -\left(\frac{1}{m} + \frac{l_a^2}{J_y} \right) K_p, \quad A_{c23} = \left(\frac{1}{m} - \frac{l_a l_b}{J_y} \right) K_p, \quad A_{c26} = -\frac{J_z w l_a}{J_y \left(l_a + l_b \right)},$$

$$A_{c28} = \frac{J_z w l_a}{J_y \left(l_a + l_b \right)}, \quad A_{c41} = \left(\frac{1}{m} - \frac{l_a l_b}{J_y} \right) K_p, \quad A_{c43} = -\left(\frac{1}{m} + \frac{l_b^2}{J_y} \right) K_p, \quad A_{c46} = \frac{J_z w l_b}{J_y \left(l_a + l_b \right)},$$

$$A_{c48} = -\frac{J_z w l_b}{J_y \left(l_a + l_b \right)}, \quad A_{c62} = \frac{J_z w l_a}{J_x \left(l_a + l_b \right)}, \quad A_{c64} = -\frac{J_z w l_a}{J_x \left(l_a + l_b \right)}, \quad A_{c65} = -\left(\frac{1}{m} + \frac{l_a^2}{J_x} \right) K_p,$$

$$A_{c67} = \left(\frac{1}{m} - \frac{l_a l_b}{J_x}\right)K_p, \qquad A_{c82} = -\frac{J_z w l_b}{J_x(l_a + l_b)}, \qquad A_{c84} = \frac{J_z w l_b}{J_x(l_a + l_b)}, \qquad A_{c85} = \left(\frac{1}{m} - \frac{l_a l_b}{J_x}\right)K_p,$$

$$A_{c87} = \left(\frac{1}{m} + \frac{l_b^2}{J_x}\right)K_p, \qquad B_{c21} = \left(\frac{1}{m} + \frac{l_a^2}{J_y}\right)K_c, \qquad B_{c22} = \left(\frac{1}{m} - \frac{l_a l_b}{J_y}\right)K_c, \qquad B_{c41} = \left(\frac{1}{m} - \frac{l_a l_b}{J_y}\right)K_c,$$

$$B_{c42} = \left(\frac{1}{m} + \frac{l_b^2}{J_y}\right)K_c, \qquad B_{c63} = \left(\frac{1}{m} + \frac{l_a^2}{J_x}\right)K_c, \qquad B_{c64} = \left(\frac{1}{m} - \frac{l_a l_b}{J_x}\right)K_c, \qquad B_{c83} = \left(\frac{1}{m} - \frac{l_a l_b}{J_x}\right)K_c,$$

$$B_{c84} = \left(\frac{1}{m} + \frac{l_b^2}{J_x}\right)K_c.$$

During a closed-loop control phase, the position and rate of the shaft are constantly monitored by contactless sensors, and are processed in a controller, so that a control current to the coils of electromagnets which attract or repel the shaft is amplified and fed back.

4. Controller design

Let he discrete-time model of (3) be described by

$$\begin{cases} X(k+1) = AX(k) + Bu(k) \\ Z(k) = CX(k) \end{cases} \tag{4}$$

where k denotes the discrete time. Note that the disturbance term is ignored.

By introducing the following synchronization errors,

$$\delta(k) = LCX(k) \tag{5}$$

where

$$\delta(k) = \begin{bmatrix} x_a - x_b \\ x_b - y_a \\ y_a - y_b \\ y_b - x_a \end{bmatrix}, L = \begin{bmatrix} 1 & -1 & 0 & 0 \\ 0 & 1 & -1 & 0 \\ 0 & 0 & 1 & -1 \\ -1 & 0 & 0 & 1 \end{bmatrix},$$

it has the modified cost function,

$$J(k) = \sum_{i=1}^{H_p} \hat{Z}^T(k+i\,|\,k)\hat{Z}(k+i\,|\,k) + \lambda \sum_{i=1}^{H_c} \hat{u}^T(k+i-1\,|\,k)\hat{u}(k+i-1\,|\,k)$$

$$+ v\sum_{i=1}^{H_p} \hat{\delta}^T(k+i\,|\,k)\hat{\Delta}(k+i\,|\,k) \tag{6}$$

where $\hat{Z}(k+i|k)$ is the future output vector, $\hat{u}(k+i-1|k)$ is the future control input vector, $\hat{\delta}(k+i|k)$ is the future synchronization errors, H_p is the prediction horizon, H_c is the control horizon, λ is the positive weighting factor used to adjust the control action, v is the non-negative weighting factor for the synchronization error.

Rewrite (6) as,

$$J(k) = \sum_{j=0}^{H_p-1} \Theta(\hat{X},\hat{u}) + \hat{X}^T\left(k+H_p|k\right)P_0\hat{X}^T\left(k+H_p|k\right) \tag{7}$$

where

$$\Theta\left(\hat{X},u\right) = \hat{X}^T\left(k+H_p-1+j|k\right)Q_j\hat{X}(k+H_p-1+j|k)$$
$$+\hat{u}^T\left(k+H_p-1+j|k\right)R_j\hat{u}\left(k+H_p-1+j|k\right) \tag{8}$$

$$P_0 = C^TC + vC^TL^TLC \tag{9}$$

$$Q_j = \begin{cases} C^TC + vC^TL^TLC & if \quad j=0,1,...,H_p-2 \\ 0 & if \quad j=H_p-1 \end{cases} \tag{10}$$

$$R_j = \begin{cases} \infty I, & if \quad j=0,1...,H_p-H_c-1 \\ \lambda I, & if \quad j=H_p-H_c,...,H_p-1' \end{cases} \tag{11}$$

where I is the unit matrix with appropriate dimension.

Hence, minimization of the cost function (7) results in the synchronization control law,

$$u(k) = -K_{H_p-1}X(k) \tag{12}$$

where

$$K_{H_p-1} = (B^TP_{H_p-1}B + \lambda I)^{-1}B^TP_{H_p-1}A \tag{13}$$

$$P_{j+1} = A^TP_jA - A^TP_jB(B^TP_jB + R_j)^{-1}B^TP_jA + Q_j \tag{14}$$

from the initial condition P_0.

Indeed, as receding horizon LQG control is a stationery feedback strategy, over an infinite interval, questions of stability naturally arise while solutions are slow to emerge. On the other hand, the stability of the proposed controller (12) can sometimes be guaranteed with finite horizons, even if there is no explicit terminal constraint. The finite horizon predictive control problem is normally associated with a time-varying RDE, which is related to the optimal value of the cost function. Attempts at producing stability result for MPC on the basics of its explicit input–output description have been remarkably unsuccessful, usually necessitating the abandonment of a specific control performance.

5. Stability analysis

Lemma 1. Consider the following ARE with an infinite-horizon linear quadratic control (Souza et al., 1996),

$$P = A^T PA - A^T PB(B^T PB + R)^{-1} B^T PA + Q \tag{15}$$

where

- $[A,B]$ is stabilizable,
- $[A,Q^{1/2}]$ is detectable,
- $Q \geq 0$ and $R > 0$.

Then

- there exists a unique, maximal, non-negative definite symmetric solution $\bar{P}$.
- $\bar{P}$ is a unique stabilizing solution, *i.e.*, $A - B(B^T \bar{P} B + R)^{-1} B^T \bar{P} A$ has all the eigenvalues strictly within the unit circle.

Rewrite (15) as

$$P = (A - BK)^T P(A - BK) + K^T RK + Q \tag{16}$$

In order to connect the RDE (14) to the ARE (15), the Fake Algebraic Riccati Technique (FART) is used as follows:

$$P_j = (A - BK_j)^T P_j(A - BK_j) + K_j^T R_j K_j + \bar{Q}_j \tag{17}$$

where $\bar{Q}_j = Q_j - (P_{j+1} - P_j)$. Clearly, while one has not altered the RDE in viewing it as a masquerading ARE, the immediate result from Lemma 1 and (17) can be obtained.

Theorem 1. Consider (17) with $\bar{Q}_j$. If

- $[A,B]$ is stabilizable,
- $[A,Q_j^{1/2}]$ is detectable,
- $\bar{Q}_j \geq 0$ and $R_j > 0$.

then P_j is stabilizing, *i.e.* the closed-loop transition matrix

$$\bar{A}_j = A - B\left(B^T P_j B + R_j\right)^{-1} B^T P_j A \tag{18}$$

has all its eigenvalues strictly within the unit circle.

Regarding the receding horizon strategy, only P_j with $j = H_p - 1$ will be applied. This leads to

$$P_{H_p-1} = A^T P_{H_p-1} A - A^T P_{H_p-1} B\left(B^T P_{H_p-1} B + \lambda I\right)^{-1} B^T P_{H_p-1} A + \bar{Q}_{H_p-1} \tag{19}$$

where $\bar{Q}_{H_p-1} = P_{H_p-1} - P_{H_p}$. Then, the stability result of the control system can be given by the following theorem.

Theorem 2. Consider (19) with the weighting matrix $\bar{Q}_{H_p-1}$. If

- $\begin{bmatrix} A, B \end{bmatrix}$ is stabilizable,
- $\begin{bmatrix} A, \bar{Q}_{H_p-1}^{1/2} \end{bmatrix}$ is detectable,
- P_{H_p-1} is non-increasing, $\lambda > 0$,

then the controller (12) is stabilizing, i.e., the closed-loop transition matrix $\bar{A}_{H_p-1} = A - B\left(B^T P_{H_p-1} B + \lambda I\right)^{-1} B^T P_{H_p-1} A$ has all its eigenvalues strictly within the unit circle.

Proof. The proof is completed by setting $j = H_p - 1$ in Theorem 1.

It can be seen from the above theorem that the prediction horizon H_p is a key parameter for stability, and an increasing H_p is always favorable. This was the main motivation to extend the one-step-ahead control to long range predictive control. However, a stable linear feedback controller may not remain stable for a real system $P(z)$ with model uncertainty, which is normally related to stability robustness of the system. The most common specification of model uncertainty is norm-bounded, and the frequency response of a nominal model (3) can be obtained by evaluating:

$$\hat{P}(z) = C\left(zI - A\right)^{-1} B \tag{20}$$

Then, the real system $P(z)$ is given by a 'norm-bounded' description:

$$P(z) = \hat{P}(z) + \Delta_A, \text{ for additive model uncertainties} \tag{21}$$

where Δ_A is stable bounded operator, and $P(z)$ is often normalized in such a way that $\|\Delta\| \le 1$.

Because one does not know exactly what Δ is, various assumptions can be made about the nature of Δ: nonlinear, linear time-varying, linear parameter-varying and linear time-invariant being the most common ones. Also, various norms can be used, and the most commonly used one is the 'H-infinity' norm $\|\Delta\|_\infty$, which is defined as the worst-case 'energy gain' of an operator even for nonlinear systems. It then follows from the small-gain theorem that the feedback combination of this system with the uncertainty block Δ_A will remain stable if

$$\bar{\sigma}\left[K_{H_p-1}\left(e^{j\omega T_s}\right)S\left(e^{j\omega T_s}\right)\right]\|\Delta_A\|_\infty < 1 \tag{22}$$

where $\bar{\sigma}[\cdot]$ denotes the largest singular value, $S(z)=\left[I+\hat{P}(z)K_{H_p-1}(z)\right]^{-1}$ is the sensitivity function. Note that (22) is only a sufficient condition for robust stability; if it is not satisfied, robust stability may nevertheless have been obtained. In practice, when tuning a controller, one can try to influence the frequency response properties in such a way as to make (22) hold.

6. Simulation study

Stability robustness with respect to variable control parameters will first be carried out. The y-axis of each graph indicates the maximum singular value of $\bar{\sigma}\left[K_{H_p-1}\left(e^{j\omega T_s}\right)S\left(e^{j\omega T_s}\right)\right]$, and the x-axis is the frequency range, $\omega = 10^{-2} \sim 1\,\text{Hz}$. Then, the performance of the proposed controller will be demonstrated in the presence of external disturbances and model uncertainties.

Consider the flywheel system with parameters given in (Zhu & Xiao, 2009), and assume that the rotor is spinning at a constant speed. As the eigenvalues of A_c are: $\pm 2.0353i$, $\pm 10.4i$, ± 149.3, ± 149.3, the open-loop continuous system is obviously unstable. With appropriate control parameters for the discrete-time model (sampling period $T_s = 0.008$ s), such as $H_p = 6$, $H_c = 1$, $\lambda = 0.01$, $v = 10$, all of the eigenvalues of the closed-loop transition matrix $\bar{A}_{Hp-1}$ are within the unit circle, which are: $-0.782 \pm 0.555i$, -0.379, 0.481, 0.378, $0.127 \pm 0.195i$ and -0.027 respectively. In another word, the system can be stabilized with this feedback controller.

6.1. Stability robustness against control parameters

The prediction and control horizons are closely related to the stability of the closed-loop system. In the case of additive uncertainties, the maximum singular value $\bar{\sigma}\left[K_{H_p-1}\left(e^{j\omega T_s}\right)S\left(e^{j\omega T_s}\right)\right]$ against variation of prediction horizon is illustrated in Fig. 2, while $H_c = 1$, $\lambda = 0.01$ and $v = 0$ are set. It can be seen that a larger prediction horizon results in a smaller singular value, which means that the stability robustness of the control system can be improved. As a rule of thumb, H_p can be chosen according to $H_p = \text{int}(2\omega_s / \omega_b)$, where ω_s is the sampling frequency and ω_b is the bandwidth of the process. Fig. 3 shows the singular value when the control horizon is varying. Clearly, a smaller control horizon H_c may enhance the stability robustness of the control system. However, if the nominal model of the process is accurate enough, and the influence of model uncertainties is negligible, then $H_c > 1$ is preferred for faster system responses.

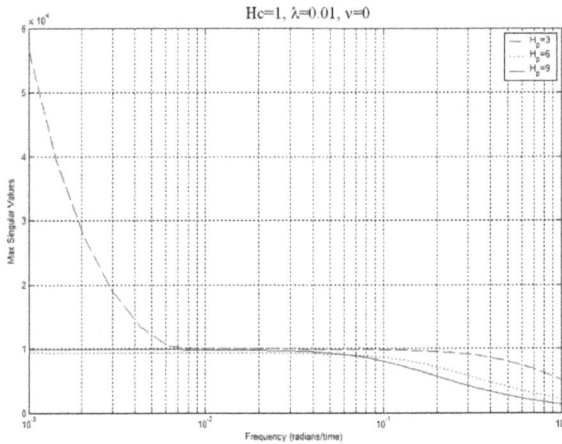

Figure 2. Maximum singular value $\bar{\sigma}\left[K_{H_p-1}\left(e^{j\omega T_s}\right)S\left(e^{j\omega T_s}\right)\right]$ against prediction horizon H_p

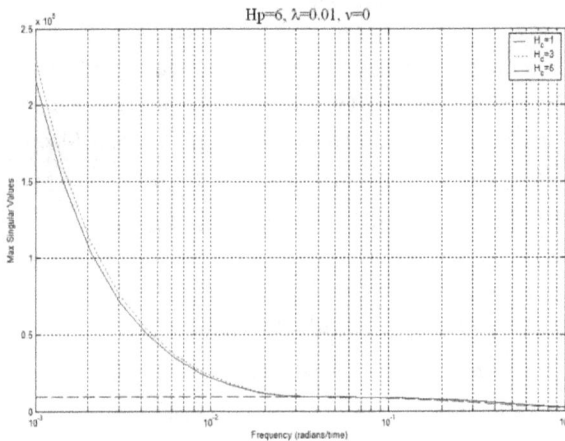

Figure 3. Maximum singular value $\bar{\sigma}\left[K_{H_p-1}\left(e^{j\omega T_s}\right)S\left(e^{j\omega T_s}\right)\right]$ against control horizon H_c

The stability robustness bounds shown in Fig. 4 is obtained by varying λ, while $H_p = 6$, $H_c = 1$ and $v = 0$ are set. Clearly, a larger value of λ can improve the stability robustness of the control system. This is because that the increasing λ will reduce the control action and the influence of the model uncertainties on system stability will become less important. Consequently, the stability robustness can be enhanced. If $\lambda \to \infty$, the feedback action disappears and the closed loop is broken. In general, a larger λ should be chosen when the system stability might be degraded due to significant model uncertainty. However, if the model uncertainty is insignificant, a smaller λ would then be expected as the system response can be improved in this case, i.e., a decrease in the response time. In practice, a careful choice of λ is necessary as it may have a large range of the values and is difficult to predetermine it.

The synchronization factor v is introduced to compensate the synchronization error of the rotor in radial direction. Fig. 5 shows that the influence of v on stability robustness is not consistent over frequency. In particular, a lower value of v can enhance the stability robustness at certain frequencies, but the performance will be degraded at higher frequencies. Another interesting observation is that the two boundaries for $v = 5$ and $v = 10$ are almost overlapping. It means that the stability robustness of the control system will not be affected if a further increase of v is applied. In general, one can increase the prediction horizon and the synchronization control weighting factor so that the stability of the control system is maintained while the synchronization performance can be improved.

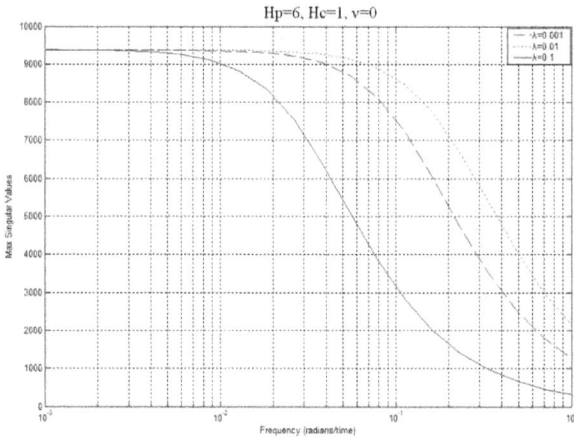

Figure 4. Maximum singular value $\bar{\sigma}\left[K_{H_p-1}\left(e^{j\omega T_s}\right) S\left(e^{j\omega T_s}\right) \right]$ against weighting factor λ

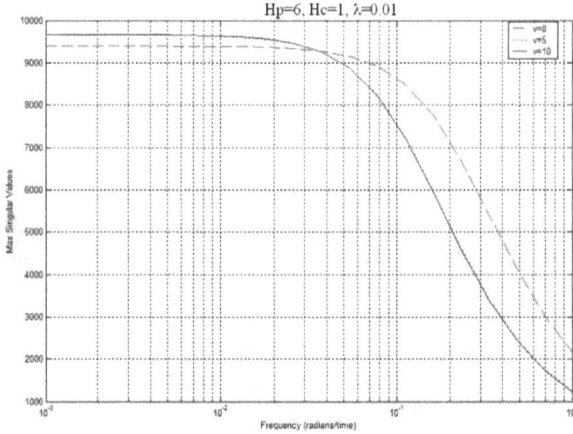

Figure 5. Maximum singular value $\bar{\sigma}\left[K_{H_p-1}\left(e^{j\omega T_s}\right)S\left(e^{j\omega T_s}\right)\right]$ against synchronization factor v

6.2. Disturbances on magnetic forces

In this simulation, force disturbances are introduced to the bearings of the rotor at different time instants, and amplitudes are 0.5N, -0.5N, 0.5N and -0.5N on xa-axis, xb-axis, ya-axis and yb-axis respectively. The duration of 0.2 seconds for each disturbance is assumed. Figs. 6-11 show the numerical results of the control algorithm when $H_p = 10$, $H_c = 1$, $\lambda = 0.01$ are set for the two cases: with $v = 0$, and $v = 10$. Clearly, without cross-coupling control action due to $v = 0$, evident synchronization errors and a conical whirl mode during the transient responses are resulted. However, when $v = 10$ is introduced, the synchronization performance can be improved significantly, especially in terms of the rolling angles, as shown in Fig. 11. Therefore, with adequately selected control parameters the improved synchronization performance as well as guaranteed stability of the FESS can be obtained, and in consequence, the whirling rotor in the presence of disturbances would be suppressed near the nominal position.

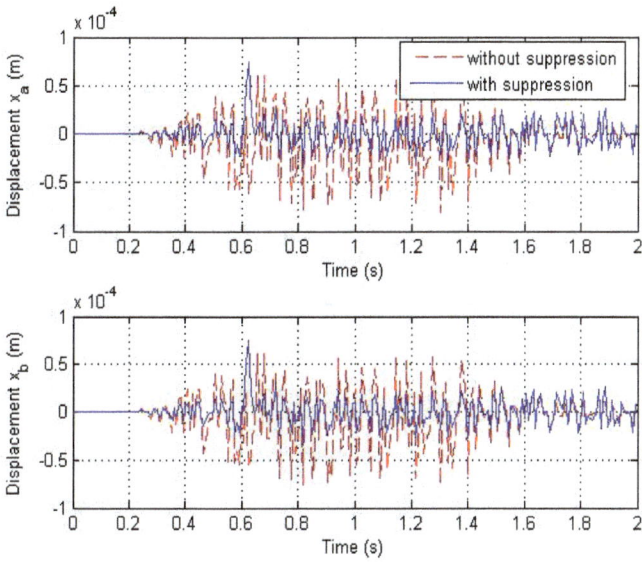

Figure 6. Radial displacements of the rotor along x-axis

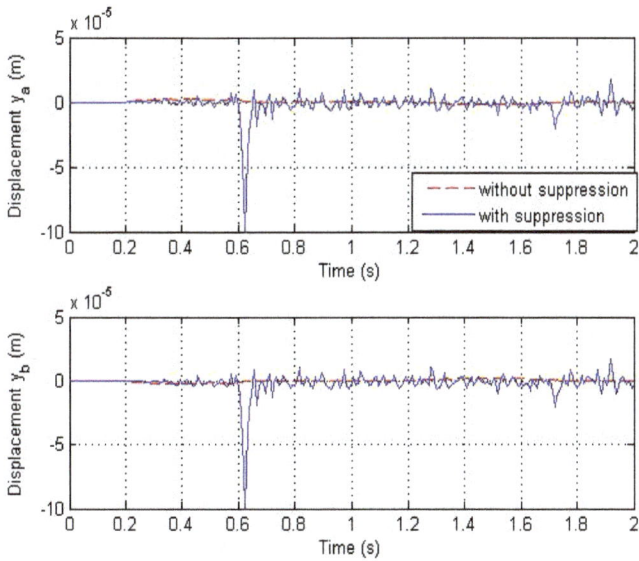

Figure 7. Radial displacements of the rotor along y-axis

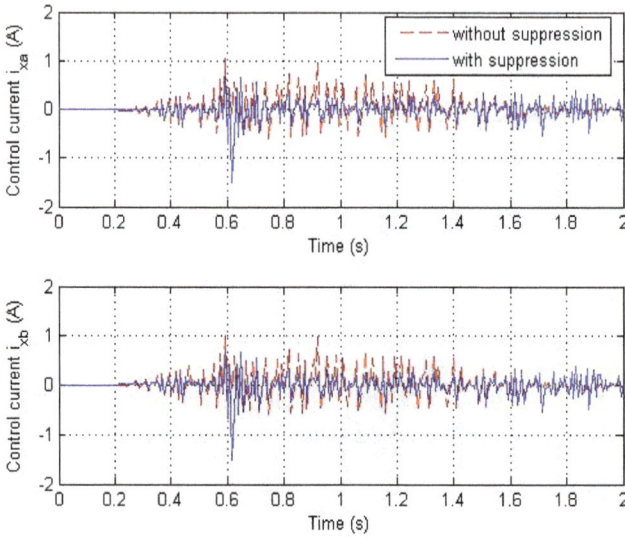

Figure 8. Control currents to bearings along x-axis

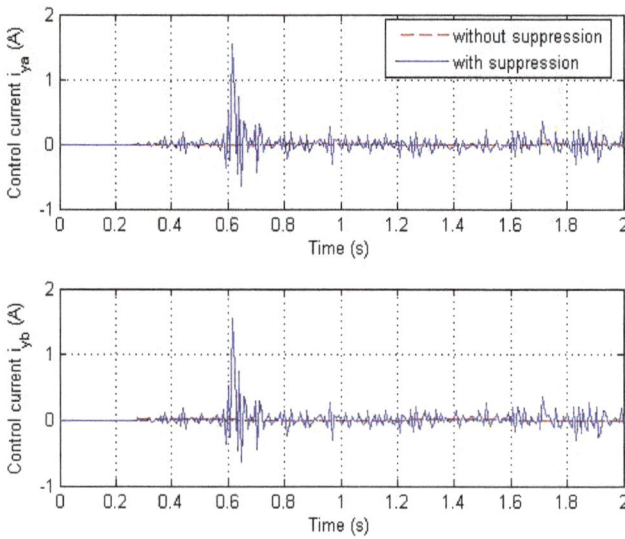

Figure 9. Control currents to bearings along y-axis

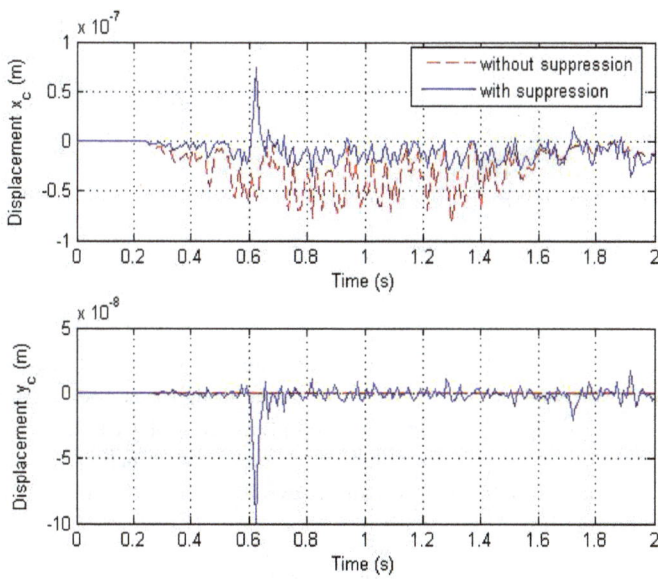

Figure 10. Displacements of rotor mass center

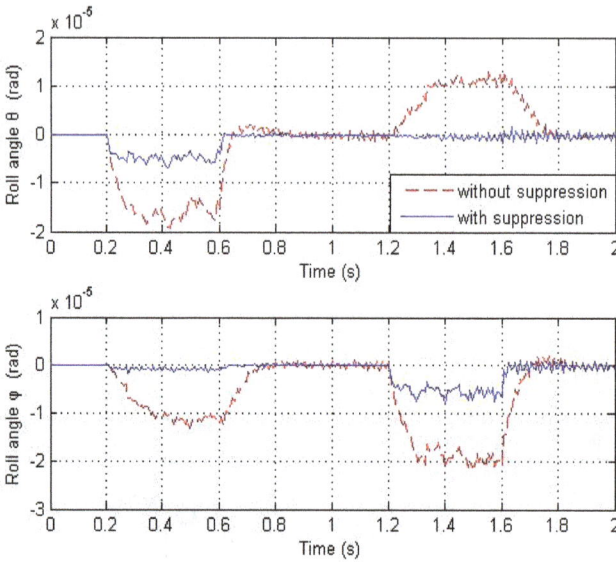

Figure 11. Rolling angles of rotor mass center

7. Conclusion

In this chapter, stability problem of magnetic bearings for a flywheel energy storage system has been formulated, and a synchronization design has been presented by incorporating cross-coupling technology into the optimal control architecture. The basic idea of the control strategy is to minimize a new cost function in which the synchronization errors are embedded, so that the gyro-dynamic rotation of the rotor can be effectively suppressed.

However, as optimal control, using receding horizon idea, is a feedback control, there is a risk that the resulting closed-loop system might be unstable. Then, stability of the control system based on the solution of the Riccati Difference Equation has also been analyzed, and some results are summarized. The illustrative example reveals that with adequately adjusted control parameters the resulting control system is very effective in recovering the unstable rotor and suppressing the coupling effects of the gyroscopic rotation at high spinning speeds as well as under external disturbances and model uncertainties.

Author details

Yong Xiao, Xiaoyu Ge and Zhe Zheng
College of Information Engineering, Shenyang University of Chemical Technology, China

8. References

Subkhan, M. & Komori, M. (2011). New concept for flywheel energy storage system using SMB and PMB, *IEEE Transactions on Applied Superconductivity*, 21(3): 1485-1488.

Samineni, S., Johnson, B. Hess, H. & Law, J. (2006). Modelling and analysis of a flywheel energy storage system for voltage sag correction, *IEEE Transactions on Industry Applications*, 42(1): 42-52.

Bitterly, J. (1998). Flywheel technology: past, present, and 21st century projections, *IEEE AES Systems Magazine*, 13(8): 13-16.

Beach, R. & Christopher, D. (1998). Flywheel technology development program for aerospace applications, *IEEE AES System Magazine*, 15(6): 9-14.

Suvire, G. & Mercado, P. (2012). Active power control of a flywheel energy storage system for wind energy applications, *IET Renewable Power Generation*, 6(1): 9-16.

Cimuca, G, Saudemont, C, Robyns, B & Radulescu, M. (2006). Control and performance evaluation of a flywheel energy storage system associated to a variable-speed wind generator, *IEEE Transactions on Industrial Electronics*, 53(4): 1074-1085.

Park, J., Kalev, C. & Hofmann, H. (2008). Modelling and control of solid-rotor synchronous reluctance machines based on rotor flux dynamics, *IEEE Transactions on Magnetics*, 44(12): 4639-4647.

Okada Y., Nagai B. & Shimane T. (1992). Cross feedback stabilization of the digitally controlled magnetic bearing. *ASME Journal of Vibration and Acoustics*, 114: 54-59.

Williams, R., Keith, F. & Allaire, P. (1990). Digital control of active magnetic bearings, IEEE Transactions on Industrial Electronics, 37(1): 19-27.

Tomizuka M., Hu J., Chiu T. & Kamano T. (1992). Synchronization of two motion control axes under adaptive feedforward control. *ASME Journal of Dynamic Systems, Measurement and Control*, 114:196-203.

Tsao, J., Sheu, L. & Yang, L. (2000). Adaptive synchronization control of the magnetically suspended rotor system. Dynamics and Control, 10: 239-53.

Yang L. & Chang W. (1996). Synchronization of twin-gyro precession under cross-coupled adaptive feedforward control. *AIAA Journal of Guidance, Control, and Dynamics*, 19: 534-539.

Summers, S., Jones, C., Lygeros, J. & Morari, M. (2011). A multiresolution approximation method for fast explicit model predictive control, *IEEE Transactions on Automatic Control*, 56(11): 2530-2541.

Rawlings, J., Bonne, D., Jorgensen, J. & Venkat, A. (2008). Unreachable setpoints in odel predictive control, *IEEE Transactions on Automatic Control*, 53(9): 2209-2215.

Mayne, D., Rawlings, J., Rao, C. & Scokaert P. (2000). Constrained model predictive control: stability and optimality, *Automatica*, 36: 789-814.

Zhu K. & Chen B. (2001). Cross-coupling design of generalized predictive control with reference models. *Proc. IMechE Part I: Journal of Systems Control Engineering*, 215: 375-384.

Xiao Y. & Zhu K. (2006). Optimal synchronization control of high-precision motion systems, *IEEE Transactions on Industrial Electronics*, 53(4): 1160-1169.

Zhang C. & Tseng K. (2007). A novel flywheel energy storage system with partially-self-bearing flywheel-rotor, *IEEE Transactions on Energy Conversion*, 22(2): 477-487.

Xiao Y., Zhu K., Zhang C., Tseng K. & Ling K. (2005). Stabilizing synchronization control of rotor-magnetic bearing system, *Proc IMechE Part I: Journal of Systems Control Engineering*, 219: 499-510.

Zhu K., Xiao Y. & Rajendra A. (2009). Optimal control of the magnetic bearings for a flywheel energy storage system, *Mechatronics*, 19: 1221–1235.

Souza, C., Gevers, M., Goodwin, G. (1986). Riccati equations in optimal filtering of nonstabilizable systems having singular state transition matrices. *IEEE Transactions on Automatic Control*, AC-31:831-838.

Chapter 7

Single- and Double-Switch Cell Voltage Equalizers for Series-Connected Lithium-Ion Cells and Supercapacitors

Masatoshi Uno

Additional information is available at the end of the chapter

1. Introduction

As demands for energy-efficient electrical devices and equipment continue to increase, the role of energy storage devices and systems becomes more and more important. Applications of such energy storage devices range from portable electronic devices, where a single cell is sufficient to provide adequate run time, to electric vehicles that require more than 100 cells in series to produce a sufficient high voltage to drive motors. Lithium-ion batteries (LIBs) are the most prevalent and promising because of their highest specific energy among commercially available secondary battery technologies.

Supercapacitor (SC) technologies, including traditional electric double-layer capacitors and lithium-ion capacitors (hybrid capacitors that combine features of double-layer capacitors and LIBs) are also drawing significant attention, because of their outstanding service life over a wide temperature range, high-power capability, and high-energy efficiency performance. The use of such SC technologies has traditionally been limited to high-power applications such as hybrid electric vehicles and regenerative systems in industries, where high-power energy buffers are needed to meet short-term large power demands. But it is found that SC technologies also have a great potential to be alternative energy storage sources to traditional secondary batteries once their superior life performance over a wide temperature range is factored in (Uno, 2011; Uno & Tanaka, 2011).

The voltage of single cells is inherently low, typically lower than 4.2, 2.7, and 3.8 V for lithium-ion cells, traditional electric double-layer capacitors, and lithium-ion capacitors, respectively. Hence in most practical uses, a number of single cells need to be connected in series to produce a high voltage level to meet the load voltage requirement. Voltages of series-connected cells are gradually imbalanced because their individual properties, such as capacity/capacitance, self-discharge rate, and internal impedance, are different from each other. Nonuniform temperature gradient among cells in a battery pack/module also lead to nonuniform self-discharging that accelerates voltage imbalance. In a voltage-imbalanced battery/module, some cells in the series connection may be overcharged and over-discharged during the charging and discharging processes, respectively, even though the average voltage of the series-connected cells is within the safety boundary. Using LIBs/SCs beyond the safety boundary not only curtails their operational life but also undermines their electrical characteristics. Overcharging must be prevented especially for LIBs since it may result in fire or even an explosion in the worst situation.

In addition to the safety issues mentioned above, the voltage imbalance also reduces the available energies of cells. When charging the cells in series, charging processes must be halted as soon as the most charged cell reaches the upper voltage limit, above which accelerated irreversible deterioration is very likely. Similarly, in order to avoid over-discharging during discharging processes, the least charged cell in the series connection limits the discharging time as a whole. Thus, voltage imbalance should be minimized in order to prolong life time as well as to maximize the available energies.

Various kind of equalization techniques have been proposed, demonstrated, and implemented for LIBs and SCs (Cao et al., 2008; Guo et al., 2006). However, conventional equalization techniques have one of the following major drawbacks:

1. Low energy efficiency because of the dissipative equalization mechanism.
2. Complex circuitry and control because of high switch count.
3. Design difficulty and poor modularity because of the need for a multi-winding transformer that imposes strict parameter matching among multiple secondary windings.

This chapter presents single- and double-switch cell voltage equalizers for series-connected lithium-ion cells and SCs. The equalization process of the equalizers is nondissipative, and a multi-winding transformer is not necessary. Hence, all the issues underlying the conventional equalizers listed above can be addressed by the presented equalizers. In Section 2, the above-mentioned issues are discussed in detail, and conventional cell voltage equalizers are briefly reviewed. In Section 3, single-switch cell voltage equalizers based on multi-stacked buck–boost converters are presented. In the single-switch equalizers, although multiple inductors are required, the circuitry can be very simple because of the single-switch configuration. Section 4 introduces double-switch cell voltage equalizers using a resonant inverter and a voltage multiplier. Although the circuitry is slightly more complex than the single-switch equalizers, its single-magnetic configuration minimizes circuit size and cost. Detailed operation analyses are mathematically made, and experimental equalization tests performed for series-connected SCs and lithium-ion cells using the prototype of the single- and double-switch equalizers are shown. Finally, in Section 5, the presented single- and double-switch equalizers are compared with conventional equalizers in terms of the required number of circuit components.

2. Conventional equalization techniques

2.1. Dissipative equalizers

The most common and traditional approach involves the use of dissipative equalizers, which do not require high-frequency switching operations. With dissipative equalizers, the voltage of series-connected cells can be equalized by removing stored energy or by shunting charge current from the cells with higher voltage. During the equalization process, the excess energy or current is inevitably dissipated in the form of heat, negatively influencing the thermal system of batteries/modules. Dissipative equalizers introduce advantages over nondissipative equalizers in terms of circuit simplicity and cost. The dissipative equalizers can be categorized into two groups: passive and active equalizers.

Fig. 1(a) shows the simplest solution using passive resistors. Voltage imbalance gradually decreases because of different self-discharge through resistors depending on cell voltages. Although simple, the relentless power loss in resistors reduces the energy efficiency, depending on the resistance values, and hence, this equalizer is rarely used in practice.

Another concept of a passive dissipative equalizer is the use of Zener diodes, as shown in Fig. 1(b). Cell voltages exceeding a Zener voltage level are cramped by Zener diodes. The power loss in the Zener diodes during the rest period is negligibly low depending on their leakage current. However, these diodes must be chosen to be capable of the largest possible charge current because the charge current flows through them when the cell voltage reaches or exceeds the Zener voltage level. In addition, a great temperature dependency of the Zener voltage, which may not be acceptable in most applications, should be factored in.

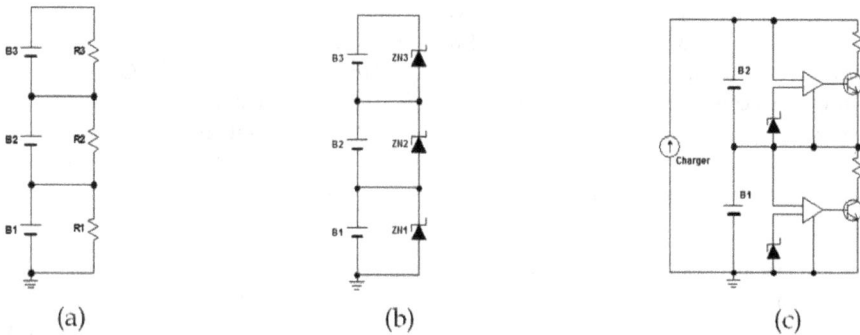

(a)   (b)   (c)

Figure 1. Passive dissipative equalizers using (a) resistors and (b) Zener diodes, and (c) active shunting equalizer.

Fig. 1(c) shows a schematic drawing of shunting equalizers (Isaacson et al., 2000; Uno, 2009). Cell voltages are monitored and compared with a preset voltage level (shunt voltage level). When the cell voltage reaches or exceeds the shunt voltage level, the charge current is bypassed through a transistor to reduce the net charge current. The product of cell voltage and shunt current is the power dissipation in the equalizer. The shunting equalizer needs as many switches, voltage sensors, and comparators as the number of series connection of cells. In addition, this equalizer inevitably causes energy loss in the form of heat generation during the equalization process. The operation flexibility is also poor because cells are equalized only during the charging process, especially at fully charged states.

Although dissipative equalization techniques seem less effective compared with nondissipative equalizers, which are reviewed in the following subsection, the shunting equalizers are widely used in various applications, and a number of battery management ICs that include shunting equalizers are available because of their simplicity, good modularity (or extendibility), and cost effectiveness.

2.2. Nondissipative equalizers

Nondissipative equalizers that transfer charges or energies among series-connected cells are considered more suitable and promising than dissipative equalizers in terms of energy efficiency and thermal management. In addition, nondissipative equalizers (including single- and double-switch equalizers presented in this chapter) are usually operational during both charging and discharging, and hence, operation flexibility can be improved compared with dissipative equalizers. Numerous nondissipative equalization techniques have been proposed and demonstrated. Representative nondissipative equalizer topologies are reviewed in the following subsections.

2.2.1. Individual cell equalizer

Fig. 2(a) depicts a schematic drawing of the individual cell equalizer (ICE) (Lee & Cheng, 2005). ICEs are typically based on individual bidirectional dc–dc converters such as switched capacitor converters (Pascual & Krein, 1997; Uno & Tanaka, 2011) and buck–boost converters (Nishijima et al., 2000), as shown in Figs. 2(b) and (c), respectively. Other types of bidirectional converters, such as resonant switched capacitor converters and Ćuk converters (Lee & Cheng, 2005), can also be used for improving equalization efficiencies. In ICE topologies, the charges or energies of the series-connected cells can be transferred between adjacent cells to eliminate cell voltage imbalance. The number of series connection of cells can be arbitrary extended by adding the number of ICEs.

Since these ICE topologies are derived from multiple individual bidirectional dc–dc converters, numerous switches, sensors, and switch drivers are required in proportion to the number of series-connected energy storage cells. Therefore, their circuit complexity and cost are prone to increase, especially for applications needing a large number of series connections, and their reliability decreases as the number of series connections increases.

2.2.2. Equalizers using a multi-winding transformer

In cell voltage equalizers using a multi-winding transformer based on flyback and forward converters, as shown in Figs. 3(a) and (b), respectively, the energies of series-connected cells can be redistributed via a multi-winding transformer to the cell(s) having the lowest voltage (Kutkut, et al., 1995). The required number of switches in the multi-winding transformer-based equalizers is significantly less than those required in ICE topologies. However, these topologies need a multi-winding transformer that must be customized depending on the number of series connections, and hence, the modularity is not good. In addition, since parameter mismatching among multiple secondary windings results in voltage imbalance that can never be compensated by control, multi-winding transformers must be designed and made with great care (Cao et al., 2008; Guo et al., 2006). In general, the difficulty of parameter matching significantly increases with the number of windings. Therefore, their applications are limited to modules/batteries with a few series connections.

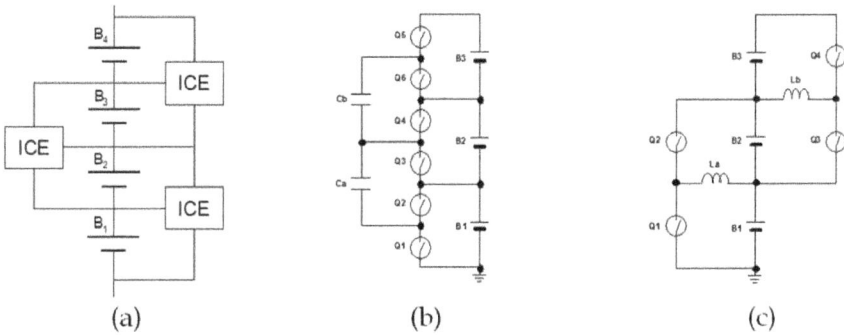

Figure 2. (a) Generic configuration of individual cell equalizer, (b) switched capacitor-based equalizer, and (c) buck–boost converter-based equalizer.

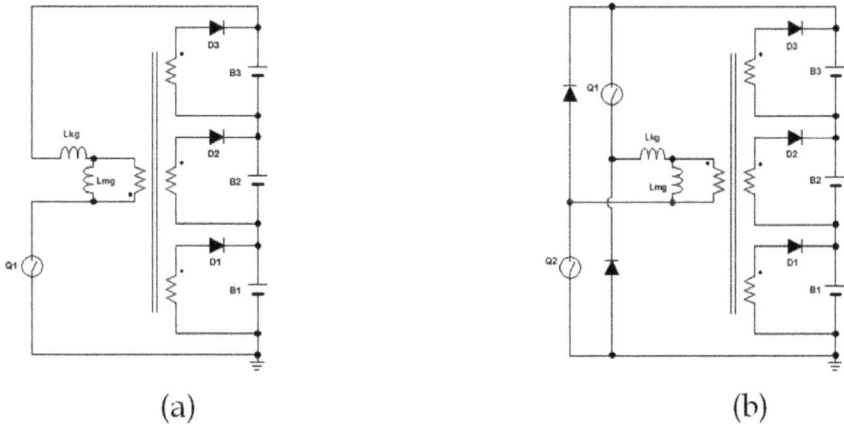

Figure 3. Equalizers using a multi-winding transformer: (a) flyback converter-based and (b) forward converter-based equalizers.

2.2.3. Equalizers using a single converter with selection switches

Figs. 4(a) and (b) show equalizers using a single converter and selection switches based on the flying capacitor and the flyback converter, respectively (Kim et al., 2011). In these topologies, individual cell voltages are monitored, and the cell(s) with the lowest and/or highest voltages are determined. In the flying capacitor-based equalizer, shown in Fig. 4(a), the energy of the most charged cell is delivered to the least charged cell via the flying capacitor C by selecting proper switches. In the flyback converter-based equalizer, shown in Fig. 4(b), the energy of series-connected cells is redistributed to the least charged cell via the flyback converter and properly selected switches.

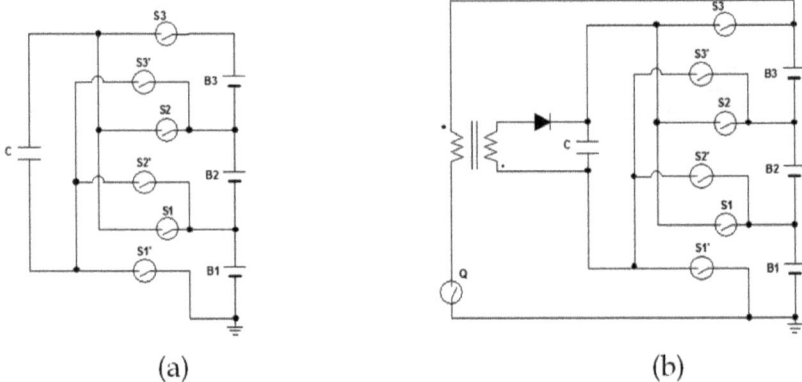

(a)   (b)

Figure 4. Equalizers using a single converter with selection switches based on (a) flying capacitor and (b) flyback converter.

These topologies can reduce the number of passive components significantly when compared with ICE topologies shown in Fig. 2, and do not need a multi-winding transformer. However, the required number of switches is still large (proportional to the number of series connections), and furthermore, microcontroller- or DSP-based intelligent management is mandatory because target cell(s) (i.e., the least and/or most charged cells) must be determined for equalization to be implemented.

3. Single-switch cell voltage equalizer using multi-stacked buck–boost converters

3.1. Circuit description and major benefits

The single-switch cell voltage equalizers are derived by multi-stacking traditional buck–boost converters that consist of two inductors and one coupling capacitor (Uno & Tanaka, 2011). Candidate topologies, which can be used as a basic topology, are shown in Fig. 5. The single-ended primary inductor converter (SEPIC) and Zeta converters can be simply adapted as the basic topology. On the other hand, the Ćuk topology cannot be used without a transformer because of its inverting property. Out of the three candidates shown in Fig. 5, the Ćuk converter is the best in terms of current pulsation at its input and output because inductors Lin and Lout are connected in series to the input and output, respectively. Although the SEPIC and Zeta converter can be adapted without a transformer, transformer-less operation may suffer from duty cycle limitations when the number of series connection is large, as will be discussed in the following subsection. Therefore, the need for the transformer in the Ćuk converter is justified, although comparative analysis is necessary to determine the best topology to achieve sufficient performance at reasonable size for a given application. The isolated Ćuk converter-based equalizer is focused in the following sections.

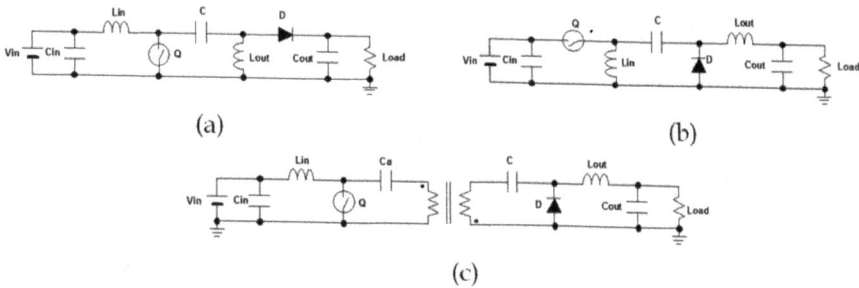

Figure 5. Traditional buck–boost converters that can be used as the basic topology for single-switch cell voltage equalizers. (a) SEPIC , (b) Zeta, and (c) isolated Ćuk converters.

The derived isolated Ćuk converter-based single-switch equalizer for four cells connected in series is shown in Fig. 6. The circuit consisting of C_{in}, L_{in}, Q, C_a, transformer, C_3, D_3, and L_3 is identical to the circuit shown in Fig. 5(c), while the circuit consisting of C_i–D_i–L_i (i = 1...4) is multi-stacked to the isolated Ćuk converter. Hence, the derived equalizer can be regarded as multi-stacked Ćuk converters.

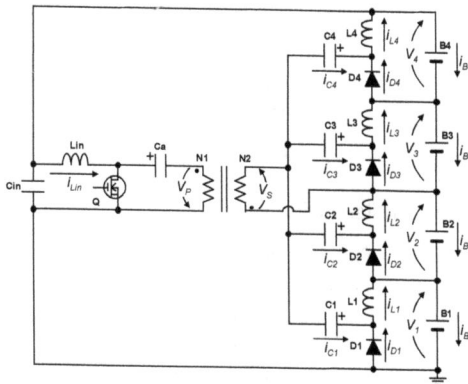

Figure 6. Single-switch cell voltage equalizer based on isolated Ćuk converter.

The required number of switches is only one; thus, reducing the circuit complexity significantly when compared with conventional equalizers that need numerous switches proportional to the number of series connections, as explained in the previous section. In addition, a multi-winding transformer is not necessary, and the number of series connections can be arbitrarily extended by stacking the circuit consisting of C_i–D_i–L_i. Therefore, in addition to the reduced circuit complexity, the single-switch equalizer offers a good modularity as well. Furthermore, as will be mathematically indicated, feedback control is not necessary when it is operated in discontinuous conduction mode (DCM), further simplifying the circuitry by removing the feedback control loop.

3.2. Operation analysis

3.2.1. Operation under voltage-balanced condition

Traditional buck–boost converters, including the isolated Ćuk converter, operate in either continuous conduction mode (CCM) or DCM. The boundary between CCM and DCM is the discontinuity of the diode current during the off-period. Although ripple currents of inductors tend to be large in DCM, currents in the circuit can be limited under desired levels without feedback control, as will be mathematically indicated later. The following analysis focuses on DCM operation. Key operation waveforms and current flow directions in DCM under the voltage-balanced condition are shown in Figs. 7 and 8, respectively. The fundamental operation is similar to the traditional isolated Ćuk converter, and the DCM operation can be divided into three periods: T_{on}, T_{off-a}, and T_{off-b}.

Under a steady-state condition, average voltages of inductors and transformer windings are zero, and hence, the average voltages of C_1–C_4 and C_a, V_{C1}–V_{C4} and V_{Ca}, can be expressed as

$$\begin{cases} V_{C1} = -V_2 \\ V_{C2} = 0 \\ V_{C3} = V_3 \\ V_{C4} = V_3 + V_4 \\ V_{Ca} = V_1 + V_2 + V_3 + V_4 \end{cases} , \qquad (1)$$

where V_1–V_4 are the voltages designated in Fig. 6.

During T_{on} period, as shown in Fig. 8(a), all the inductors are energized and their currents increase linearly. When Q is turned off, T_{off-a} period begins and diodes start to conduct, as shown in Fig. 8(b). As the inductors release stored energies, the inductor currents as well as the diode currents decrease linearly. When the diode currents fall to zero, period T_{off-b} begins. In this period, all the currents in the equalizer are constant because the voltages across inductors are zero.

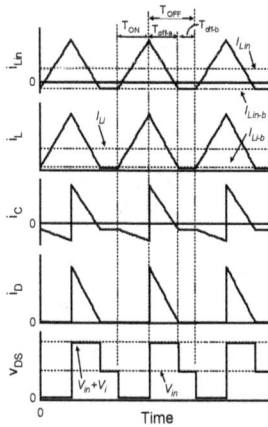

Figure 7. Key operation waveforms of isolated Ćuk converter-based single-switch equalizer under a voltage-balanced condition.

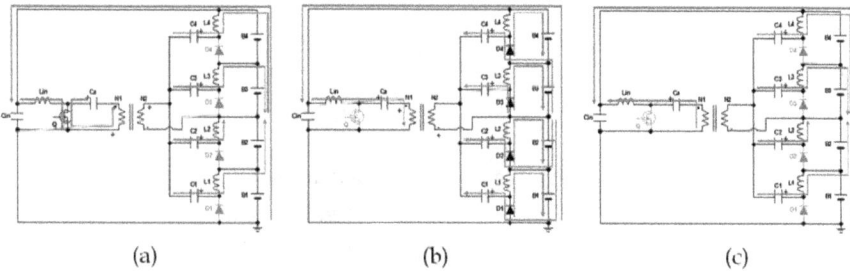

(a)	(b)	(c)

Figure 8. Current flow directions during periods (a) T_{on}, (b) $T_{off\text{-}a}$, and (c) $T_{off\text{-}b}$ under a voltage-balanced condition.

The voltage-time product of inductors under a steady-state condition is zero, yielding

$$\begin{cases} D\left(V_S + V_{C1} + V_2\right) & = D_a\left(V_1 + V_D\right) \\ D\left(V_S + V_{C2}\right) & = D_a\left(V_2 + V_D\right) \\ D\left(V_S + V_{C3} - V_3\right) & = D_a\left(V_3 + V_D\right) \\ D\left(V_S + V_{C4} - V_3 - V_4\right) = D_a\left(V_4 + V_D\right) \end{cases} \tag{2}$$

where D and D_a are the duty ratio of T_{on} and $T_{off\text{-}a}$, respectively, V_D is the forward voltage of diodes, and V_S is the transformer secondary voltage designated in Fig. 6 expressed as

$$V_S = \frac{V_P}{N} = \frac{V_1 + V_2 + V_3 + V_4}{N} = \frac{V_{in}}{N}, \tag{3}$$

where N is the transformer turn ratio. From Eqs. (1) and (2),

$$V_i = \frac{D}{D_a} V_S - V_D , \tag{4}$$

where $i = 1...4$. This equation means that the equalizer produces the uniform output voltages to the cells, and all the cell voltages can eventually become uniform.

In order for the equalizer to operate in DCM, $T_{off\text{-}b}$ period must exist, meaning $D_a < (1 - D)$. From Eq. (4), the critical duty cycle to ensure DCM operation, $D_{critical}$, is given by

$$D_{critical} < \frac{V_i + V_D}{V_S + V_i + V_D} . \tag{5}$$

Eqs. (3) and (5) imply that without the variable N, as V_{in} increases, D must be lowered for a given value of V_i. In other words, the duty cycle limitation confronts when the number of series connection is large. On the other hand, with the introduction of the transformer, the issue on the duty cycle limitation can be overcome by properly determining N.

According to Fig. 7, the average currents of L_i and L_{in}, I_{Li} and I_{Lin}, are expressed as

$$\begin{cases} I_{Li} = (D + D_a) \dfrac{V_S D T_S}{2L_i} + I_{Li\text{-}b} \\ I_{Lin} = (D + D_a) \dfrac{V_S D T_S}{2L_{in}} + I_{Lin\text{-}b} \end{cases} , \tag{6}$$

where T_S is the switching period, and $I_{Li\text{-}b}$ and $I_{Lin\text{-}b}$ are the currents flowing through L_i and L_{in}, respectively, during period $T_{off\text{-}b}$ as designated in Fig. 7. Assuming that impedances of C_1–C_4 are equal, i_{C1}–i_{C4} as well as $I_{L1\text{-}b}$–$I_{L4\text{-}b}$ can be uniform, as expressed by

$$NI_{Lin\text{-}b} = -I_{L1\text{-}b} - I_{L2\text{-}b} - I_{L3\text{-}b} - I_{L4\text{-}b} = -4I_{Li\text{-}b} . \tag{7}$$

The average current of C_i, I_{Ci}, is expressed as

$$I_{Ci} = I_{Li\text{-}b} + D \frac{V_S D T_S}{2L_i} - D_a \frac{N^2 V_S D T_S}{2 \cdot 4L_{in}} = 0 . \tag{8}$$

From Eqs. (6)–(8),

$$\frac{I_{Li}}{I_{Lin}} = \frac{ND_a}{4D} . \tag{9}$$

With $I_{Ci} = 0$, Kirchhoff's current law in Fig. 6 yields

$$I_{Li} = I_{Di} , \tag{10}$$

where I_{Di} is the average current of D_i. From Figs. 7 and 8(b), I_{Di} is expressed as

$$I_{Di} = D_a \left(\frac{V_S D T_S}{2 L_i} + \frac{N^2 V_S D T_S}{2 \cdot 4 L_{in}} \right) = \frac{V_S D D_a T_S}{2} \left(\frac{4 L_{in} + N^2 L_i}{4 L_{in} L_i} \right). \tag{11}$$

Substituting Eqs. (10) and (11) into Eq. (9) produces

$$I_{Lin} = \frac{V_S D^2 T_S}{2N} \left(\frac{4 L_{in} + N^2 L_i}{L_{in} L_i} \right). \tag{12}$$

Eqs. (10)–(12) imply that currents in the equalizer under a voltage-balanced condition can be limited under a desired level as long as a variation range of V_S is known. In Eq. (12), for example, V_S is variable and D is determinable, while others are fixed values, and hence, with a known variation range of V_S, I_{Lin} can be designed limited under the desired level by properly determining D. I_{Li} and I_{Di} (that can be expressed by Eqs. (10) and (11)) can be similarly designed because D_a is a predictable variable given by Eq. (4). Thus, the currents in the equalizer operating in DCM can be limited under desired levels even in fixed duty cycle operations, and feedback control is not necessary for the single-switch equalizer, further simplifying the circuit by eliminating the feedback control loop.

3.2.2. Operation under voltage-imbalanced condition

As expressed by Eq. (4), the single-switch equalizer inherently produces the uniform output voltages to the cells. This characteristic implies that in the case where voltages of cells are imbalanced, the currents from the equalizer tend to concentrate to a cell having the lowest voltage. Fig. 9 shows the key operation waveforms under a voltage-imbalanced condition. Asterisks added to the symbols in Fig. 9 correspond to the cell with the lowest voltage, B*. As shown in Fig. 9, when there is voltage imbalance, only i_{D*} flows, whereas the other diode currents (i_{Di}) are zero for the entire periods. Since all the currents concentrate to D*, the average current of D* is obtained by transforming Eq. (11) as

$$I_{D*} = D_a \left(\frac{4 V_S D T_S}{2 L_i} + \frac{N^2 V_S D T_S}{2 L_{in}} \right) = \frac{V_S D D_a T_S}{2} \left(\frac{4 L_{in} + N^2 L_i}{L_{in} L_i} \right), \tag{13}$$

which is fourfold larger than that of Eq. (11). Since I_{Di} is zero under the voltage-imbalanced condition, I_{Li} is also zero according to Eq. (10), although ripples exist. I_{Lin} under the voltage-imbalanced condition is identical to that under the voltage-balanced condition because I_{Lin} is independent of cell voltages, as expressed by Eq. (12).

3.3. Experimental

3.3.1. Prototype and its fundamental performance

A 5-W prototype of the isolated Ćuk converter-based single-switch equalizer was built for 12 cells connected in series, as shown in Fig. 10. Component values are listed in Table 1.

C_{out1}–C_{out12}, which were not depicted in figures for the sake of simplicity, are smoothing capacitors connected to the cells in parallel. The RCD snubber was added at the primary winding in order to protect the switch from surge voltages generated by the transformer leakage inductance. The prototype was operated with a fixed $D = 0.3$ at $f = 150$ kHz.

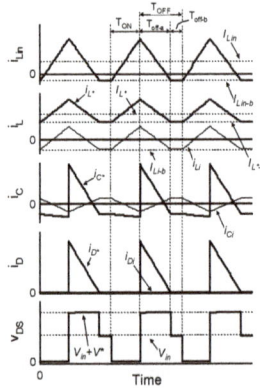

Figure 9. Key operation waveforms under a voltage-imbalanced condition.

The experimental setup for power conversion efficiency measurement is shown in Fig. 11. The tap Y and X in Fig. 11 were selected to emulate the voltage-balanced and -imbalanced ($V_1 < V_i$ ($i = 2...12$)) conditions, respectively. The external power supply, V_{ext}, was used, and the input and output of the equalizer were broken at point Z in order to measure efficiencies. The efficiencies were measured by changing the ratio of V_1/V_{in} between approximately 1/12 and 1/16. During the efficiency measurement, cells were removed and only smoothing capacitors (C_{out1}–C_{out12}) were used to sustain the voltages of V_1–V_{12}.

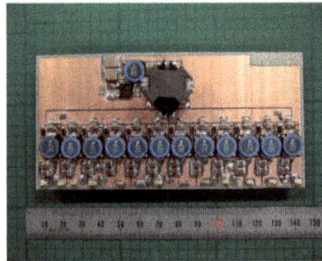

Figure 10. Photograph of a 5-W prototype of the isolated Ćuk converter-based single-switch cell voltage equalizer for 12 cells connected in series.

The measured power conversion efficiencies and output power characteristics as a function of V_1 are shown in Fig. 12. The efficiencies increased with V_1 because the diode voltage drop represented a lesser portion of the output voltage (i.e., V_1). The efficiencies under the voltage-balanced condition were higher than those under the voltage-imbalanced condition;

the peak efficiencies under the voltage-balanced and -imbalanced conditions were approximately 70% and 65%, respectively. The lower efficiencies under the voltage-imbalanced condition were due to the current concentration to C_1, D_1, and L_1, which caused increased Joule losses in resistive components.

Component	Value
C_{in}	Ceramic Capacitor, 20 μF
C_a	Ceramic Capacitor, 22 μF
$C_{out\,1}$–$C_{out\,12}$	Ceramic Capacitor, 200 μF
C_1–C_{12}	Ceramic Capacitor, 22 μF
L_{in}	1 mH
L_1–L_{12}	47 μH
Transformer	$N_1{:}N_2 = 24{:}4$, $L_{kg} = 8.7$ μH, $L_{mg} = 2.03$ mH
Q	N-Ch MOSFET, IRFR13N20D, $R_{on} = 235$ mΩ
D_1–D_{12}	Schottky Diode, CRS08, $V_D = 0.36$ V
RCD Snubber	$R = 2.2$ kΩ, $C = 470$ pF

Table 1. Component values used for the prototype.

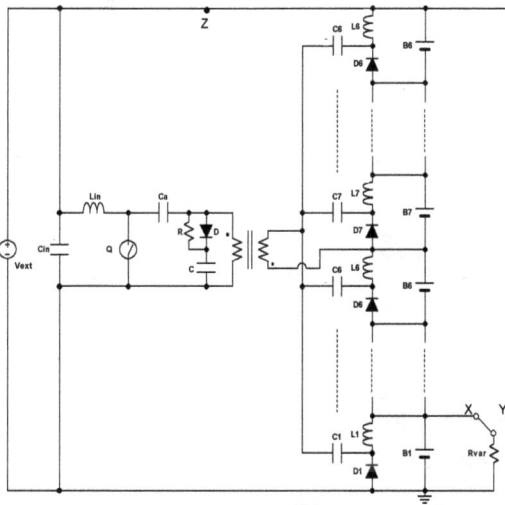

Figure 11. Experimental setup for efficiency measurement.

Figs. 13(a) and (b) show typical operation waveforms measured under the voltage-balanced and -imbalanced conditions, respectively. Under the voltage-balanced condition, as shown in Fig. 13(a), all the inductor currents were uniform, although the oscillations caused by interactions between inductors and parasitic capacitance of the MOSFET were observed. Under the voltage-imbalanced condition, i_{L2}–i_{L12} were uniform and their averages were zero, whereas only i_{L1} showed an average higher than zero because of the current concentration.

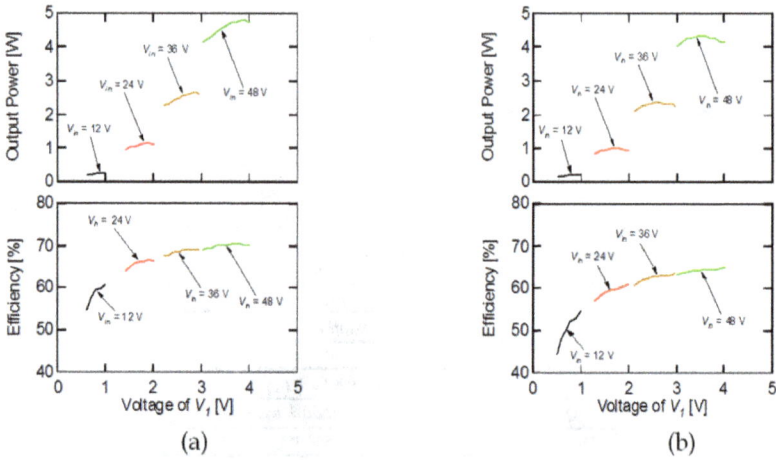

Figure 12. Measured power conversion efficiencies and output powers as a function of V_1 under (a) voltage-balanced and (b) -imbalanced conditions.

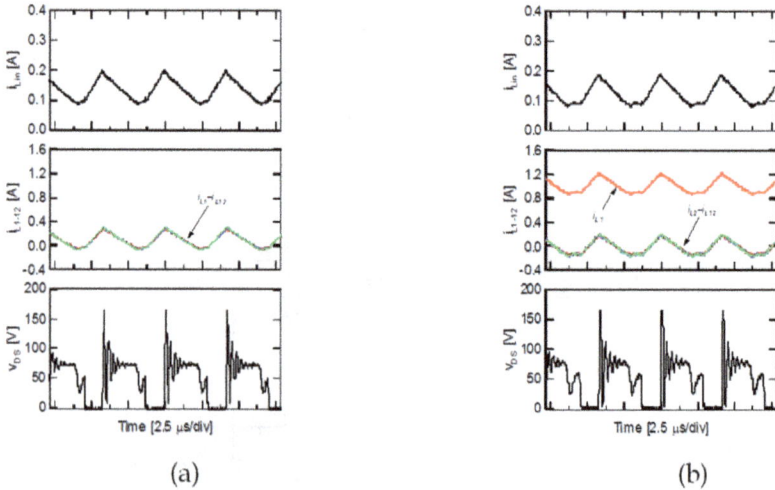

Figure 13. Measured waveforms under (a) voltage-balanced and (b) -imbalanced conditions.

3.3.2. Equalization

The experimental equalization test was performed for 12-series SCs with a capacitance of 500 F at a rated charge voltage of 2.5 V. The voltages of SCs were initially imbalanced between 0.85–2.5 V. The resultant equalization profiles are shown in Fig. 14(a). As the equalizer redistributed energies from the series connection to cells with low voltages, voltages of cells with low initial voltages increased while those with high initial voltages decreased. The voltage imbalance was gradually eliminated as time elapsed, and the standard deviation of cell voltages eventually decreased down to approximately 7 mV at the end of the experiment; thus, demonstrating the equalizer's equalization performance. The cell voltages kept decreasing even after the voltage imbalance disappeared. This decrease was due to the power conversion loss in the equalizer. After the cell voltages were balanced, the energies of the cells were meaninglessly circulated by the equalizer, and therefore, the equalizer should be disabled after cell voltages are sufficiently balanced in order not to waste the stored energies of cells.

Another experimental equalization was performed for 12-series lithium-ion cells with a capacity of 2200 mAh at a rated charge voltage of 4.2 V. The state of charges (SOCs) of the cells were initially imbalanced between 0%–100%. The experimental results are shown in Fig. 14(b). Although the resultant profiles were somewhat elusive because of the nonlinear characteristics of the lithium-ion chemistry, the voltage imbalance was successfully eliminated, and all the cell voltage converged to a uniform voltage level.

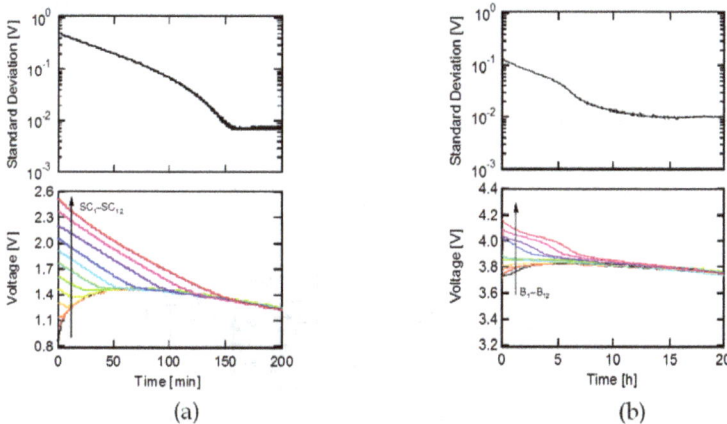

Figure 14. Experimental profiles of 12 series-connected (a) SCs and (b) lithium-ion cells equalized by the isolated Ćuk converter-based single-switch equalizer.

4. Double-switch resonant cell voltage equalizer using a voltage multiplier

4.1. Circuit description and major benefits

The double-switch resonant equalizer is essentially a combination of a conventional series resonant inverter and a voltage multiplier, shown in Figs. 15(a) and (b), respectively. The voltage multiplier shown in Fig. 15(b) is an example circuit that can produce a 4 times higher voltage than the amplitude of the input. The voltages of the stationary capacitors C'_1–C'_4 automatically become uniform as the amplitude of the input square wave under a steady-state condition (when diode voltage drops are neglected). Detailed operation analyses on both the resonant inverter and the voltage multiplier are separately made in the following section.

By combining the series-resonant inverter and the voltage multiplier, the double-switch equalizer can be synthesized as shown in Fig. 16. The leakage inductance of the transformer is used as the resonant inductor, L_r. The magnetizing inductance of the transformer is not depicted in Fig. 16 for the sake of simplicity. The stationary capacitors C'_1–C'_4 in the voltage multiplier in Fig. 15(b) are replaced with energy storage cells B_1–B_4.

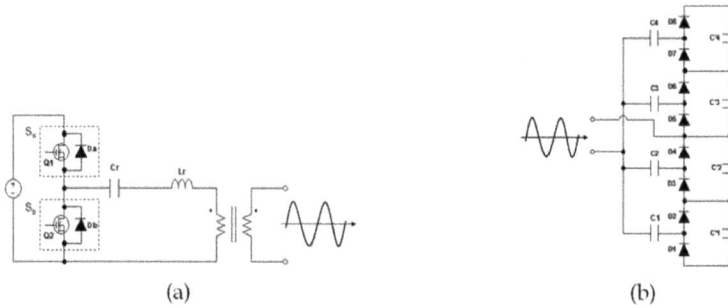

(a)   (b)

Figure 15. (a) Series-resonant inverter and (b) 4x-voltage multiplier.

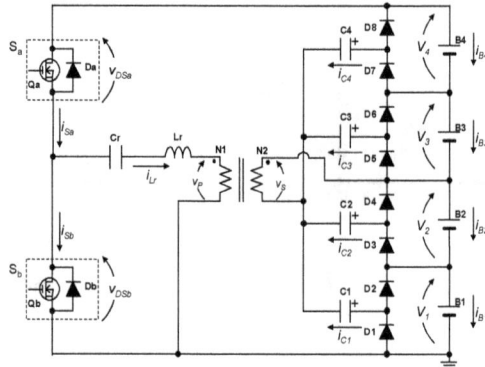

Figure 16. Double-switch series-resonant equalizer using voltage multiplier.

The required number of switches and magnetic components are only two and one, respectively, and hence, the circuit complexity as well as the size and cost of the circuit can be significantly reduced when compared with conventional equalizers, which need multiple switches and/or magnetic components in proportion to the number of series connections. In addition, the proposed resonant equalizer can be configured without a multi-winding transformer, and the number of series connections can be readily extended by stacking a capacitor and diodes, thus offering a good modularity (or extendibility). As mathematically explained in the following analysis, feedback control is not necessarily needed to limit currents under the desired current levels, and therefore, the circuit can be further simplified by removing the feedback control loop.

Although the series-resonant inverter with a transformer is used for the proposed equalizer, other types of resonant inverters, such as parallel-, series-parallel-, and LLC-resonant inverters, can be used instead, and may even offer better performance.

4.2. Operation analysis

4.2.1. Fundamental operation

Similar to traditional resonant inverters, the double-switch resonant equalizer is operated at the switching frequency f higher than the resonant frequency, f_r, above which the series-resonant circuit consisting of Lr and Cr represents an inductive load. The key operation waveforms at $f > f_r$ are shown in Fig. 17, and f_r is given by

$$f_r = \frac{1}{2\pi\sqrt{L_r C_r}},$$ (14)

where L_r is the inductance of the resonant inductor Lr, and C_r is the capacitance of the resonant capacitor Cr. MOSFETs Qa and Qb are complementarily operated with a fixed duty cycle slightly less than 50% in order to provide adequate dead times to prevent a short-through current. Above f_r, the current of the resonant circuit i_{Lr} lags behind the fundamental component of the voltage v_{DSb}, which corresponds to the input voltage for the resonant circuit, and Qa and Qb are turned on at zero-voltage.

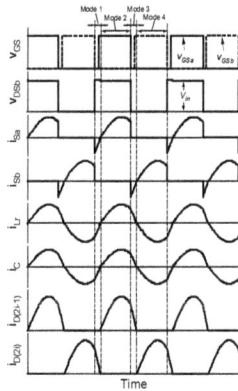

Figure 17. Key operation waveforms of resonant equalizer above resonant frequency ($f > f_r$).

The operation of the series-resonant equalizer can be divided into four modes: Mode 1–4, and the current flow directions under a voltage-balanced condition during each mode are shown in Fig. 18. In Mode 1, the current of the resonant circuit, i_{Lr}, flows through the anti-parallel diode of Sa, Da, toward the series connection of B1–B4, and the energy transfer capacitors in the voltage multiplier, C_i (i = 1...4), discharge through even-numbered diodes, $D_{(2i)}$, as shown in Fig. 18(a). Before i_{Sa} reaches zero, the gate signal for Q_a, v_{GSa}, is applied. Since the voltage across Q_a, v_{DSa}, is zero at this moment, Q_a is turned on at zero-voltage. After i_{Sa} is reversed, as shown in Fig. 17, Q_a starts to conduct, and Mode 2 begins. In Mode 2, the resonant circuit is energized by the series connection of B_1–B_4, and C_i is charged via odd-numbered diodes, $D_{(2i-1)}$. As Q_a is turned off, the current is diverted from Q_a to the anti-parallel diode of S_b, D_b, and Mode 3 begins. C_i is still being charged. The gate signal for Q_b, v_{GSb}, is applied and Q_b is turned on at zero-voltage, before the current of S_b, i_{Sb}, is reversed. As i_{Sb} reaches zero, the operation shifts to Mode 4, in which C_i discharges through $D_{(2i)}$. When Q_b is turned off, the current is diverted from Q_b to D_a, and the operation returns to Mode 1. Thus, similar to the conventional resonant inverters, the double-switch resonant equalizer achieves zero-voltage switching (ZVS) operation when Q_a and Q_b are turned on.

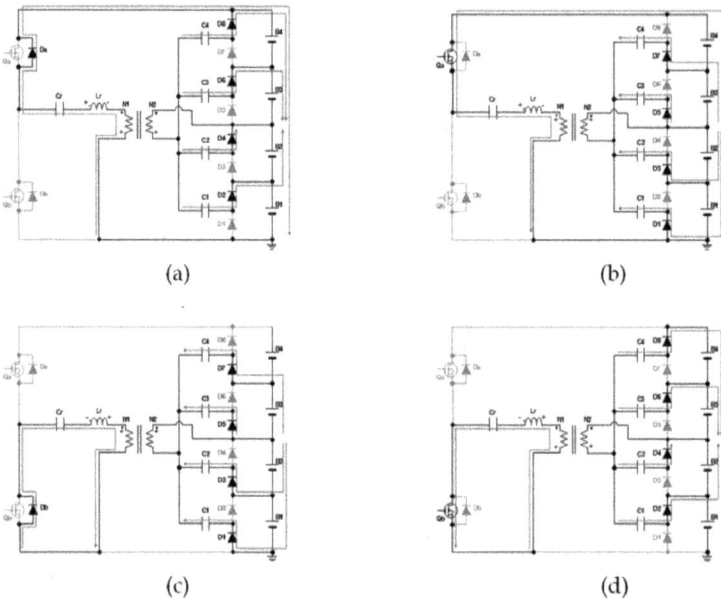

(a)

(b)

(c)

(d)

Figure 18. Current flow directions during Mode (a) 1, (b) 2, (c) 3, and (d) 4.

Repeating the above sequence, energies of the series connection of B_1–B_4 are supplied to the resonant inverter, and then are transferred to the voltage multiplier that redistributes the energies to B_1–B_4. Thus, the energies of B_1–B_4 are redistributed via the resonant inverter and voltage multiplier. Throughout a single switching cycle, C_i as well as B_1–B_4 are charged and discharged via $D_{(2i-1)}$ and $D_{(2i)}$, and consequently, voltages of B_1–B_4, V_1–V_4, become automatically uniform. The voltage equalization mechanism by the voltage multiplier is discussed in detail in the following subsection.

4.2.2. Voltage multiplier

The peak voltage of C_i during the time $D_{(2i-1)}$ is on, V_{CiO}, can be expressed as

$$\begin{cases} V_{C1O} = V_{S-O} - V_D - V_1 - V_2 \\ V_{C2O} = V_{S-O} - V_D - V_2 \\ V_{C3O} = V_{S-O} - V_D \\ V_{C4O} = V_{S-O} - V_D + V_3 \end{cases} , \qquad (15)$$

where V_{S-O} is the peak voltage of the transformer secondary winding when $D_{(2i-1)}$ is on, and V_D is the forward voltage drop of the diodes. Similarly, the bottom voltages of C_i when $D_{(2i)}$ is on, V_{CiE}, are

$$\begin{cases} V_{C1E} = -V_{S-E} + V_D - V_2 \\ V_{C2E} = -V_{S-E} + V_D \\ V_{C3E} = -V_{S-E} + V_D + V_3 \\ V_{C4E} = -V_{S-E} + V_D + V_3 + V_4 \end{cases} , \qquad (16)$$

where V_{S-E} is the bottom voltage of the transformer secondary winding when $D_{(2i)}$ is on.

Subtracting Eq. (16) from Eq. (15) yields the voltage variation of C_i during a single switching cycle, ΔV_{Ci}:

$$\Delta V_{Ci} = \left(V_{S-O} + V_{S-E} \right) - V_i - 2V_D . \qquad (17)$$

Generally, an amount of charge delivered via a capacitor having a capacitance of C, and an equivalent resistance for the charge transfer, R_{eq}, are given by

$$\begin{cases} Q = It = C\Delta V \\ \Delta V = \dfrac{It}{C} = \dfrac{I}{Cf} = IR_{eq} \end{cases} , \qquad (18)$$

where ΔV is the voltage variation caused by charging/discharging. Substitution of Eq. (18) into Eq. (17) produces

$$I_{Ci}R_{eqi} = \left(V_{S-O} + V_{S-E} \right) - 2V_D - V_i , \qquad (19)$$

where I_{Ci} is the average current flowing via C_i.

Eq. (19) yields a dc equivalent circuit of the voltage multiplier, as shown in Fig. 19. All cells, B_1–B_4, are tied to a common dc-source V_{dc}, which provides a voltage of $(V_{S-O} + V_{S-E})$, via two diodes and one equivalent resistor, R_{eqi}. When V_1–V_4 are balanced, I_{C1}–I_{C4} can be uniform as long as C_1–C_4 are designed so that all the equivalent resistances, R_{eq1}–R_{eq4}, are uniform. In the case of voltage imbalance, the current preferentially flows to the cell(s) having the lowest voltage, and its voltage increases more quickly than the others. Eventually, voltages V_1–V_4 automatically reach a uniform voltage level.

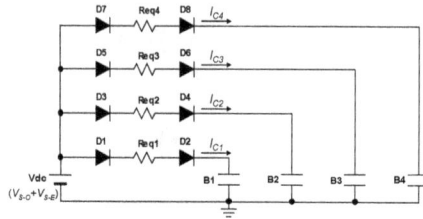

Figure 19. DC equivalent circuit for voltage multiplier.

Current flows in the dc equivalent circuit in Fig. 19 can then be transformed to those in the original circuit shown in Figs. 16 and 18. We consider the case that V_1 is the lowest and when no currents flow through R_{eq2}–R_{eq4}, as a simple example. Since any current flowing through R_{eqi} represents charging/discharging the capacitor C_i, as indicated by Eq. (18), no currents in R_{eq2}–R_{eq4} mean that no currents flow through C_2–C_4 as well as D_3–D_8 in the original circuit. Meanwhile, the current from the transformer secondary winding concentrates to C_1 and D_1–D_2 in the original circuit. Thus, under a voltage-imbalanced condition, currents flow through only the capacitor(s) and diodes that are connected to the cell(s) having the lowest voltage, although practical current distribution tendencies are dependent on R_{eqi} as well as the voltage conditions of V_i.

4.2.3. Series-resonant circuit

The average voltage across a transformer winding throughout a single switching cycle is zero, and the on-duties of $D_{(2i-1)}$ and $D_{(2i)}$ are both 50%. Therefore, the average voltages of C_1–C_4, V_{C1}–V_{C4}, can be obtained from Fig. 18, and are expressed as

$$\begin{cases} V_{C1} = -\dfrac{V_1}{2} - V_2 \\[2mm] V_{C2} = -\dfrac{V_2}{2} \\[2mm] V_{C3} = \dfrac{V_3}{2} \\[2mm] V_{C4} = V_3 + \dfrac{V_4}{2} \end{cases} \tag{20}$$

The Square voltage waves in the resonant circuit are approximated to the sinusoidal fundamental components, as shown in Fig. 20, in which key waveforms of the series-resonant inverter and their fundamental components are sketched. By assuming that the voltage of C_i is constant as V_{Ci} throughout a single switching cycle, the voltage of the transformer secondary winding, v_s, is

$$v_S = \begin{cases} V_{C1} + V_1 + V_2 + V_D = V_{C2} + V_2 + V_D = V_{C3} + V_D = V_{C4} - V_3 + V_D \quad \left(D_{(2i-1)}areon\right) \\ V_{C1} + V_2 - V_D = V_{C2} - V_D = V_{C3} - V_3 - V_D = V_{C4} - V_3 - V_4 - V_D \quad \left(D_{(2i)}areon\right) \end{cases}. \quad (21)$$

v_S is a square wave with an amplitude of V_S, which is obtained from Eqs. (20) and (21) as

$$V_S = V_i + 2V_D. \quad (22)$$

The amplitude of the fundamental component of the transformer primary winding, $V_{m\text{-}P}$, can be obtained from Eq. (22) with the Fourier transfer,

$$V_{m-P} = \frac{2}{\pi}N\left(V_i + 2V_D\right). \quad (23)$$

Similarly, the amplitude of the fundamental component of v_{DSb}, $V_{m\text{-}in}$, is

$$V_{m-in} = \frac{2}{\pi}V_{in} = \frac{2}{\pi}\left(V_1 + V_2 + V_3 + V_4\right). \quad (24)$$

The amplitude of i_{Lr}, I_m, is obtained as

$$I_m = \frac{V_{m-in} - V_{m-P}}{|Z|} = \frac{V_{m-in} - V_{m-P}}{\sqrt{\left(\omega L_r - \frac{1}{\omega C_r}\right)^2}} \frac{V_{m-in} - V_{m-P}}{Z_0\sqrt{\left(\frac{\omega}{\omega_r} - \frac{\omega_r}{\omega}\right)^2}}, \quad (25)$$

where Z_0 is the characteristic impedance of the resonant circuit given by

$$Z_0 = \omega_r L_r = \frac{1}{\omega_r C_r}. \quad (26)$$

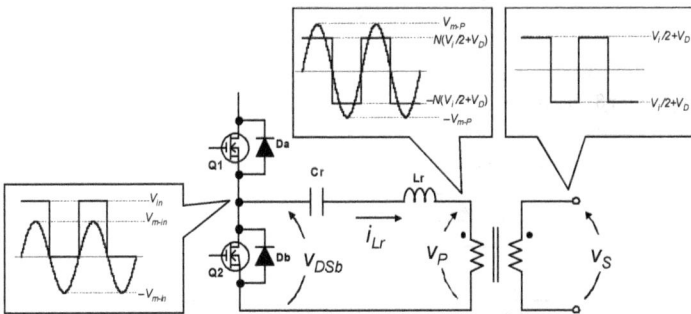

Figure 20. Key waveforms and their fundamental components.

In order for the series-resonant inverter to transfer energies to the voltage multiplier connected to the secondary winding, $V_{m\text{-}in}$ must be higher than $V_{m\text{-}P}$. Assuming that the number of series connections is four and V_1–V_4 are balanced as V_i, the criterion of N is obtained from Eqs. (23) and (24), as

$$N < \frac{4V_i}{V_i + 2V_D}. \qquad (27)$$

Eqs. (23) and (25) indicate that the smaller N is, the larger I_m will be, resulting in the larger power transfer from the resonant circuit to the voltage multiplier. Thus, with small N, an equalization speed can be accelerated, although it tends to cause increased losses in resistive components in the resonant circuit as well as in the voltage multiplier.

4.3. Experimental

4.3.1. Prototype and its fundamental performance

A 10-W prototype of the double-switch series-resonant equalizer was built for 8 cells connected in series, as shown in Fig. 21. Table 2 lists the component values used for the prototype. C_{out1}–C_{out8} are smoothing capacitors connected to cells in parallel (not shown in Fig. 16 for the sake of simplicity). The transformer leakage inductance, L_{kg}, was used as the resonant inductor L_r, whereas the magnetizing inductance, L_{mg}, was designed to be large enough not to influence the series-resonant operation. The prototype equalizer was operated with a fixed $D = 0.48$ at a switching frequency of 220 kHz.

Figure 21. Photograph of a 10-W prototype of the double-switch series-resonant equalizer using a voltage multiplier for 8 cells connected in series.

Component	Value
C_1–C_8	Tantalum Capacitor, 47 µF, 80 mΩ
C_{out1}–C_{out8}	Ceramic Capacitor, 200 µF
C_r	Film Capacitor, 100 nF
Q_a, Q_b	N-Ch MOSFET, HAT2266H, R_{on} = 9.2 mΩ
D_1–D_{16}	Schottky Diode, CRS08, V_D = 0.36 V
Transformer	N_1:N_2 = 30:5, L_{kg} = 4.7 µH, L_{mg} = 496 µH

Table 2. Component values used for the prototype of the series-resonant equalizer.

The experimental setup for the efficiency measurement for the resonant equalizer is shown in Fig. 22. The efficiency measurement was performed using the intermediate tap and the variable resistor in order to emulate the voltage-balanced and -imbalanced conditions. With the tap Y selected, the current flow paths under the voltage-balanced condition are emulated, whereas those under the voltage-imbalanced condition of $V_1 < V_i$ ($i = 2...12$) can be emulated by selecting the tap X. The input and the output of the equalizer were separated at the point Z to measure efficiencies. The efficiencies were measured by changing the ratio of V_1/V_{in} between approximately 1/8 and 1/20. Cells were disconnected and only the smoothing capacitors (C_{out1}–C_{out8}) were used to sustain the voltages of V_1–V_8 in the efficiency measurement.

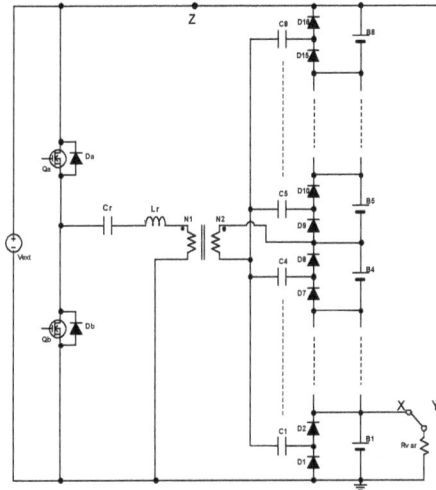

Figure 22. Experimental setup for efficiency measurement for the resonant equalizer.

The measured power conversion efficiencies and output power characteristics as a function of V_1 are shown in Fig. 23. As V_1 increased, the efficiencies significantly increased because the diode voltage drop accounted for a lesser portion of the output voltage (i.e., V_1). The measured peak efficiencies under the voltage-balanced and -imbalanced conditions were 73% and 68%, respectively. The efficiency trends under the voltage-balanced condition were higher than those under the voltage-imbalanced condition. The lower efficiency trend under the voltage-balanced condition can be attributed to increased Joule losses in resistive components in the series-resonant inverter and the voltage multiplier.

Measured waveforms of i_{Lr} and v_{DSb} at V_{in} = 32 V and V_1 = 4 V under the voltage-balanced and -imbalanced conditions are shown in Figs. 24(a) and (b), respectively. The amplitude of i_{Lr} under the voltage-balanced condition was slightly greater than that under the voltage-imbalanced condition. In the operation analysis made in Section 4.2.3, the voltage across C_i was assumed constant and the voltages of transformer windings were treated as square waves. However, in practice, the voltage across C_i varies as current flows, and transformer winding voltages are not ideal square waves. Under the voltage-imbalanced condition, as currents in the voltage multiplier concentrated to C_1, V_{m-P} tended to increase because of an increased voltage variation of C_i. Consequently, the amplitude of i_{Lr}, I_m, decreased as V_{m-P} increased, as can be understood from Eq. (25).

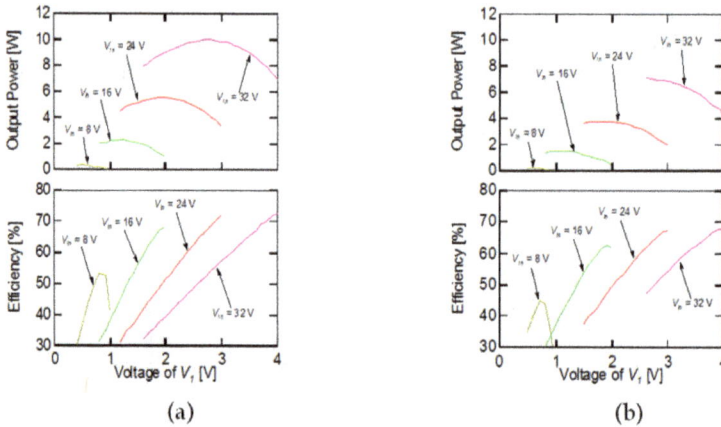

Figure 23. Measured power conversion efficiencies and output powers of the series-resonant equalizer as a function of V_1 under (a) voltage-balanced and (b) -imbalanced conditions.

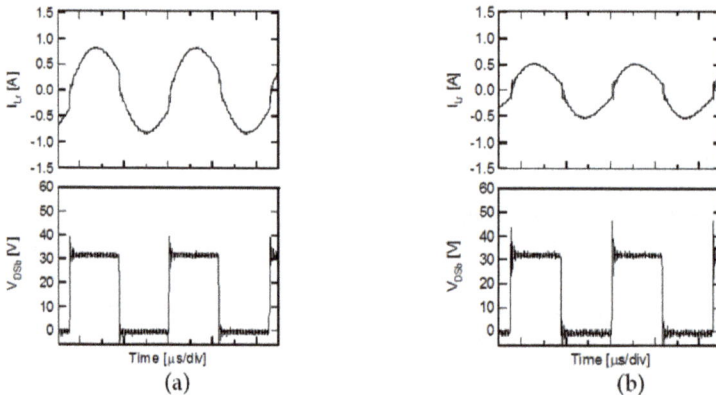

Figure 24. Measured waveforms under (a) voltage-balanced and (b) -imbalanced conditions.

4.3.2. Equalization

The experimental equalization test using the prototype of the series-resonant equalizer was performed for 8-series SCs with a capacitance of 500 F at a rated charge voltage of 2.5 V. The initial voltages of SCs were imbalanced in the range 1.8–2.5 V. The results of the equalization test are shown in Fig. 25(a). The cell voltages with a high initial voltage decreased, while those with a low initial voltage increased by the energy redistribution mechanism. The voltage imbalance gradually disappeared, and the standard deviation of the cell voltages eventually decreased down to approximately 4 mV at the end of the equalization test. The cell voltages were almost completely balanced, and hence, the equalization performance of the series-resonant equalizer was demonstrated.

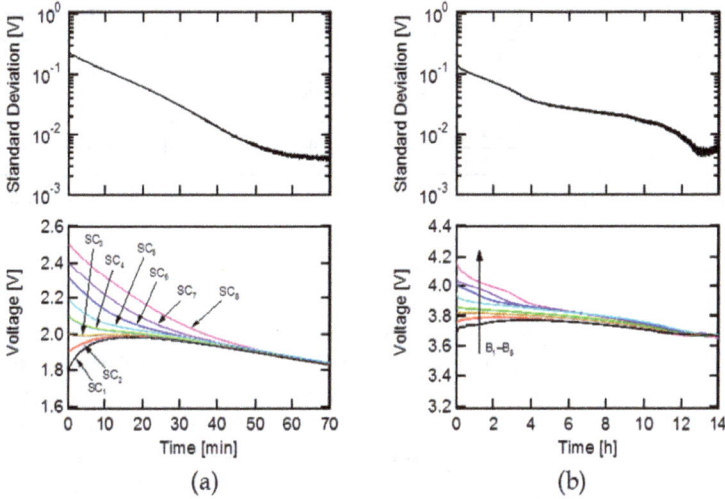

Figure 25. Experimental profiles of 8 series-connected (a) SCs and (b) lithium-ion cells equalized by the series-resonant equalizer.

A similar experimental equalization was performed for 8-series lithium-ion cells with a capacity of 2200 mAh at a rated charge voltage of 4.2 V. The initial SOCs of the cells were imbalanced between 0%–100%. The measured equalization profiles are shown in Fig. 25(b). Despite the nonlinear characteristics of lithium-ion cells, the standard deviation of the cell voltages gradually decreased, and the voltage imbalance was successfully eliminated.

5. Comparison with conventional equalizers

The presented single- and double-switch equalizers are compared with conventional equalizers in terms of the number of required power components in Table 3, where n is the number of series connections of cells. Obviously, the passive dissipative equalizers using resistors or diodes are the simplest topology, although they are neither efficient nor practical as mentioned in Section 2.1. Except for the equalizers using a multi-winding transformer, the required number of switches in conventional nondissipative equalizers is proportional to the number of series connections, and therefore, the circuit complexity tends to significantly increase for applications requiring a large number of series connections. Although the equalizers using a multi-winding transformer need only one or two switches, the need of a multi-winding transformer is considered as their major drawback because of the requirement for strict parameter matching among multiple secondary windings, resulting in design difficulty and poor modularity.

Topology		Switch	Resistor	Inductor	Capacitor	Diode	Transformer
Dissipative Equalizer	Resistor	-	n	-	-	-	-
	Zener Diode	-	-	-	-	n	-
	Active Shunting	n	n	-	-	-	-
Induvidual Cell Equalizer	Switched Capacitor	$2n$	-	-	$n-1$	-	-
	Buck-Boost Converter	$2(n-1)$	-	$n-1$	-	-	-
Multi-Winding Transformer-Based	Flyback Converter	1	-	-	-	n	1 ($n+1$ windings)
	Forward Converter	2	-	-	-	$n+2$	1 ($n+1$ windings)
Single Converter with Selection Switches	Flying Capacitor	$2n$	-	-	1	-	-
	Flyback Converter	$2n+1$	-	-	1	1	1
Single-Switch Equalizer (Isolated Ćuk-Based)		1	-	$n+1$	$n+1$	n	1
Double-Switch Series-Resonant Equalizer		2	-	$(1)^*$	$n+1$	$2n$	1

* Inductor is not necessary when leakage inductance of the transformer is used as a resonant inductor

(Smoothing Capacitors are excluded)

Table 3. Comparison in terms of required number of power components.

On the other hand, the single-switch equalizer using multi-stacked buck–boost converters (isolated Ćuk converter-based) can operate with a single switch, and therefore, the circuit complexity can be significantly reduced when compared with conventional equalizers that require many switches in proportion to the number of series connections. In addition, since a multi-winding transformer is not necessary and the number of series connections can be readily arbitrary extended by stacking a circuit consisting of L, C, and D. A drawback of this single-switch equalizer is the need of multiple inductors, with which the equalizer is prone to be bulky and costly as the number of the series connections increases.

The double-switch resonant equalizer using a voltage multiplier is able to operate with two switches and a single transformer (in the case that the leakage inductance of the transformer is used as a resonant inductor). In addition to the reduced number of switches, the required number of magnetic components is only one, and hence, the resonant equalizer achieves simplified circuitry coupled with a reduction in size and cost when compared with those requiring multiple magnetic components. The modularity of the resonant equalizer is also good; by adding C and D, the number of series connections can be arbitrarily extended.

6. Conclusions

Cell voltage equalizers are necessary in order to ensure years of safe operation of energy storage cells, such as SCs and lithium-ion cells, as well as to maximize available energies of cells. Although various kinds of equalization techniques have been proposed, demonstrated, and implemented, the requirement of multiple switches and/or a multi-winding transformer in conventional equalizers is not desirable; the circuit complexity tends to significantly increase with the number of switches, and the strict parameter matching among multiple secondary windings of a multi-winding transformer is a serious issue resulting in design difficulty and poor modularity. Two novel equalizers, (a) the single-switch equalizer using multi-stacked buck–boost converters and (b) the double-switch equalizer using a resonant inverter and voltage multiplier are presented in this chapter in order to address the above issues.

The single-switch equalizer using multi-stacked buck–boost converters can be derived by multi-stacking any of the traditional buck–boost converters: SEPIC, Zeta, or Ćuk converters. In addition to the single-switch configuration, a multi-winding transformer is not necessary, and therefore, the circuit complexity can be significantly reduced as well as improving the modularity when compared with conventional equalizers, which require multiple switches and/or a multi-winding transformer. The detailed operation analysis was mathematically made for the isolated Ćuk converter-based topology.

The double-switch equalizer using a resonant inverter and voltage multiplier can be synthesized by, namely, combining a resonant inverter and a voltage multiplier. Although two switches are necessary, the required number of switches is sufficiently small to achieve a reduced circuit complexity. Since the number of required magnetic components is only one (i.e., a transformer), the size and cost of the equalizer are considered to be minimal when compared with equalizers requiring multiple magnetic components. The series-resonant inverter was used as a resonant inverter, and a detailed operation analysis was separately made for the voltage multiplier and the series-resonant inverter.

The prototypes of the single- and double-switch equalizers were built for series-connected cells, and experimental equalization tests were performed for series-connected SCs and lithium-ion cells from initially-voltage-imbalanced conditions. The energies of cells with a high initial voltage are redistributed to the cells with a low initial voltage, and eventually, voltage imbalance of SCs and lithium-ion cells were almost perfectly eliminated by the equalizers after sufficient time elapsed.

Author details

Masatoshi Uno
Japan Aerospace Exploration Agency, Japan

7. References

Cao, J., Schofield, N. & Emadi, A. (2008). Battery Balancing Methods: A Comprehensive Review, *Proceedings of IEEE Vehicle Power and Propulsion Conference*, ISBN 978-1-4244-1848-0, Harbin, China, September 3-5, 2008

Guo, K. Z., Bo, Z. C., Gui, L. R. & Kang, C. S. (2006). Comparison and Evaluation of Charge Equalization Technique for Series Connected Batteries, *Proceedings of IEEE Applied Power Electronics Conference and Exposition*, ISBN 0-7803-9716-9, Jeju, South Korea, June 18-22, 2006

Isaacson, M. J., Hollandsworth, R. P., Giampaoli, P. J., Linkowaky, F. A., Salim, A. & Teofilo, V. L. (2000). Advanced Lithium Ion Battery Charger, *Proceedings of Battery Conference on Applications and Advances*, ISBN 0-7803-5924-0, Long Beach, California, USA, January 11-14, 2000

Kim, C. H., Kim. M. Y., Kim. Y. D. & Moon, G. W. (2011). A Modularized Charge Equalizer Using Battery Monitoring IC for Series Connected Li-Ion Battery String in an Electric

Vehicle, *Proceedings of IEEE International Power Electronics Conference*, ISBN 978-1-61284-956-0, Jeju, South Korea, May 30-June 3, 2011

Kutkut, N. H., Divan, D. M. & Novotny, D. W. (1995). Charge Equalization for Series Connected Battery Strings. *IEEE Transaction on Industry Applications*, Vol. 31, No. 3, (May & June 1995), pp. 562-568, ISSN 0093-9994

Lee, Y. S. & Cheng, M. W. (2005). Intelligent Control Battery Equalization for Series Connected Lithium-Ion Battery Strings. *IEEE Transaction on Industrial Electronics*, Vol. 52, No. 5, (October 2005), pp. 1297-1307, ISSN 0278-0046

Nishijima, K., Sakamoto, H. & Harada, K. (2000). A PWM Controlled Simple and High Performance Battery Balancing System, *Proceedings of IEEE Power Electronics Specialist Conference*, ISBN 0-7803-5692-6, Galway, Ireland, June 18-23, 2009

Pascual, C. & Krein, P. T. (1997). Switched Capacitor System for Automatic Series Battery Equalization, *Proceedings of IEEE Applied Power Electronics Conference and Exposition*, ISBN 0-7803-3704-2, Atlanta, Georgia, USA, February 23-27, 1997

Uno, M. (2009). Interactive Charging Performance of a Series Connected Battery with Shunting Equalizer, *Proceedings of IEEE International Telecommunications Energy Conference*, ISBN 978-1-4244-2491-7, Incheon, South Korea, October 18-22, 2009

Uno, M. & Tanaka, K. (2011). Single-Switch Cell Voltage Equalizer Using Multi-Stacked Buck-Boost Converters Operating in Discontinuous Conduction Mode for Series-Connected Energy Storage Cells. *IEEE Transaction on Vehicular Technology*, Vol. 60, No. 8, (October 2011), pp. 3635-3645, ISSN 0018-9545

Uno, M. & Tanaka, K. (2011). Influence of High-Frequency Charge-Discharge Cycling Induced by Cell Voltage Equalizers on Life Performance of Lithium-Ion Cells. *IEEE Transaction on Vehicular Technology*, Vol. 60, No. 4, (May 2011), pp. 1505-1515, ISSN 0018-9545

Uno, M. (2011). Supercapacitor-Based Electrical Energy Storage System, In: *Energy Storage in the Emerging Era of Smart Grids*, Carbone, R., pp. 21-40, InTech, ISBN 978-953-307-269-2, Rijeka, Croatia

Uno, M. & Tanaka, K. (2011). Accelerated Ageing Testing and Cycle Life Prediction of Supercapacitors for Alternative Battery Applications, *Proceedings of IEEE International Telecommunications Energy Conference*, ISBN 978-1-4577-1248-7, Amsterdam, Netherland, October 9-13, 2011

Low Voltage DC System with Storage and Distributed Generation Interfaced Systems

George Cristian Lazaroiu and Sonia Leva

Additional information is available at the end of the chapter

1. Introduction

The complexity of the problems related to the generation, transport and utilization of energy increased in the last decades, with the intensification of the global problems regarding environment protection, climatic changes and the exhaust of the natural resources. In addition, the European Union is facing some specific problems, the most important being the one linked to the nowadays high dependency of the imported energy resources. Placed under the pressure of the agreements assumed through Kyoto protocol, The European Union launched in 2000 the third Green Paper "Towards an European strategy for security of supply". The necessity that the renewable sources to become an important part of the energy generation sector it is highlighted. An important increase of their share it is planned. In particular the place of the new sources in a liberalized energy market is discussed, as well as their purpose as main promoters of the "distributed generation"(DG) concept. The interconnection of the storage systems and distributed generation units in the existing power system affects the classical principles of operation for this latter. From the utility point of view, the operation of the these sources in parallel with the power system presents a high interest, as leads to the diminution of the transport capacity, permits the voltage regulation, maintains the systems stability, increases the equipments lifetime. Moreover, the actual trend of increasing installation of these units implies the establishment of their impact on the operation of the power system and on the power quality.

In the last years, the electrical industry sector is suffering important changes: besides the structural changes induced by the deregulated energy market, important developments of the customers installations and devices are taking place. In parallel to these aspects, energy and environmental considerations encourages the spread use of the renewable energy sources and the utilization of higher energy efficiency levels, recurring at distributed generation (DG) [1].

These bring to the question if the present distribution networks are still the most adequate to satisfy the nowadays demands. The major part of the DGs and storage systems are generating dc power or require an intermediate dc stage before power injection in a possible ac network. These considerations brought to the possibility to use dc distribution networks, in the presence of sensitive loads and distributed generation. The low voltage dc system ensures a stable voltage level for the supplied customers, connected at the dc bus through ac-dc or dc-dc converters. The choice of the most suitable equipments, which respond to the requirements of an optimum operation of the entire system, is necessary. For guaranteeing the correct and robust operation of the system is essential to adopt adequate control logic for obtaining the best system performances.

2. DC system layout

The profusion of dc power (internally used by various customers facilities based on electronics, present in the conversion state of the UPSs and generated or used in the energy conversion stage by some distributed energy sources) has opened the door for the consideration of a dc distribution system, where all the converters and distributed energy sources are connected. The possible use of a dc distribution network for residential customers has been analyzed in [2], which illustrated the suitability to directly supply with dc power some specific customers. Low voltage dc distribution systems, where the various converters are connected to the main dc bus, were proposed for navy applications, or, for the industrial sector, the supply of variable speed drives.

The main purpose of the dc distribution system is to ensure a high degree of power quality and supply reliability for the customers supplied by it. Also, the network is thought to facilitate the interconnection of the distributed generators and of the storage systems [3]. The dc distribution system, illustrated in Fig. 1, presents a series of advantages with respect to the ac network. The distributed generation units can be interconnected directly with the dc system or through only one ac/dc converter, avoiding in this way the many conversion stages that reduce the system reliability and are responsible for higher power losses. Storage energy systems, for the power quality improvement, can also be interconnected directly or through inverters with the dc distribution system. Also, is not required to realize the synchronization between the generators and the dc network and the control of the system is simpler, as it is not necessary frequency regulation for the islanding operation [4].

The dc bus is the only common point for all the converters. Hence, the control of every converter is based on the dc voltage feedback that is compared with a reference value, in correspondence with every device converter.

The low voltage dc distribution systems must fulfill the following requirements [5], [6]:

- the dc system, with distributed generators and storage systems interconnected, must have a stable operation during ac grid-connected and islanding functioning;
- should deliver premium voltage quality and should ensure high supply continuity for the ac and dc loads;
- the dc system should be expandable: by adding supplementary loads and energy sources, the control strategy must be redesigned in a small manner;
- the electric safety should be ensured respecting the existing standards.

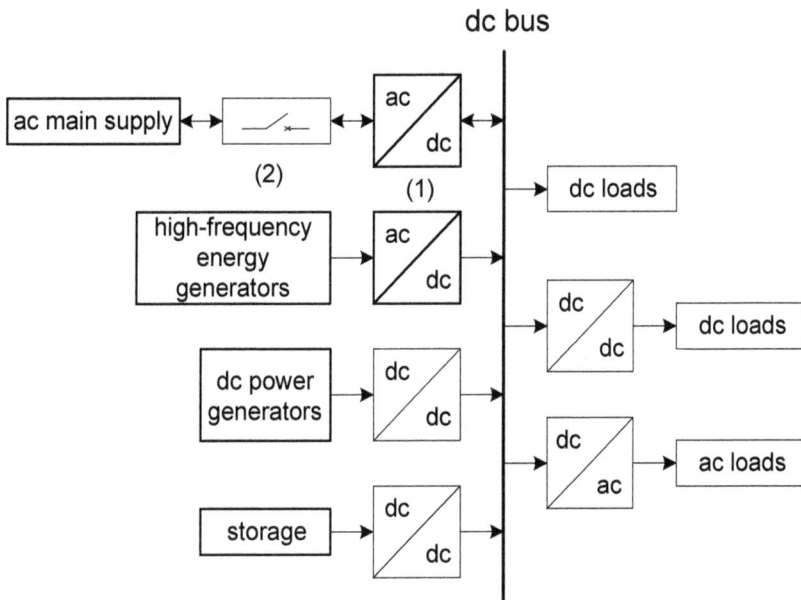

Figure 1. Layout of dc system with storage and distributed generation interfaced systems.

The design process of the low voltage dc distribution system requires the selection of the most suitable combination of energy sources, power-conditioning devices, and energy-storage systems for responding to the necessities and requirements of the dc low voltage dc distribution system, together with the implementation of an efficient energy dispatch strategy.

In Fig. 2, the layout of the dc distribution network with distributed generators, storage energy systems and sensitive loads, is illustrated.

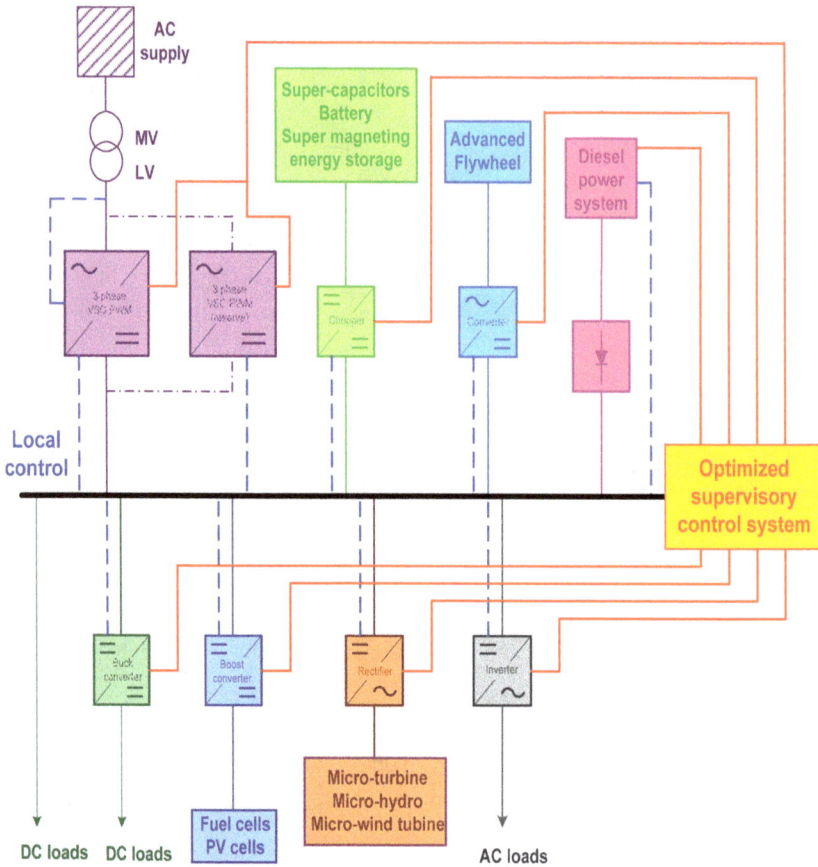

Figure 2. Layout of the dc distribution system

2.1. AC/DC interface

The voltage source converter is a PWM current-controlled one, forcing the instantaneous phase current to follow a sinusoidal current reference template; also, presents some advantages with respect to the voltage-controlled method, like major control stability and it is easier to implement [7]. Using the known PWM techniques, the ac input current drawn by the converter is controlled to become a sine wave, which is in phase with the input voltage and thus the resulting input power factor is maintained at, or very close to, unity. Since the ac input current is forced by the implementation of PWM methods to be sinusoidal, the resulting current waveform will also have a significantly lower harmonic distortion. Even if the current going through the boost inductors will be sinusoidal, there will be a high-frequency component on top of the fundamental due to the PWM switching. Therefore, a small passive filter is required on the source-side of the boost inductors to eliminate this harmonic.

In order to obtain the benefits mentioned above, proper control of the power converter semiconductor switches is required. The operation of the different topologies becomes even more complicated when perturbations in ac supply system occur. The use of the forced commutated converter imposes the constraint that the output dc voltage must have a higher level than the peak value of the maximum ac input voltage. This implies a step-up or boost type of ac-to-dc power conversion, with the converter more commonly known as a boost converter.

2.2. DC network devices

2.2.1. Battery system

The high power quality degree of the low voltage dc distribution system is ensured with the help of storage energy systems. The storage energy systems must operate each time the ac/dc interface converter is not able to cover the difference between the load requested and the power generated by the distributed generators, case that can appear during voltage sags or short interruptions of the ac network. The storage energy system sizing is determined in correlation with the operations that has to fulfill. An over sizing of the storage energy system allows to realize the peak shaving function, storing the energy produced by the distributed generators during light load and using it during peak load, avoiding in this way to take it from the ac network. Still, considering the high costs associated with the storage energy systems, it is not considered the peak shaving function and is avoided the over sizing of the storage system.

The existing technologies for the storage energy systems are:

a. lead-acid batteries;
b. super-capacitors;
c. compressed air;
d. flywheels.

The lead-acid batteries are most spread and offer the best ratio cost/performances. By connecting more cells in series or parallel, the capacity desired at the required values of voltage, respectively current are obtained.

Super-capacitors, from a constructive point of view, are similar to electrochemical batteries in that each of the two electrodes is immersed in an electrolyte and they are separated by an ion permeable membrane. The main difference, compared to electrochemical batteries, is that no electrochemical reactions or phase changes take place and all energy is stored in an electrostatic field; for this reason the process is highly reversible and the charge discharge cycle can be repeated frequently and virtually without limit. The charge separation process, which requires a voltage difference across the electrodes, takes places on the two interfaces electrode electrolyte; thus, super-capacitors are usually known as double layer capacitors.

Each electrode electrolyte interface represents a capacitor; therefore, the complete cell comprises two capacitors in series. The thickness of the double layer depends on the concentration of the electrolyte and on the size of the ions.

The voltage regulation function is strictly dependent of the storage system capacity and availability. The bidirectional chopper allows the recharge of the storage system, if it is necessary, when the power is flowing from the ac network to the dc system or when in the dc network an excess energy is available. During the discharge process of the storage system, the control system maintains and stabilizes the dc voltage during islanding operation of the dc network. The chopper is behaving as a boost chopper during the discharge process of the storage system and as a buck chopper during the recharge cycle. The dynamic behavior of super-capacitors is strongly related to the ion mobility of the electrolyte used and to the porosity effects of the porous electrodes; the storage process in the double layer is a superficial effect, consequently the electrode surface behaviors play an important role. The most common super-capacitors for industrial application are based on carbon for the electrode materials and an organic solution for the electrolyte.

2.2.2. Diesel power system

The Diesel power system is designated to supply the dc loads during sustained interruptions in the ac network. The Diesel power system is constituted from a Diesel engine with a synchronous generator and a three phase diode bridge. The diode bridge, of high reliability, has to prevent the power flow through the Diesel power system during the islanding operation of the dc system. In this case, the Diesel power system regulates the voltage of the dc distribution system by acting on the synchronous generator excitation system, while the load demand is achieved regulating the output of the Diesel engine. When the dc power system is grid connected with the ac main supply, the Diesel group is not turned on and has to step in during the loss of the ac grid.

2.2.3. Distributed generators

The diffusion of the distributed energy sources and the storage systems will increase in the power systems all over the world. Most of the small power distributed energy resources and storage energy systems located close to the point of consumption of the customer are generating dc power or require a dc intermediate stage. Hence many of these sources can be connected to a dc distribution system. These energy sources, in correlation with the type of the technology implied and depending on their operating characteristics, are generating powers at different voltage levels. In order to connect them to the same dc bus, these sources are interfaced with the help of power electronic converters, like inverters or choppers.

The wind turbine is interconnected with the help of an ac/dc converter. In order to control the wind speed, within the laboratory implementation, the wind speed is generated by an ac motor drive connected to the wind turbine. The wind turbine is a permanent magnet brushless generator.

The photovoltaic generators allow to directly convert the solar radiations into electrical energy without producing pollutants [8]. The photovoltaic cells are simple in design and require reduced maintenance operations. A key advantage of the photovoltaic cells is the fact that can operate interconnected with the public network or in remote areas. Another important characteristic is their modularity that allows reaching new panels with the purpose to increase the power generated. The diffusion of the photovoltaic plants and the utilization of this energy are limited by the costs of the power generated. Even if this cost is independent on the type of used photovoltaic plant, is directly related to the efficiency of the plant. In the last years it can be noted a progress of the photovoltaic cells, both from the point of view of cost reduction and the improvement of the efficiencies obtained with the technology multilayer. Nowadays, the cost of the electrical energy produced by the photovoltaic plants cannot compete with the energy produced by other technology of distributed generators. The photovoltaic plants are sized with respect to the local loads, where the energy excess is sold to the network or is stored with the help of storage systems. During the design of the photovoltaic plants, the aspects related to electrical safety have to be considered. The photovoltaic panels are ideal for remote applications that require power between watts to hundreds of kilowatts of electrical power. Also in the areas where there is a public network, some applications that require non-interruptible power or standby power can use the photovoltaic power. The photovoltaic plants, as consequence of the incentives proposed by various governments, are the most diffused form of distributed generation in the low voltage distribution systems. To this contributed also the possibility to integrate the photovoltaic panels within the existing buildings. The photovoltaic panel consists of 72 series connected mono-crystal silica cells interconnected to the dc system with the help of a dc chopper.

The fuel cell is a proton exchange membrane fuel cell supplied with pressurized hydrogen and ambient air. The fuel cell is interconnected to the dc system with the help of a dc chopper.

The distributed generators contribution is to allow the injection in the ac network of the power excess, during light load, and to supply, in parallel with the storage system, the dc power system during islanding operation.

2.2.4. Loads

The loads interconnected to the dc distribution system can be dc loads or ac loads. In the first case, the supply can be realized directly, if the loads operation permits, or by using a buck converter, when the loads operating voltage is lower than the dc voltage level. In the second case, an inverter is the interface device.

3. Logic of control

The use of the optimized control is based on the supervision of the sources and converters states that has to determine the power requests such that to realize the load sharing [9]. The optimized control strategy needs to ensure high reliability of the system.

The supervisory control allows avoiding the interaction between the electric devices controllers and obviating the occurring high or sudden transients (Fig. 3). The input commands and the limiting values to these controllers come from the supervisory control that produces the required references for the system devices. The supervisory control system determines the behavior of each component: only one device is operating as a voltage source (when the component is directly responsible for the regulation of the dc voltage) while others are operating as current sources (when the component is injecting or absorbing its power available at that instant of time) [10].

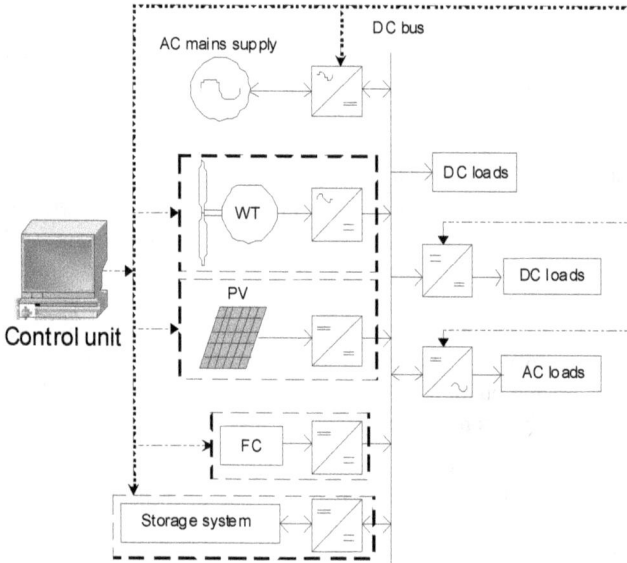

Figure 3. Layout of control implementation

The supervisory control is based on the fact the power systems are current-controlled devices, even when these are behaving as voltage sources. Hence, the optimized control strategy imposes the current of the various power systems, and in this way also the power injected by these devices, for maintaining the dc voltage at the desired value. The philosophy of supervisory controlling the sources and loads is important to be correctly established such that the interaction between the various devices, operating in parallel, must not lead to the instability of the system. The supervisory control has to manage the sources present in the dc distribution system in order to be guaranteed the load sharing.

4. Network components modelling and simulation

The energy sources, the dc loads and the power electronic devices interconnected to the dc bus are modeled within the ATP/EMTP software package. The control strategies are implemented using the Models subroutine.

4.1. Storage system

The charge and discharge states are commanded by a control loop that compares the measured dc voltage V_{DC} with the thresholds V_{stc} (voltage state of charge) and V_{std} (voltage state of discharge). The control strategy of the chopper is realized taking into consideration the various charge states of the storage system, respectively:

- state of charge: absorbed current ($I_{bat}<0$) and increasing voltage (V_{bat}). This state occurs if the dc voltage exceeds the threshold V_{stc};
- state of discharge: injected current ($I_{bat}>0$) and decreasing voltage (V_{bat}). This state occurs if the dc voltage is below the threshold V_{std};
- inert state: no power flow between the storage and the dc power system. This state occurs if the dc voltage is between the thresholds V_{stc} and V_{std}.

During the charge state, the current absorbed by the battery is imposed in conformity with the characteristics of the storage system and must not be influenced by the value of the dc voltage. The threshold V_{stc} is a reference for the charge state and does not represent a value at which the dc voltage has to be maintained. On the contrary, the reference V_{std} is the dc voltage value that the storage energy system imposes during the discharge state, caused by sags or short interruptions of the ac network. In this case the injected current is the result of the control strategy, such that the dc voltage is stabilized and maintained by the storage energy system at the reference value V_{std}.

If the voltage at storage system terminals is between the maximum and minimum limits is realized the second part of the control strategy, bordered in Fig. 4. Otherwise, the chopper is shut down and the storage system is maintained in its state of charge (fully charged or fully discharged). The dc voltage is filtered with a low pass filter for reducing the high frequency ripple of the dc voltage. The output of the low pass filter, represented by V_{mis}, is then compared in the upper and lower loops with the thresholds V_{stc}, respectively V_{std}. In the upper control loop of Fig. 4 the recharge process of the storage system is conditioned by the constraint that the dc network voltage has to exceed the threshold V_{stc} for a certain duration Δt, e.g. 1 s. It must be noted that during storage system recharge process, the reference current value is negative as the storage system is absorbing current.

During the discharge process, lower control loop in Fig. 4 the voltage V_{mis} is compared with the threshold value V_{std} and the resulting error is the input of a proportional-integral (PI) regulator that stabilizes the dc network voltage at the reference V_{std}. The output signal is limited (IF condition) such that the requested current does not exceed the maximum injecting current of the battery I_{dmax}.

In order to verify the control strategy of the interface bidirectional chopper, the voltage values of the two thresholds were chosen V_{stc} = 1.06 p.u. and V_{std} = 1 p.u.; the reference value has to be maintained by the voltage control loop even in the case of sags and short interruptions of the ac network. The current references associated to the charge and discharge processes are strictly dependent of the maximum current that the storage system can inject.

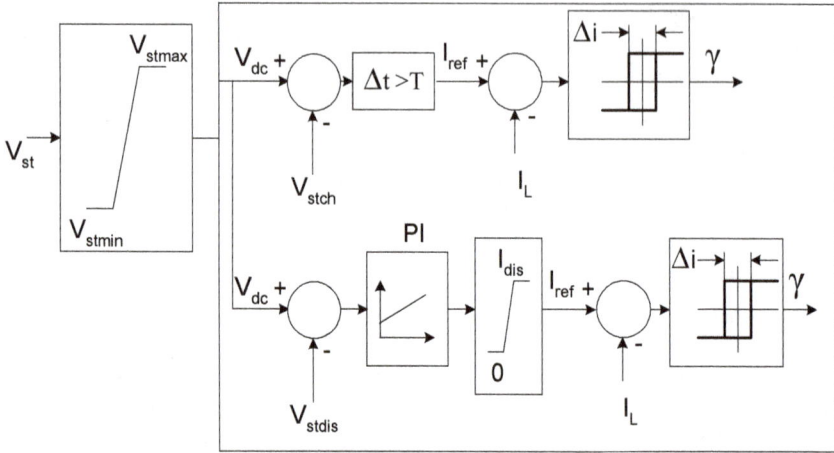

Figure 4. Control loop of the storage energy system

In a first simulation a 50% load reduction is considered, occurred at the time instant 0.07 s. Fig. 5 shows the dc voltage waveform and the two thresholds V_{stc} and V_{std}. Fig. 6 illustrates the storage system current waveform and the reference values I_{stc} and I_{std}. It can be seen that, after the load diminution, the storage system current is null as long as the dc voltage is below the threshold V_{stc}.

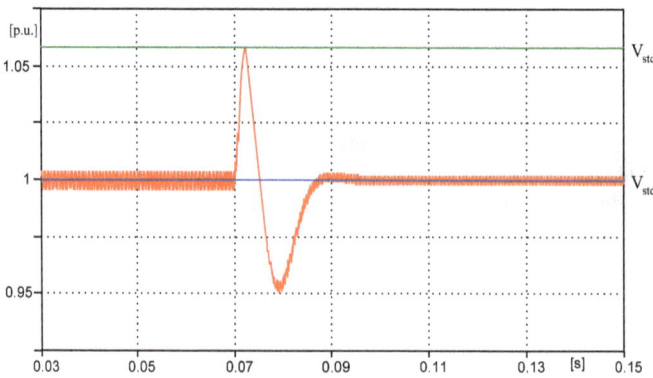

Figure 5. Dc voltage waveform and the reference thresholds V_{stc} and V_{std}.

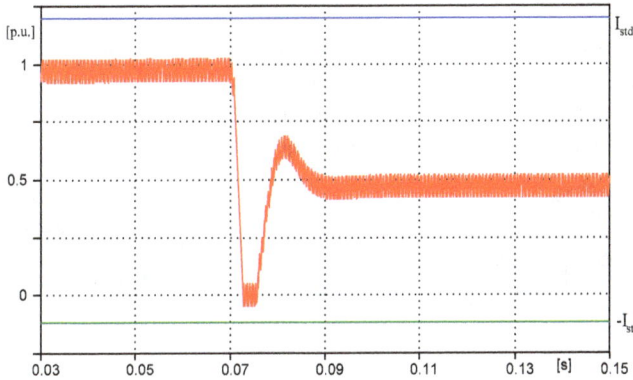

Figure 6. Waveform of the current flowing through the inductor and thresholds I_{std} and $-I_{stc}$.

4.2. AC/DC converter

The ac/dc interface converter is realizing the bidirectional power transfer between the ac and the dc network in correspondence with the dc voltage behavior. As the control strategy is based on the dc voltage feedback, a reference value V_{DCref} has been assigned; this value is maintained and stabilized by the interface converters in any operating condition. This is achieved using the control strategy of Fig. 7; the effective dc voltage, ripple filtered, is compared with V_{DCref}. The VSC converter is absorbing from the ac network three sinusoidal currents and in phase with the ac voltage, avoiding in this way the reactive power flow. The control strategy illustrated in Fig. 7 shows the hysteresis current controller for phase a, forcing the phase currents to follow the reference template. Identical controllers are used in phase b and c. The control strategy imposes the template of the reference current i_{refl} that is obtained multiplying the ac voltages v_a for a gain G. This one results from a PI controller that guarantees null steady state error of the dc voltage. The gain G imposes that the currents taken from the ac network are always in phase with the voltage (positive gain) or in opposition (negative gain), if the phase shift introduced by the LC filter is not considered. The phase shift is determined by the reactive power absorbed by the LC filter that at fundamental frequency is behaving as a capacitance. The phase shift can be significant, even if the LC filter is correctly sized.

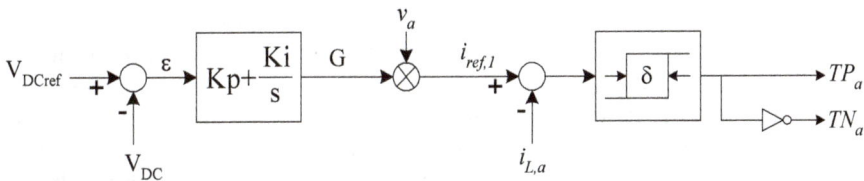

Figure 7. Control strategy of the voltage source converter.

4.2.1. PV system

The PV generator is interfaced with the dc network with the help of a boost chopper. The control system is shown in Fig. 8. The reference current is obtained using the signal I_{MPP} resulting from the MPPT system of the PV generator. This system is not implemented as its dynamics is much slower with respect to the one of the boost converter.

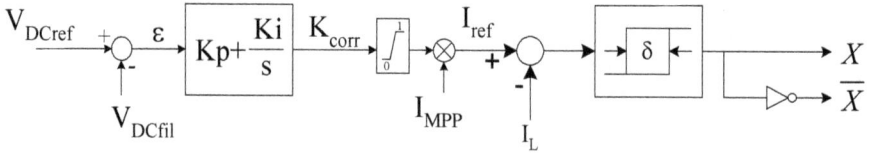

Figure 8. Control system of the boost chopper.

Thus, I_{MPP} is considered constant and corresponding to the MPP in standard environmental conditions of solar radiation and temperature. The control system produces null reference current in case the dc voltage V_{DCfil} exceeds the V_{DCref}. In this case the boost chopper limits the power produced by the PV generator such that to maintain the dc voltage at the reference value assigned.

4.2.2. DC loads

The dc loads are interfaced through a buck converter. The load control strategy requires a feedback loop. The current reference template is generated using a PI regulator. A hysteresis band modulation method is used to produce the PWM pattern for the power valves.

5. Case studies

5.1. Fault the mains supply

Initially, the system has been tested both for showing the transition between the voltage source converter and the battery. A three phase fault occurs at time instant t=0.4 s and it lasts for 250 ms. The ac voltages in this case study are illustrated in Fig. 9.

The dc voltage, maintained before the fault by the interface converter at the nominal value, will decrease until it reaches the threshold value V_{BS} = 0.96 p.u. of the storage energy system reference voltage (Fig. 10). This will determine the intervention of battery that comes in and supplies the load demand. In correspondence, the battery current will decrease until becomes zero, as in Fig. 11. After a first transient when the battery is injecting the maximum power for compensating the dc voltage drop, the control regulator of the battery system is stabilizing the dc bus voltage at the reference value V_{BS} and the current injected by the battery will reach the regime value.

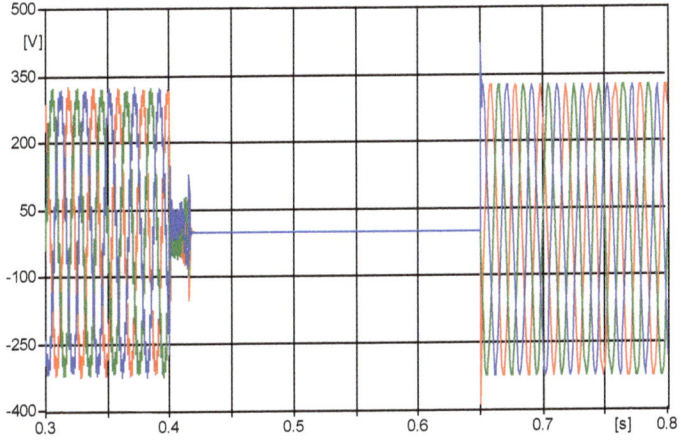

Figure 9. Voltages upstream the ac/dc interface converters during a three phase fault in the ac system.

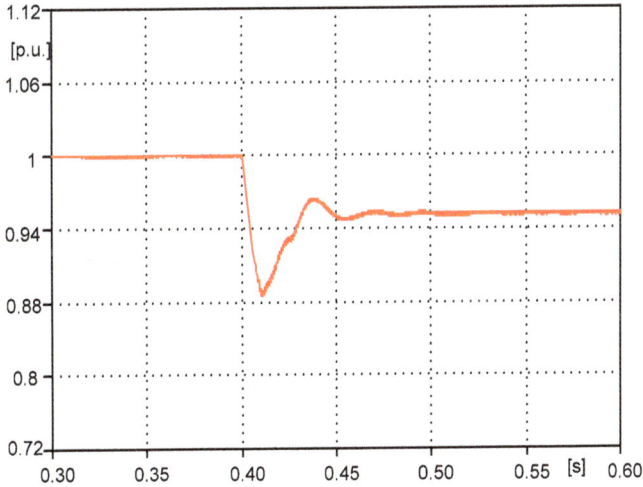

Figure 10. DC voltage waveform during a three phase fault in the ac grid.

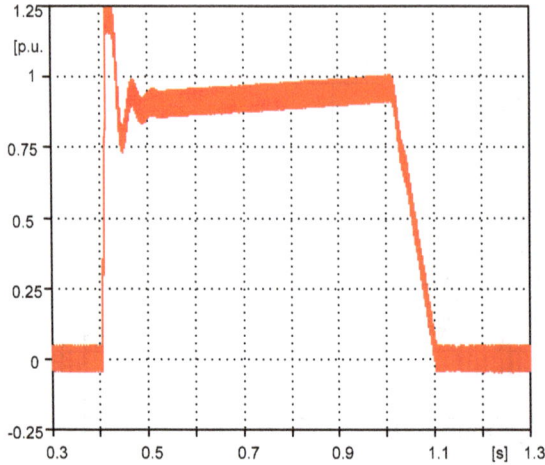

Figure 11. Waveform of the battery current.

After the fault clearance, the synchronization of the ac/dc interface converter with the ac network is required for having a proper operation of the power system.

5.2. Sustained inperruption in the mains supply

The interruption of the ac main supply is occurring at time instant t=0.4 s and lasts for 600 ms. Initially the dc voltage is maintained by the interface converters at the rated value, but after the interruption occurrence the storage system controls the dc voltage at the reference value. At instant t=0.5 s, the Diesel power system is started and the output voltage follows a ramp increase until it reaches the reference value of 0.97 p.u.. The transition between the two devices is illustrated in Fig. 10, where the interaction instant is highlighted.

The currents injected by the Diesel group and the output voltages of the generator of Diesel power system are illustrated in Fig. 14. As it can be observed, when the Diesel output voltage intersects the dc bus voltage, the Diesel group starts injecting power in the dc grid.

When the Diesel group starts injecting power into the dc power system, the voltage regulator of the storage energy system initiates decreasing the current flowing through the battery chopper inductor until it becomes null.

The ac grid is restored at time instant t=0.95 s but the ac/dc interface converter cannot start supplying the load until the synchronization with the ac grid is achieved. The currents flowing through the ac/dc interface inductors increase and the dc voltage, maintained by the Diesel generator at the reference value assigned, is brought by the interface converter to the initial state, respectively 1 p.u., as depicted in Fig. 14.

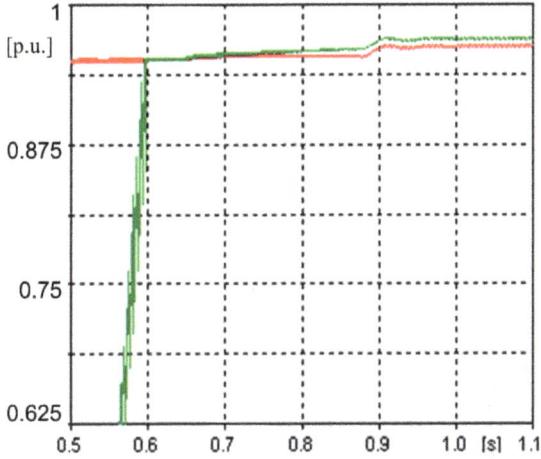

Figure 12. The transition between storage system (red) and Diesel system (green).

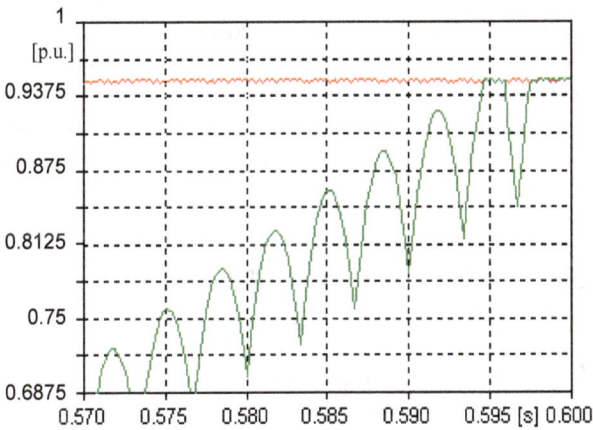

Figure 13. Magnified representation of the transition between storage system (red) and Diesel system (green).

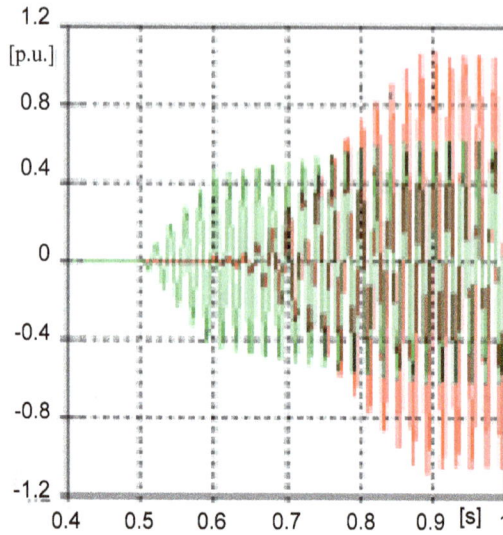

Figure 14. Characteristic current (red) and voltages (green) of the Diesel group.

5.3. DC faults

A short-circuit on the dc busbar leads to the interruption of the supply and each device interconnected to the dc busbar (load, generator, interface converter) is affected. An IGBT electronically breaking system is used for the interruption of the dc circuit. Two electronically interrupters were used, one for each pole, controlled by an over-current relay. The choice for two interrupters is justified by the necessity to interrupt the pole-earth fault. The relay measures the currents of the two poles and commands their disconnection even when one current exceeds the established threshold. This choice protects the ac/dc converter in case of unsymmetrical dc pole-earth faults. The layout of the dc system, in case of a pole-pole fault, used for the protection system investigation is illustrated in Fig. 15.

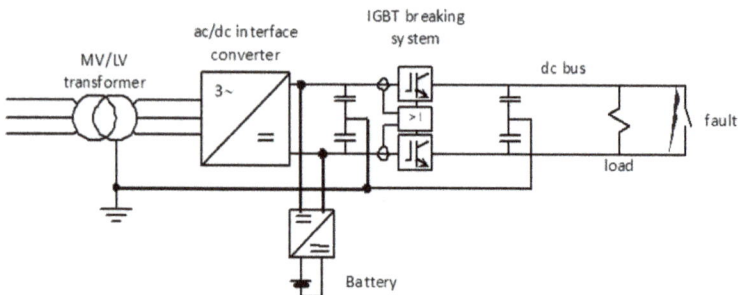

Figure 15. Layout of the dc system protection investigation

The voltage of the bus where the dc fault occurs is shown in Fig. 16. The voltage quickly drops to zero in a time period depending on the system time constants, allowing the complete discharge of the dc capacitors installed across the dc bus of the voltage source converter. In the first instants, the voltage quickly decreases in a time period shorten than the clearing time of the IGBT breaking system.

The waveform of the dc voltages upstream the IGBT breaking system are illustrated in Fig. 17. Initially, the dc voltages drop due to the dc fault, but slowly and much less with respect to the one of the dc bus due to the effect of the decoupling inductance present within the IGBT breaking system. Afterwards the breaking system intervention, the voltages grow as in the first instances after the fault occurrence, the converter control system maintains unchanged the power exchanged with the ac system which is entirely stored in the dc capacitors installed across the dc bus of the VSC.

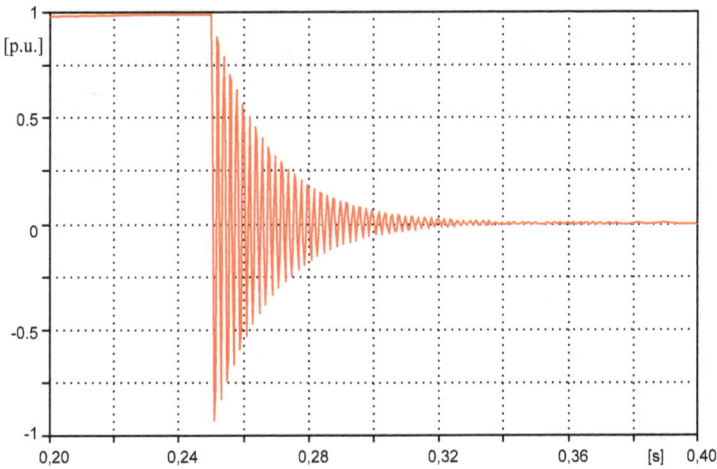

Figure 16. DC voltage of the low voltage distribution system

Then, the converter control system maintains the dc voltages at the rated value. The overvoltages depend on the value of the dc capacitors installed across the dc bus of the VSC and on the fastness of the controller intervention. The current flowing through the IGBT breaking system is illustrated in Fig. 18. When the current reaches the 2 p.u. threshold, the relay commands the breaking system opening, within milliseconds. The IGBT breaking system has the advantage to immediately limit the current, which is very important for the VSCs, which are difficultly supporting current spikes even of short duration. The protection of the interface converter is guaranteed both in case of overvoltages and overcurrents, due to the dc faults occurrence. The dc fault occurrence does not lead to overcurrents in the ac system, due to the decoupling between the ac and dc networks, as seen in Fig. 19.

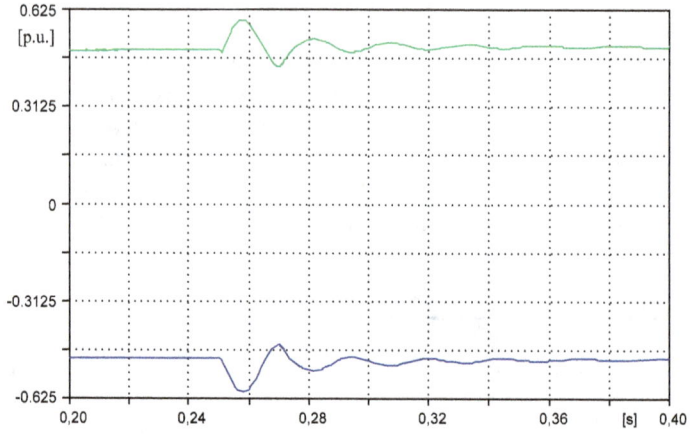

Figure 17. DC voltages upstream the IGBT breaking systems

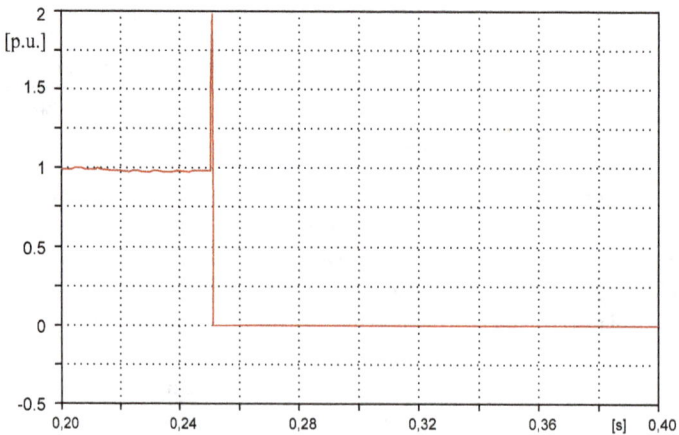

Figure 18. Current flowing within the IGBT breaking system

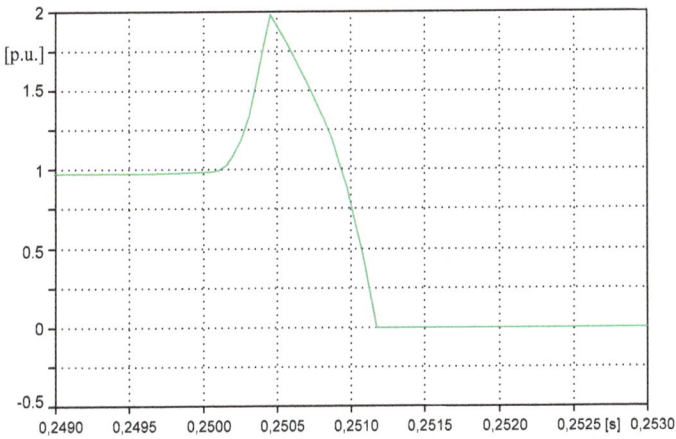

Figure 19. Interruption moment of the dc fault current in the IGBT breaking system

The ac phase voltages are not highly influenced by the dc fault transients. A small increase of the ac voltage, due to the small voltage drop, is occurring.

6. Conclusions

The integration of renewable sources and storage systems within the existing power system affects its traditional principles of operation, the utilization of these alternative sources presenting advantages and disadvantages. The existing trend of installing more and more small capacity sources implies the establishment as accurate as possible of their impact on power system operation. The realization of a low voltage dc distribution grid is technologically feasible and the use of direct current may lead to less power losses and more transmissible power with respect to the ac one. The design of a dc distribution grid, where distributed generation sources and storing energy devices are interconnected, is a difficult task. The choice of the most suitable equipments, which respond to the requirements of an optimum operation of the entire system, is necessary. For guaranteeing the correct and robust operation of the system is essential to adopt adequate control logic for obtaining the best system performances. As the ac fault protection system does not represent such a difficult task, the dc circuit breaking has been covered. A protection strategy using an IGBT breaking system is used. The fast and safe isolation of the dc fault, such that the dc distribution system equipments are not damages, is the most challenging task.

Author details

George Cristian Lazaroiu
Department of Electrical Engineering, University Politehnica from Bucharest, Bucharest, Romania

Sonia Leva
Department of Energy of the Politecnico di Milano, Milan, Italy

7. References

[1] IEEE Standard for interconnecting distributed resources with electric power systems, IEEE Standard 1547, 2003.

[2] M. Baran, N.R. Mahajan, "DC distribution for industrial systems: opportunities and challenges" in IEEE Transactions on Industry Applications, vol. 39, no. 6, Nov.-Dec. 2003, pp.1596 – 1601.

[3] M. Brenna, G. C. Lazaroiu, E. Tironi, "High Power Quality and DG Integrated Low Voltage dc Distribution System", in Proc. IEEE Power Engineering Society General Meeting 2006, June 18-22, Montreal, Canada, pp. 6

[4] M. Brenna, G.C. Lazaroiu, G. Superti_Furga, E. Tironi, "Bidirectional front-end converter for DG with disturbance insensitivity and islanding detection capability," IEEE Trans. Power Del., vol. 23, no. 2, pp. 907–914, July 2008.

[5] S. Leva, et al., Hybrid Renewable Energy-Fuel Cell System: design and performance evaluation, Electric Power Systems Research, Vol. 79, 2009, pp.316-324

[6] S. Leva, Dynamic Stability of Isolated System in presence of PQ Disturbances, IEEE Transactions on Power Delivery, Vol. 23, No.2, April 2008, pp. 831-840

[7] A. Yazdani, et al., Modeling Guidelines and a Benchmark for Power System Simulation Studies of Three-Phase Single-Stage Photovoltaic Systems, IEEE Transactions on Power Delivery, Vol. 26, No. 2, 2011, pp. 1247 - 1264

[8] M. Berrera, R. Faranda, S. Leva, Experimental test of seven widely-adopted MPPT algorithms, IEEE Power Tech 2009, Bucarest, Romania, 28 June - 2 July 2009

[9] N. Mohan; T. M. Undeland; W.P. Robbins: "Power electronics: Converters, applications and design", 2nd Edition New York, Wiley & Sons, 1995.

[10] M. Brenna, F. Foiadelli, D. Zaninelli, "New Stability Analysis for Tuning PI Controller of Power Converters in Railway Application," IEEE Trans. Ind. El., vol. 58, no. 2, pp. 533–543, Feb 2011.

Chapter 9

Hybrid Energy Storage and Applications Based on High Power Pulse Transformer Charging

Yu Zhang and Jinliang Liu

Additional information is available at the end of the chapter

1. Introduction

1.1. HES based on pulse transformer charging

In the fields of electrical discipline, power electronics and pulsed power technology, the common used modes of energy transferring and energy storage include mechanical energy storage (MES), chemical energy storage (CHES), capacitive energy storage (CES), inductive energy storage (IES) and the hybrid energy storage (HES) [1-3]. The MES and CHES are important ways for energy storage employed by people since the early times. The MES transfers mechanical energy to pulse electromagnetic energy, and the typical MES devices include the generator for electricity. The CHES devices, such as batteries, transfer the chemical energy to electrical energy. The energy storage modes aforementioned usually combine with each other to form an HES mode. In our daily life, the MES and CHES usually need the help of other modes to deliver or transfer energy to drive the terminal loads. As a result, CES, IES and HES become the most important common used energy storage modes for users. So, these three energy storage modes are analyzed in detail as the central topics in this chapter.

The CES is an energy storage mode employing capacitors to store electrical energy [3-5]. As Fig. 1(a) shows, C_0 is the energy storage component in CES, and the load of C_0 can be inductors, capacitors and resistors respectively. Define the permittivity of dielectric in capacitor C_0 as ε, the electric field intensity of the stored electrical energy in C_0 as E. The energy density W_E of CES is as

$$W_E = \frac{1}{2}\varepsilon E^2 \ .$$

(1)

Usually, W_E which is restricted to ε and the breakdown electric field intensity of C_0 is about $10^4 \sim 10^5 \, J/m^3$. The traditional Marx generators are in the CES mode [4-5].

The IES is another energy storage mode using inductive coils to generate magnetic fields for energy storage. As shown in Fig. 1(b), the basic IES cell needs matched operations of the opening switch (S_{open}) and the closing switch (S_{close}) [6-7], while L_0 is as the energy storage component. When the charging current of L_0 reaches its peak, S_{open} becomes open and S_{close} becomes closed at the same time. As the instantaneously induced voltage on L_0 grows fast, the previously stored magnetic energy in the magnetic field is delivered fast to the load through S_{close}. The load of L_0 also can respectively be inductors, capacitors and resistors. The explosive magnetic flux compression generator is a kind of typical IES device [7]. The coil winding of pulse transformer which has been used in Tokamak facility is another kind of important IES device [8]. Define the permeability of the medium inside the coil windings as μ, the magnetic induction intensity of the stored magnetic energy as B. The energy density W_B of IES is as

$$W_B = \frac{1}{2}\frac{B^2}{\mu} \ . \tag{2}$$

Usually, W_B restricted by μ and B is about $10^7\ J/m^3$. IES has many advanced qualities such as high density of energy storage, compactness, light weight and small volume in contrast to CES. However, disadvantages of IES are also obvious, such as requirement of high power opening switches, low efficiency of energy transferring and disability of repetitive operations.

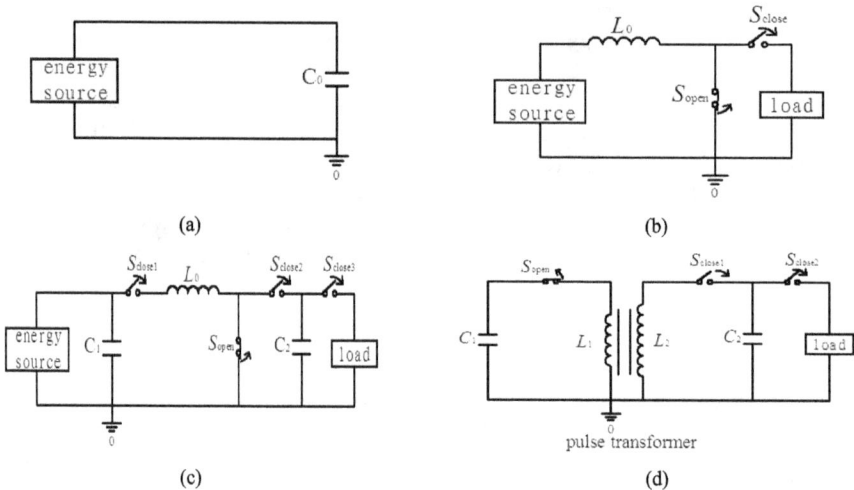

Figure 1. Schematics of three kinds of common-used energy storage modes. (a) Capacitive energy storage mode; (b) Inductive energy storage mode; (c) Typical hybrid energy storage mode; (d) Hybrid energy storage based on pulse transformer.

In many applications, CES combining with IES is adopted for energy storage as a mode of HES. Fig. 1(c) shows a typical HES mode based on CES and IES. Firstly, the energy source charges C_1 in CES mode. Secondly, S_{close1} closes and the energy stored in C_1 transfers to L_0

through the resonant circuit in IES mode. Thirdly, the previously closed switch S_{open} opens, and S_{close2} closes at the same time. The accumulated magnetic energy in L_0 transfers fast to capacitor C_2 in CES mode again. Finally, S_{close3} closes and the energy stored in C_2 is delivered to the terminal load. So, in the HES mode shown in Fig. 1(c), the HES cell orderly operates in CES, IES and CES mode to obtain high power pulse energy. Furthermore, the often used HES mode based on CES and IES shown in Fig. 1(d) is a derivative from the mode in Fig. 1(c). In this HES mode, pulse transformer is employed and the transformer windings play as IES components. In Fig. 1(d), if S_{open} and S_{close1} operate in order, the HES cell also orderly operates in CES, IES and CES mode. Of course, switch S_{close1} in Fig. 1(d) also can be ignored in many applications for simplification.

Generally speaking, a system can be called as HES module if two or more than two energy storage modes are included in the system. In this chapter, the centre topics just focus on CES, IES and the HES based on the CES and IES, as they have broad applications in our daily life. The CES and IES both have their own advantages and defects, but the HES mode based on these two achieves those individual advantages at the same time. In applications, a lot of facilities can be simplified as the HES module including two capacitors and a transformer shown in Fig. 2 [9-16]. Switch S_1 has ability of closing and opening at different time. This kind of HES module based on transformer charging can orderly operate in CES, IES and CES mode. And it has many improved features for application at the same time, such as high efficiency of energy transferring, high density of energy storage and compactness.

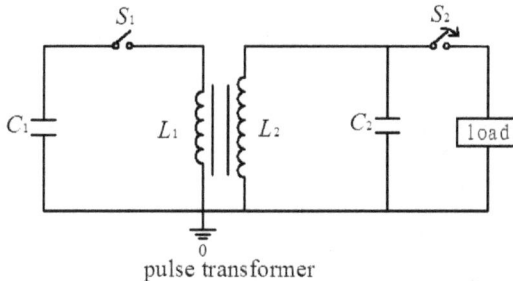

Figure 2. Schematic of the common used hybrid energy storage mode based on capacitors and pulse transformer

1.2. Applications of HES based on pulse transformer charging

The HES based on pulse transformer charging is an important technology for high-voltage boosting, high-power pulse compression, pulse modification, high-power pulse trigger, intense electron beam accelerator and plasma source. The HES cell has broad applications in the fields such as defense, industry, environmental protection, medical care, physics, cell biology and pulsed power technology.

The HES based on pulse transformer charging is an important way for high-power pulse compression. Fig. 3(a) shows a high-power pulse compression facility based on HES in

Nagaoka University of Technology in Japan [9], and its structure is shown in Fig. 3(b). The Blumlein pulse forming line plays as the load capacitor in the HES cell, and two magnetic switches respectively control the energy transferring. The pulse compression system can compress the low voltage pulse from millisecond range to form high voltage pulse at 50ns/480kV range.

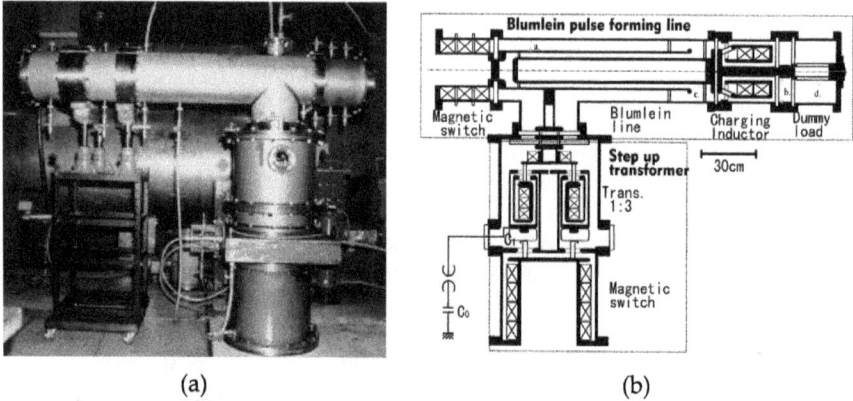

(a)   (b)

Figure 3. Typical high power pulse compressor with a transformer-based HES module. (a) The pulse compressor system; (b) the diagram and schematic of the pulse compressor system

The HES based on pulse transformer charging is an important way for high-power pulse trigger. Fig. 4(a) shows a solid state pulse trigger with semiconductor opening switches (SOS) in the Institute of Electrophysics Russian Academy of Science [10-11]. Fig. 4(b) presents the schematic of the pulse trigger, which shows a typical HES mode based on pulse transformer charging. SOS switch and IGBT are employed as the switches controlling energy transferring. The pulse trigger delivers high-voltage trigger pulse with pulse width at 70ns and voltage ranging from 20 to 80kV under the 100Hz repetition. And the average power delivered is about 50kW.

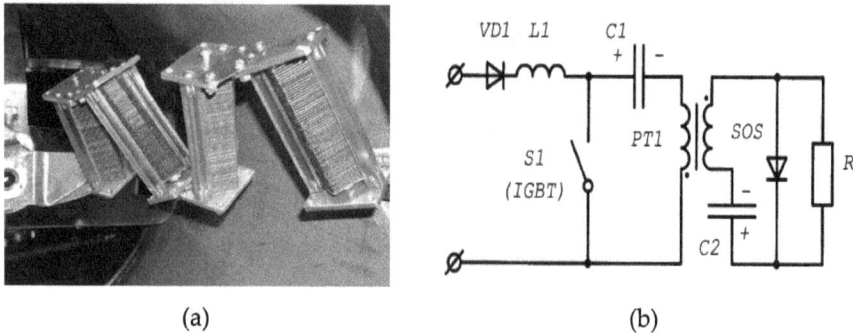

(a)   (b)

Figure 4. Typical high-voltage narrow pulse trigger with the transformer-based HES module. (a) The pulse trigger with the SOS switches; (b) The schematic of the high-power pulse trigger system

The HES cell based on pulse transformer charging is also an important component in intense electron beam accelerator for high-power pulse electron beams which are used in the fields of high-power microwave, plasma, high-power laser and inertial fusion energy (IFE). Fig. 5(a) shows the "Sinus" type accelerator in Russia [12], and it also corresponds to the HES mode based on transformer charging shown in Fig. 2. The pulse transformer of the accelerator is Tesla transformer with opened magnetic core, while spark gap switch controls energy transferring. The accelerator has been used to drive microwave oscillator for high-power microwave. Fig. 5(b) presents a high-power KrF laser system in Naval Research Laboratory of the U. S. A., and the important energy storage components in the system just form an HES cell based on transformer charging [13-14]. The HES cell drives the diode for pulse electron beams to pump the laser, and the laser system delivers pulse laser with peak power at 5GW/100ns to the IFE facility.

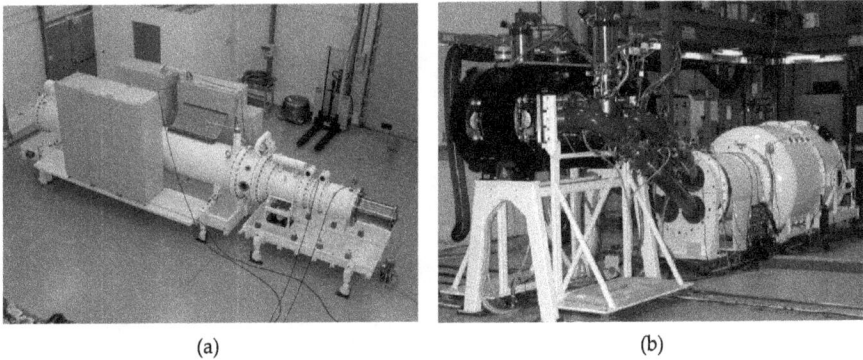

(a)   (b)

Figure 5. Typical intense electron beam accelerator with the transformer-based HES module. (a) The pulse electron beam accelerator based on HES for high-power microwave application in Russia; (b) The pulse electron beam accelerator based on HES for high-power laser application in Naval Research Laboratory, the U. S. A.

The HES based on pulse transformer charging also have important applications in ultra-wideband (UWB) electromagnetic radiation and X-ray radiography. Fig. 6 shows an ultra-wideband pulse generator based HES mode in Loughborough University of the U. K. [15]. The air-core Tesla transformer charges the pulse forming line (PFL) up to 500kV, and spark gap switch controls the energy transferring form the PFL to antenna. The "RADAN" series pulse generators shown in Fig. 7 are portable repetitive high-power pulsers made in Russia for X-ray radiography [16]. The "RADAN" pulser which consists of Tesla transformer and PFL are also based on the HES mode shown in Fig. 2. The controlling switches are thyristors and spark gap.

Besides, the HES cell is also used in shockwave generator [17], dielectric barrier discharge [18], industrial exhaust processing [19], material surface treatment [20], ozone production [21], food sterilization [22], cell treatment and cell mutation [23].

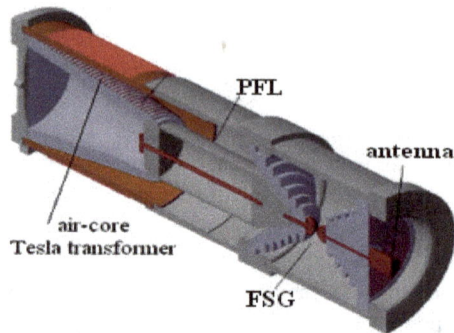

Figure 6. Compact 500kV pulse generator based on HES for UWB radiation in Loughborough University, U. K.

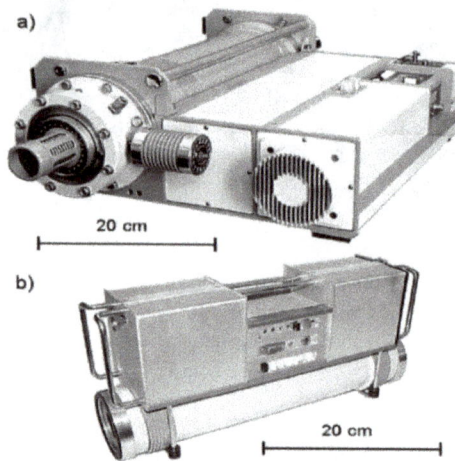

Figure 7. The compact "RADAN" pulse generators for X-ray radiography in Russia

2. Parametric analysis of pulse transformer with closed magnetic core in HES

Capacitor and inductor are basic energy storage components for CES and IES respectively, and pulse transformer charging is important to the HES mode shown in Fig. 2. So, it is essential to analyze the characteristic parameters of the common used high-power pulse transformer, and provide theoretical instructions for better understanding of the HES based on transformer charging.

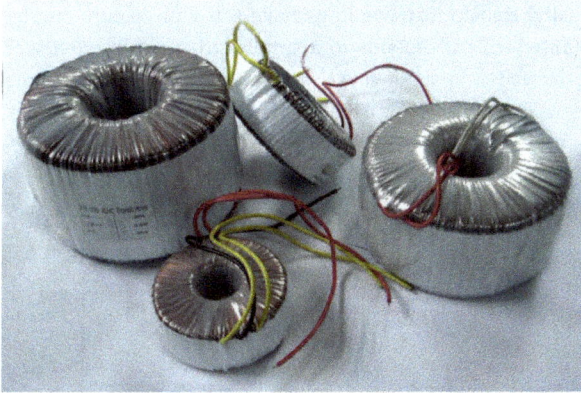

Figure 8. Common used pulse transformers with closed toroidal magnetic cores

There are many kinds of standards for categorizing the common used pulse transformers. From the perspective of magnetic core, pulse transformers can be divided into two types, such as the magnetic-core transformer [24-25] and the air-core transformer [26]. In view of the geometric structures of windings, the pulse transformer can be divided to many types, such as pulse transformer with closed magnetic core, solenoid-winding transformer, curled spiral strip transformer [26], the cone-winding Tesla transformer [16, 27], and so on. The transformer with magnetic core is preferred in many applications due to its advantages such as low leakage inductance, high coupling coefficient, high step-up ratio and high efficiency of energy transferring. Russian researchers produced a kind of Tesla transformer with cone-like windings and opened magnetic core, and the transformer with high coupling coefficient can deliver high voltage at MV range in repetitive operations [27]. Usually, pulse transformer with closed magnetic core, as shown in Fig.8, is the typical common used transformer which has larger coupling coefficient than that of Tesla transformer. The magnetic core can be made of ferrite, electrotechnical steel, iron-based amorphous alloy, nano-crystallization alloy, and so on. The magnetic core is also conductive so that the core needs to be enclosed by an insulated capsule to keep insulation from transformer windings.

Paper [28] presents a method for Calculation on leakage inductance and mutual inductance of pulse transformer. In this chapter, the common used pulse transformer with toroidal magnetic core will be analyzed in detail for theoretical reference. And a more convenient and simple method for analysis and calculation will be presented to provide better understanding of pulse transformer [24-25].

The typical geometric structure of pulse transformer with toroidal magnetic core is shown in Fig. 9(a). The transformer consists of closed magnetic core, insulated capsule of the core and transformer windings. The cross section of the core and capsule is shown in Fig. 9(b). Transformer windings are formed by high-voltage withstanding wires curling around the capsule, and turn numbers of the primary and secondary windings are N_1 and N_2, respectively. Usually, transformer windings have a layout of only one layer of wires as shown in Fig. 9(a), which corresponds to a simple structure. In other words, this simple structure can be viewed as a single-layer solenoid with a circular symmetric axis in the azimuthal direction. The transformer usually immerses in the transformer oil for good heat sink and insulation.

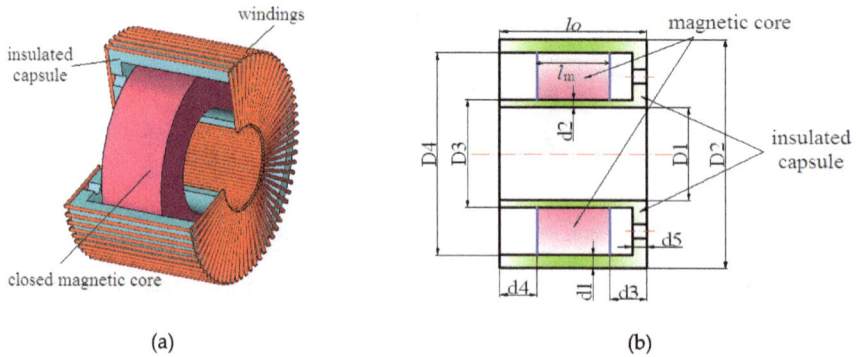

Figure 9. Typical structure of the pulse transformer with a closed magnetic core and an insulated capsule. (a) Assembly structure of the pulse transformer; (b) Geometric structure of the cross section of the pulse transformer.

Define the geometric parameters in Fig. 9(b) as follows. The height, outer diameter and inner diameter of the closed magnetic core are defined as l_m, D_4 and D_3 respectively. The height, outer diameter and inner diameter of the insulated capsule are defined as l_0, D_2 and D_1 respectively. The thicknesses of the outer wall, inner wall and side wall of insulated capsule are defined as d_1, d_2 and d_5 in order. The distances between the side surfaces of capsule and magnetic core are d_3 and d_4 shown in Fig. 9(b). Define diameters of wires of the primary windings and secondary windings as d_p and d_s respectively. The intensively wound primary windings with N_1 turns have a width about $N_1 d_p$.

2.1. Inductance analysis of pulse transformer windings with closed magnetic core

2.1.1. Calculation of magnetizing inductance

Define the permittivity and permeability of free space as ε_0 and μ_0, relative permeability of magnetic core as μ_r, the saturated magnetic induction intensity of core as B_s, residue magnetic induction intensity of core as B_r, and the filling factor of magnetic core as K_T. The cross section area S of the core is as

$$S = (D_4 - D_3)l_m / 2. \tag{3}$$

Define the inner and outer circumferences of magnetic core as l_1 and l_2, then $l_1 = \pi D_3$ and $l_2 = \pi D_4$. The primary and secondary windings tightly curl around the insulated capsule in separated areas in the azimuthal direction. In order to get high step-up ratio, the turn number N_1 of primary windings is usually small so that the single-layer layout of primary windings is in common use. Define the current flowing through the primary windings as i_p, the total magnetic flux in the magnetic core as Φ_0, and the magnetizing inductance of transformer as L_μ. According to Ampere's circuital law,

$$\Phi_0 = i_p \mu_0 \mu_r N_1^2 S K_T \ln(l_2 / l_1) / (l_2 - l_1) \tag{4}$$

As $\Phi_0 = L_\mu i_p$, L_μ is obtained as

$$L_\mu = \frac{\mu_0 \mu_r N_1^2 S K_T \ln(l_2 / l_1)}{l_2 - l_1}. \tag{5}$$

2.1.2. Leakage inductance of primary windings

The leakage inductances of primary and secondary windings also contribute to the total inductances of windings. The leakage inductance L_{lp} of the primary windings is caused by the leakage magnetic flux outside the magnetic core. If μ_r of magnetic core is large enough, the solenoid approximation can be used. Through neglecting the leakage flux in the outside space of the primary windings, the leakage magnetic energy mainly exists in two volumes. As Fig.10 shows, the first volume defined as V_1 corresponds to the insulated capsule segment only between the primary windings and the magnetic core, and the second volume defined as V_2 is occupied by the primary winding wires themselves. The leakage magnetic field in the volume enclosed by transformer windings can be viewed in uniform distribution. The leakage magnetic energy stored in V_1 and V_2 are as W_{m1} and W_{m2}, respectively.

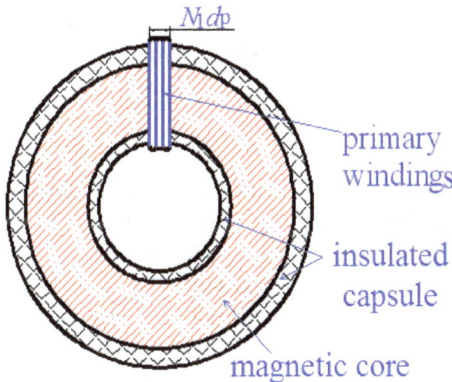

Figure 10. Primary windings structure of pulse transformer with closed magnetic core.

Define the magnetic field intensity generated by i_p from the N_1-turn primary windings as H_p in V_1. According to Ampere's circuital law, $H_p \approx i_p/d_p$. From Fig. 10,

$$V_1 = N_1 l_0 d_p (d_1 + d_2) + (D_4 - D_3) N_1 d_p (d_4 + d_3) / 2. \tag{6}$$

When the magnetic core works in the linear district of its hysteresis loop, the magnetic energy W_{m1} stored in V_1 is as

$$W_{m1} = \frac{\mu_0}{2} H_p^2 V_1 = \frac{\mu_0}{2} (\frac{i_p}{d_p})^2 V_1. \tag{7}$$

In V_2, the leakage magnetic field intensity defined as H_{px} can be estimated as

$$H_{px} = \frac{i_p}{d_p} \frac{x}{d_p}, \quad 0 \le x \le d_p. \tag{8}$$

From the geometric structure in Fig. 10, $V_2 = 2N_1 d_p^2 (l_0 + (D_2 - D_1)/2)$, the leakage magnetic energy W_{m2} stored in V_2 is as

$$W_{m2} = \frac{1}{2} \mu_0 \int_0^{d_p} d(V_2 H_{px}^2) = \frac{\mu_0 V_2}{2} \frac{i_p^2}{3 d_p^2}. \tag{9}$$

So, the total leakage magnetic energy W_{mp} stored in V_1 and V_2 is presented as

$$W_{mp} = W_{m1} + W_{m2} = L_{lp} i_p^2 / 2 \tag{10}$$

In (10), L_{lp} is the leakage inductance of the primary windings, and L_{lp} can be calculated as

$$L_{lp} = \frac{\mu_0}{3 d_p^2} (V_2 + 3V_1). \tag{11}$$

2.1.3. Leakage inductance of secondary windings

Usually, the simple and typical layout of the secondary windings of transformer is also the single layer structure as shown in Fig. 11(a). The windings are in single-layer layout both at the inner wall and outer wall of insulated capsule. As D_2 is much larger than D_1, the density of wires at the inner wall is larger than that at the outer wall. However, if the turn number N_2 becomes larger enough for higher step-up ratio, the inner wall of capsule can not provide enough space for the single-layer layout of wires while the outer wall still supports the previous layout, as shown in Fig. 11(b). We call this situation as "quasi-single-layer " layout. In the "quasi-single-layer " layout shown in Fig. 11 (b), the wires at the inner wall of capsule is in two-layer layout. After wire 2 curls in the inner layer, wire 3 curls in the outer layer next to wire 2, and wire 4 curls in the inner layer again next to wire 3, then wire 5 curls in the outer layer again next to wire 4, and so on. This kind of special layout has many advantages, such as minor voltage between adjacent coil turns, uniform voltage distribution between two layers, good insulation property and smaller distributed capacitance of windings.

In this chapter, the single-layer layout and "quasi-single-layer " layout shown in Fig. 11 (a) and (b) respectively are both analyzed to provide reference for HES module. And the multi-layer layout [29] can also be analyzed by the way introduced in this chapter.

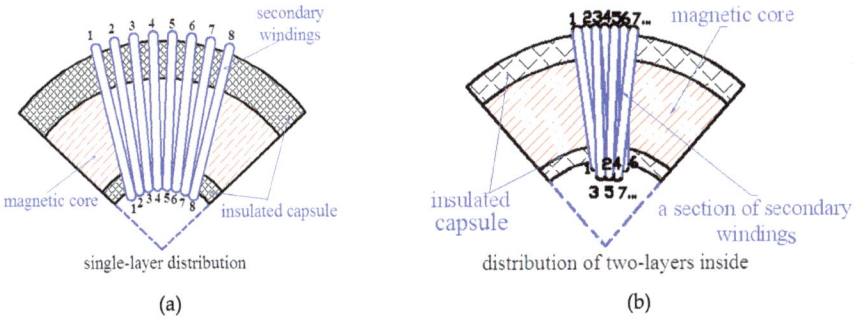

Figure 11. Secondary windings structures of pulse transformer with closed magnetic core. (a) Single-layer distribution of the secondary windings of transformer; (b) Inter-wound "quasi-single-layer" distribution of the secondary windings

Define the current flowing through the secondary windings as i_s, the two volumes storing leakage magnetic energy as V_a and V_b, the corresponding leakage magnetic energy as W_{ma} and W_{mb}, the total leakage magnetic energy as W_{ms}, wire diameter of secondary windings as d_s, and the leakage inductance of secondary windings as L_{ls}.

Firstly, the single-layer layout shown in Fig. 11 (a) is going to be analyzed. The analytical model is similar to the model analyzed in Fig. 10. If $(D_2-D_1) \ll D_1$, the length of leakage magnetic pass enclosed by the secondary windings is as $l_{ms} = N_2 d_s (D_2 + D_1) / 2(D_1 - d_s)$. The leakage magnetic field intensity defined as H_s in V_a is presented as $H_s = N_2 i_s / l_{ms}$. V_a and W_{ma} can be estimated as

$$\begin{cases} V_a = \dfrac{N_2 d_s}{4(D_1 - d_s)}[l_0(D_2^2 - D_1^2) - l_m \pi(D_4^2 - D_3^2)] \\[4mm] W_{ma} = \dfrac{\mu_0}{2}\dfrac{N_2^2 i_s^2}{l_{ms}^2} V_a \end{cases} \tag{12}$$

In volume V_b which is occupied by the secondary winding wires themselves, W_{mb} can be estimated as

$$W_{mb} = \frac{1}{2}\mu_0 \int_0^{d_s} (H_s \frac{x}{d_s})^2 l_{ms}(2l_0 + 4d_s + D_2 - D_1)dx = \frac{\mu_0 d_s N_2^2 i_s^2}{3 l_{ms}}[l_0 + 2d_s + (D_2 - D_1)/2]. \tag{13}$$

In view of that $W_{ms} = W_{ma} + W_{mb} = L_{ls}i_s^2 / 2$, the leakage inductance of single-layer layout of the secondary windings is as

$$L_{ls} = \frac{\mu_0 N_2^2 V_a}{l_{ms}^2} + \frac{2\mu_0 d_s N_2^2}{3l_{ms}}[l_0 + 2d_s + (D_2 - D_1)/2] \tag{14}$$

As to the "quasi-single-layer" layout shown in Fig. 11 (b), it also can be analyzed by calculating the leakage magnetic energy firstly. Under this condition, the length of leakage magnetic pass enclosed by the secondary windings is revised as $l_{ms} \approx N_2 d_s (D_2 + D_1) / 4(D_1 - d_s)$. The leakage magnetic energy W_{ma} and W_{mb} can be estimated as

$$\left\{ \begin{aligned} W_{ma} &= \frac{\mu_0}{2} \frac{N_2^2 i_s^2}{l_{ms}^2} V_a \\ W_{mb} &= \frac{1}{2}\mu_0 \{ \int_0^{d_s} (H_s \frac{x}{d_s})^2 [\pi(D_2 + d_s)(l_0 + 2d_s) + \pi(D_1 - d_s)(l_0 + 2d_s) + \\ &\quad + \pi(D_1 - 3d_s)(l_0 + 2d_s) + \pi(D_2^2 - D_1^2)/2]dx \} \\ &= \frac{\mu_0 \pi N_2^2 i_s^2 d_s}{6d_s^2}[(l_0 + 2d_s)(D_2 + 2D_1 - 3d_s) + (D_2^2 - D_1^2)/2] \end{aligned} \right. \tag{15}$$

Finally, the leakage inductance of the "quasi-single-layer" layout is obtained by the same way of (14) as

$$L_{ls} = \frac{\mu_0 N_2^2}{l_{ms}^2} V_a + \frac{\mu_0 \pi N_2^2 d_s}{2d_s^2}[(l_0 + 2d_s)(D_2 + 2D_1 - 3d_s) + (D_2^2 - D_1^2)/2]. \tag{16}$$

2.1.4. The winding inductances of pulse transformer

Define the total inductances of primary windings and secondary windings as L_1 and L_2 respectively, the mutual inductance of the primary and secondary windings as M, and the effective coupling coefficient of transformer as K_{eff}. From (5), (11), (14) or (16),

$$\left\{ \begin{aligned} L_1 &= L_\mu + L_{ps} \\ L_2 &= L_\mu (N_2 / N_1)^2 + L_{ss} \end{aligned} \right. \tag{17}$$

When $\mu_r \gg 1$, M and K_{eff} are presented as

$$\left\{ \begin{aligned} M &= L_\mu N_2 / N_1 \\ K_{eff} &= \frac{M}{\sqrt{L_1 L_2}} = \sqrt{1 - \frac{L_{ps} + L_{ss}(N_1 / N_2)^2}{L_\mu}} \end{aligned} \right. \tag{18}$$

2.2. Distributed capacitance analysis of pulse transformer windings

The distributed capacitances of pulse transformer include the distributed capacitances to ground [30], capacitance between adjacent turns or layers of windings [29-32], and capacitance between the primary and secondary windings [32-33]. It is very difficult to accurately calculate every distributed capacitance. Even if we can do it, the results are not liable to be analyzed so that the referential value is discounted. Under some reasonable approximations, lumped capacitances can be used to substitute the corresponding distributed capacitances for simplification, and more useful and instructive results can be obtained [29]. Of course, the electromagnetic dispersion theory can be used to analyze the lumped inductance and lumped capacitance of the single-layer solenoid under different complicated boundary conditions [34-35]. In this section, an easier way is introduced to analyze and estimate the lumped capacitances of transformer windings.

2.2.1. Distributed capacitance analysis of single-layer transformer windings

In the single-layer layout of transformer windings shown in Fig. 11(a), the equivalent schematic of transformer with distributed capacitances is shown in Fig. 12. C_{Dpi} is the distributed capacitance between two adjacent coil turns of primary windings, and C_{Dsi} is the counterpart capacitance of the secondary windings. C_{psi} is the distributed capacitance between primary and secondary windings. Common transformers have distributed capacitances to the ground, but this capacitive effect can be ignored if the distance between transformer and ground is large. If the primary windings and secondary windings are viewed as two totalities, the lumped parameters C_{Dp}, C_{Ds} and C_{ps} can be used to substitute the "sum effects" of C_{Dpi}s, C_{Dsi}s and C_{psi}s in order, respectively. And the lumped schematic of the pulse transformer is also shown in Fig. 12. C_{ps} decreases when the distance between primary and secondary windings increases. In order to retain good insulation for high-power pulse transformer, this distance is usually large so that C_{ps} also can be ignored. At last, only the lumped capacitances, such as C_{Dp} and C_{Ds}, have strong effects on pulse transformer.

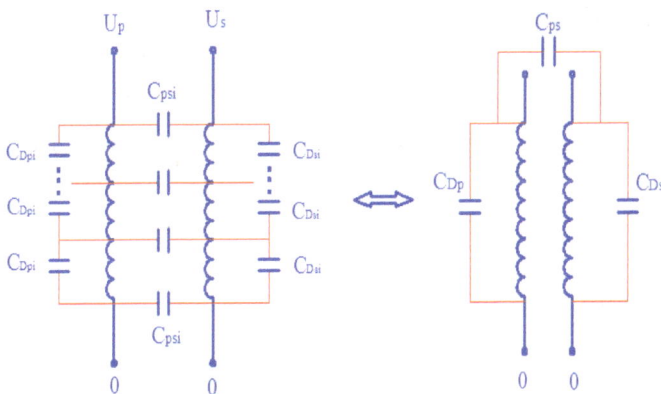

Figure 12. The distributed capacitances of single-layer wire-wound pulse transformer and the equivalent schematic with lumped parameters

In the single-layer layout shown in Fig. 11(a), define the lengths of single coil turn in primary and secondary windings as l_{s1} and l_{s2} respectively, the face-to-face areas between two adjacent coil turns in primary and secondary windings as S_{w1} and S_{w2} respectively, and the distances between two adjacent coil turns in primary and secondary windings as Δd_p and Δd_s respectively. According to the geometric structures shown in Fig. 10 and Fig. 11(a), $l_{s1}=2l_0+4d_p+D_2-D_1$, $l_{s2}=2l_0+4d_s+D_2-D_1$, $S_{w1}=d_pl_{s1}$ and $S_{w2}=d_sl_{s2}$. Because the coil windings distribute as a sector, Δd_p and Δd_s both increase when the distance from the centre point of sector increases in the radial direction. Δd_p and Δd_s can be estimated as

$$\Delta d_p(r) = \frac{2N_1 d_p r}{(N_1 - 1)(D_1 - d_p)}, \ \Delta d_s(r) = \frac{2N_2 d_s r}{(N_2 - 1)(D_1 - d_s)}, \ \frac{D_1}{2} < r < \frac{D_2}{2}. \tag{19}$$

If the relative permittivity of the dielectric between adjacent coil turns is ε_r, C_{Dpi} and C_{Dsi} can be estimated as

$$\begin{cases} C_{Dpi} = \int_{\frac{D_1}{2}}^{\frac{D_2}{2}} \frac{\varepsilon_0 \varepsilon_r S_{w1}(N_1 - 1)(D_1 - d_p)}{2N_1 d_p} \frac{dr}{r^2} = \frac{\varepsilon_0 \varepsilon_r l_{s1}(N_1 - 1)(D_1 - d_p)(D_2 - D_1)}{N_1 D_1 D_2} \\ C_{Dsi} = \int_{\frac{D_1}{2}}^{\frac{D_2}{2}} \frac{\varepsilon_0 \varepsilon_r S_{w2}(N_2 - 1)(D_1 - d_s)}{2N_2 d_s} \frac{dr}{r^2} = \frac{\varepsilon_0 \varepsilon_r l_{s2}(N_2 - 1)(D_1 - d_s)(D_2 - D_1)}{N_2 D_1 D_2} \end{cases} \tag{20}$$

Actually, the whole long coil wire which forms the primary or secondary windings of transformer can be viewed as a totality. The distributed capacitances between adjacent turns are just formed by the front surface and the back surface of the wire totality itself. In view of that, lumped capacitances C_{Dp} and C_{Ds} can be used to describe the total distributed capacitive effect. As a result, C_{Dp} and C_{Ds} are calculated as

$$\begin{cases} C_{Dp} = \int_0^{(N_1-1)l_{s1}} d_p dl_{s1} \int_{\frac{D_1}{2}}^{\frac{D_2}{2}} \frac{\varepsilon_0 \varepsilon_r (N_1 - 1)(D_1 - d_p)}{2N_1 d_p} \frac{dr}{r^2} = \frac{\varepsilon_0 \varepsilon_r l_{s1}(N_1 - 1)^2 (D_1 - d_p)(D_2 - D_1)}{N_1 D_1 D_2} \\ \quad = (N_1 - 1)C_{Dpi} \\ C_{Ds} = \int_0^{(N_2-1)l_{s2}} d_s dl_{s2} \int_{\frac{D_1}{2}}^{\frac{D_2}{2}} \frac{\varepsilon_0 \varepsilon_r (N_2 - 1)(D_1 - d_s)}{2N_2 d_s} \frac{dr}{r^2} = \frac{\varepsilon_0 \varepsilon_r l_{s2}(N_2 - 1)^2 (D_1 - d_s)(D_2 - D_1)}{N_2 D_1 D_2} \\ \quad = (N_2 - 1)C_{Dsi} \end{cases} \tag{21}$$

From (21), C_{Dp} or C_{Ds} is proportional to the wire length l_{s1} or l_{s2}, while larger turn number and smaller distance between adjacent coil turns both cause larger C_{Dp} or C_{Ds}

2.2.2. Distributed capacitance analysis of inter-wound "quasi-single-layer" windings

Usually, large turn number N_2 corresponds to the "quasi-single-layer " layout of wires shown in Fig. 13(a). In this situation, distributed capacitances between the two layers of wires at the inner wall of capsule obviously exist. Of course, lumped capacitance C_{Ls} can be used to describe the capacitive effect when the two layers are viewed as two totalities, as shown in Fig. 13(b). Define C_{Ds1} and C_{Ds2} as the lumped capacitances between adjacent coil turns of these two totalities, and C_{Ds} is the sum when C_{Ds1} and C_{Ds2} are in series. As a result, the lumped capacitances which have strong effects on pulse transformer are C_{Dp}, C_{Ds1}, C_{Ds2} and C_{Ls}.

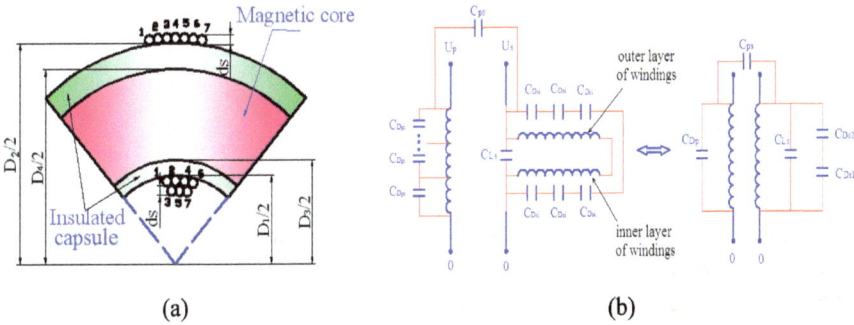

Figure 13. The distributed capacitances of "quasi-single-layer" pulse transformer and the equivalent schematic with lumped parameters. (a) Double-layer inside distribution of the secondary windings; (b) Equivalent circuit of transformer with distributed capacitances and lumped parameters

If the coil turns are tightly wound, the average distance between two adjacent coil turns is d_s. The inner layer and outer layer at the inner wall of capsule have coil numbers as $1+N_2/2$ and $N_2/2-1$ respectively. According to the same way for (21), C_{Ds1} and C_{Ds2} are obtained as

$$\begin{cases} C_{Ds1} = \dfrac{\varepsilon_0\varepsilon_r l_{s2}(N_2/2+1)^2(D_1-d_s)(D_2-D_1)}{(N_2/2)D_1 D_2} \\ C_{Ds2} = \dfrac{\varepsilon_0\varepsilon_r (l_{s2}+3d_s)(N_2/2-1)^2(D_1-d_s)(D_2-D_1)}{(N_2/2)D_1 D_2} \end{cases} \tag{22}$$

The non-adjacent coil turns have large distance so that the capacitance effects are shielded by adjacent coil turns. In the azimuthal direction of the inner layer wires, small angle $d\theta$ corresponds to the azimuthal width of wires as dl and distributed capacitance as dC_{Ls}. Then, $dC_{Ls} = \dfrac{\varepsilon_0\varepsilon_r(D_2-D_1+5d_s+2l_0)}{2d_s}dl$. If the voltage between the $(n-1)$th and nth turn of coil ($n \leq N_2$) is ΔU_0, the inter-wound method of the two layers aforementioned retains the voltage between two layers at about $2\Delta U_0$. So, the electrical energy W_{Ls} stored in C_{Ls} between the two layers is as

$$W_{Ls} = \frac{1}{2}\int_l (2\Delta U_0)^2 dC_{Ls} = \Delta U_0^2 \varepsilon_0\varepsilon_r N_2(D_2-D_1+5d_s+2l_0)/2. \tag{23}$$

In view of that $W_{Ls}=C_{Ls}(2\Delta U_0)^2/2$, C_{Ls} can be calculated as

$$C_{Ls} = \frac{\varepsilon_0 \varepsilon_r (D_2 - D_1 + 5d_s + 2l_0)N_2}{4}. \tag{24}$$

According to the equivalent lumped schematic in Fig. 13(b), the total lumped capacitance C_{Ds} can be estimated as

$$C_{Ds} = \frac{1}{1/C_{Ds1} + 1/C_{Ds2}} + C_{Ls}. \tag{25}$$

2.3. Dynamic resistance of transformer windings

Parasitic resistance and junction resistance of transformer windings cause loss in HES cell. Define the resistivity of winding wires under room temperature (20°C) as ρ_0, the work temperature as T_w, resistivity of winding wires under T_w as $\rho(T_w)$, radius of the conductive section of wire as r_w, total wire length as l_w, and the static parasitic resistance of winding wires as R_{w0}. The empirical estimation for R_{w0} is as

$$R_{w0} = \rho(T_w)\frac{l_w}{\pi r_w^2} = \rho_0[1 + 0.004(T_w - 20)]\frac{l_w}{\pi r_w^2}. \tag{26}$$

When the working frequency f is high, the "skin effect" of current flowing through the wire corss-section becomes obvious, which has great effects on R_{w0}. Define the depth of "skin effect" as Δd_w, and the dynamic parasitic resistance of winding wires as $R_w(f, T_w)$. As $\Delta d_w = (\rho / \pi f \mu_0)^{0.5}$, $R_w(f, T_w)$ is presented as

$$R_w(f,T_w) = \begin{cases} \dfrac{l_w}{(2r_w - \dfrac{\rho(T_w)}{\pi f \mu_0})\sqrt{\dfrac{\pi}{f\mu_0\rho(T_w)}}}, & r_w > \Delta d_w \\[4mm] R_{w0} = \rho(T_w)\dfrac{l_w}{\pi r_w^2}, & r_w \le \Delta d_w \end{cases} \tag{27}$$

3. Pulse response analysis of high power pulse transformer in HES

In HES cell based on pulse transformer charging, the high-frequency pulse response characteristics of transformer show great effects on the energy transferring and energy storage. Pulse response and frequency response of pulse transformer are very important issues. The distributed capacitances, leakage inductances and magnetizing inductance have great effects on the response pulse of transformer with closed magnetic core [36-39]. In this Section, important topics such as the frequency response and pulse response characteristics to square pulse, are discussed through analyzing the pulse transformer with closed magnetic core.

3.1. Frequency-response analysis of pulse transformer with closed magnetic core

The equivalent schematic of ideal pulse transformer circuit is shown in Fig. 14(a). L_{lp} and L_{ls} are the leakage inductances of primary and secondary windings of transformer calculated in (11), (14) and (16). Lumped capacitances C_{ps}, C_{DP} and C_{Ds} represent the "total effect" of the distributed capacitances of transformer, while C_{DP} and C_{Ds} are calculated in (21) and (25). L_μ is the magnetizing inductance of pulse transformer calculated in (5). Define the sum of wire resistance of primary windings and the junction resistance in primary circuit as R_P, the counterpart resistance in secondary circuit as R_s, load resistance as R_L, the equivalent loss resistance of magnetic core as R_c, and the sinusoidal/square pulse source as U_1.

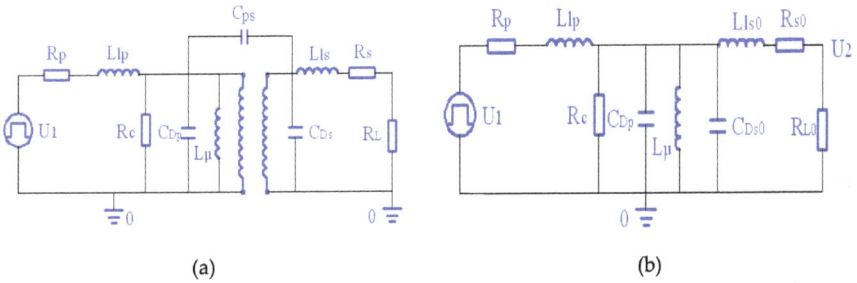

Figure 14. Equivalent schematics of pulse transformer based on magnetic core with a square pulse source and a load resistor. (a) Equivalent schematic of pulse transformer with all the distributed parameters; (b) Simplified schematic of pulse transformer when the secondary circuit is equated into the primary circuit.

Usually, C_{ps} is so small that it can be ignored due to the enough insulation distance between the primary and secondary windings. In order to simplify the transformer circuit in Fig. 14(a), the parameters in the secondary circuit such as C_{Ds}, L_{ls}, R_s and R_L, can be equated into the primary circuit as C_{Ds0}, L_{ls0}, R_{s0} and R_{L0}, respectively. And the equating law is as

$$\begin{cases} L_{ls0} = L_{ls}(\dfrac{N_1}{N_2})^2, \, C_{Ds0} = C_{Ds}(\dfrac{N_2}{N_1})^2 \\ R_{s0} = R_s(\dfrac{N_1}{N_2})^2, \, R_{L0} = R_L(\dfrac{N_1}{N_2})^2 \end{cases} . \tag{28}$$

3.1.1. Low-frequency response characteristics

Define the frequency and angular frequency of the pulse source as f and ω_0. When the transformer responds to low-frequency pulse signal ($f < 10^3$ Hz), Fig. 14(b) can also be simplified. In Fig. 14(b), C_{Dp} is in parallel with C_{Ds0}, and the parallel combination capacitance of these two is about $10^{-6} \sim 10^{-9}$F so that the reactance can reach $10\mathrm{k}\Omega \sim 1\mathrm{M}\Omega$. Meanwhile, the reactance of L_μ is small. As a result, C_{Dp} and C_{Ds0} can also be ignored. Reactances of L_{ls0} and

L_{lp} (10^{-7}H) are also small under the low-frequency condition, and they also can be ignored. Usually, the resistivity of magnetic core is much larger than common conductors to restrict eddy current. In view of that $R_{s0} \ll R_{L0} \ll R_c$, the combination of R_{s0}, R_{L0} and R_c can be substituted by R_0. Furthermore, $R_0 \cong R_{L0}$. Finally, the equivalent schematic of pulse transformer under low-frequency condition is shown in Fig.15(a).

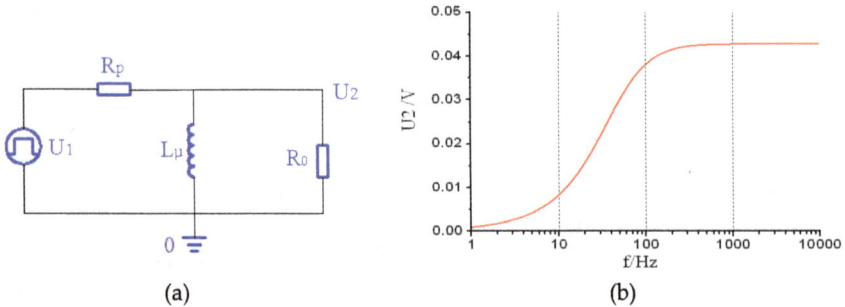

Figure 15. Simplified schematic and analytical result of transformer for low-frequency pulse response. (a) Equivalent schematic of pulse transformer under the condition of low frequency; (b) Low-frequency response results of an example of transformer

In Fig. 15(a), L_μ and R_0 are in parallel, and then in series with R_P which is at mΩ range. R_0 is usually very small due to the equating process from (28). When ω_0 of the pulse source increases, reactance of L_μ also increases so that $\omega_0 L_\mu \gg R_0$. In this case, the L_μ branch gets close to opening, and an ideal voltage divider is formed only consisting of R_P and R_0. At last, the pulse source U_1 is delivered to the load R_0 without any deformations. And the response voltage pulse signal U_2 of transformer on the load resistor is as

$$U_2 = U_1 \frac{R_0}{R_0 + R_p}. \tag{29}$$

When $R_P \ll R_0$, $U_1 = U_2$ which means the source voltage completely transfers to the load resistor. On the other hand, if $\omega_0 L_\mu \ll R_0$, L_μ shares the current from the pulse source so that the current flowing through R_0 gets close to 0. In this situation, the pulse transformer is not able to respond to the low-frequency pulse signal U_1.

An example is provided as follows to demonstrate the analysis above. In many measurements, coaxial cables and oscilloscope are used, and the corresponding terminal impedance is about $R_L = 50\Omega$. So, the R_0 may be at mΩ range when it is equated to the primary circuit. Select conditions as follows: $R_P = 0.09\Omega$, $L_\mu = 12.6\mu$H, and U_1 is the periodical sinusoidal voltage pulse with amplitude at 1V. The low-frequency response curve of pulse transformer is obtained from Pspice simulation on frequency scanning, as Fig. 15(b) shows. When f of U_1 is larger than the second inflexion frequency (100Hz), response signal U_2 is large and stable. However, when f is less than the first inflexion frequency (10Hz), response signal U_2 gets close to 0. And the cut-off frequency f_L is about 10Hz.

The conclusion is that low-frequency response capability of pulse transformer is mainly determined by L_μ, and the response capability can be improved through increasing L_μ calculated in (5).

3.1.2. High-frequency response characteristics

When the transformer responds to high-frequency pulse signal ($f > 10^6$ Hz), conditions "$\omega_0 L_\mu \gg R_0$" and "$\omega_0 L_\mu \gg R_p$" are satisfied so that the branch of L_μ seems open. In Fig. 14(b), the combination effect of R_{s0}, R_{L0} and R_c still can be substituted by R_0. Substitute L_{lp} and L_{ls0} by L_l, and combine C_{Ds0} and C_{Dp} as C_D. The simplified schematic of pulse transformer for high-frequency response is shown in Fig. 16(a).

In Fig. 16(a), when ω_0 of pulse source increases, reactance of L_l increases while reactance of C_D decreases. If ω_0 is large enough, $\omega_0 L_l \gg R_0 \gg 1/(\omega_0 C_D)$ and the response signal U_2 gets close to 0. On the other hand, condition $1/(\omega_0 C_D) \geq R_0$ is satisfied when ω_0 decreases. The pulse current mainly flows through the load resistor R_0, and the good response of transformer is obtained. Especially, when $\omega_0 L_l \ll R_p$, L_l also can be ignored. Under this situation, R_p is in series with R_0 again, and the response signal U_2 which corresponds to the best response still conforms to (29).

Select the amplitude of the periodical pulse signal U_1 at 1V. If R_p, L_l and C_D are at ranges of mΩ, 0.1μH and pF respectively, the high-frequency response curve of transformer is also obtained as shown in Fig.16(b) from Pspice simulation. When f is less than the first inflexion frequency (about 300kHz), response signal U_2 is stable. When f is larger than the second inflexion frequency (about 10MHz), response signal U_2 gets close to 0. And the cut-off frequency f_H is about 10MHz.

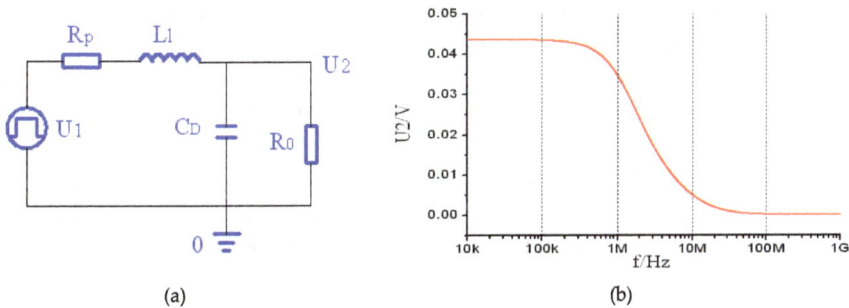

(a)   (b)

Figure 16. Simplified schematic and analytical result of transformer for high-frequency pulse response. (a) Equivalent schematic of pulse transformer under the condition of high frequency; (b) High-frequency response results of an example of transformer

The conclusion is that high-frequency response characteristics of transformer are mainly determined by distributed capacitance C_D and leakage inductance L_l. The high-frequency response characteristics can be obviously improved through restricting C_D and L_l , or increasing L_μ.

3.2. Square pulse response of pulse transformer with closed magnetic core

In Fig. 14(b), $R_{s0} \ll R_{L0} \ll R_c$, and the combination effect of R_{s0}, R_{L0} and R_c can be substituted by R_0. Combine C_{Ds0} with C_{Dp} as C_D. The simplified schematic of pulse transformer circuit for square pulse response is shown in Fig. 17. U_1 and U_2 represent the square voltage pulse source and the response voltage signal on the load respectively. The total current from the pulse source is $i(t)$, while the branch currents flowing through R_0, C_D and L_μ are as i_1, i_2 and i_3 respectively.

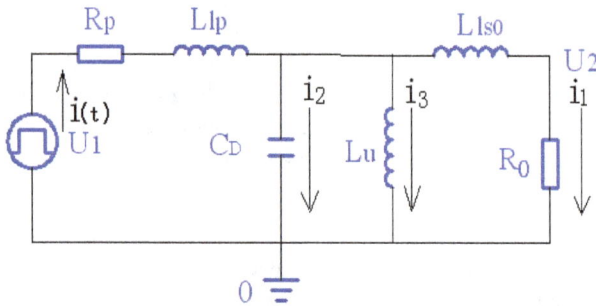

Figure 17. Equivalent schematic of transformer for square pulse response

3.2.1. Response to the front edge of square pulse

Usually, L_μ ranges from 10^{-6}H up to more than 10^{-5}H, and the square pulse has front edge and back edge both at 100ns~1μs range. So, when the fast front edge and back edge of square pulse appear, reactance of L_μ is much larger than the equated load resistor R_0. Under this condition, i_3 is so small that the effect of L_μ on the front edge response can be ignored.

Define the voltage of C_D as $U_c(t)$. As aforementioned, L_μ has little effect on the response to the front edge of square pulse. Through Ignoring the L_μ branch, the circuit equations are presented in (30) with initial conditions as $i(0)=0$, $i_1(0)=0$ and $U_c(0)=0$.

$$\begin{cases} U_1(t) = i(t)R_p + L_{lp}di(t)/dt + L_{ls0}di_1(t)/dt + i_1(t)R_0 \\ \quad L_{ls0}di_1(t)/dt + i_1(t)R_0 = \int i_2(t)dt / C_D \\ \quad\quad i(t) = i_1(t) + i_2(t) \end{cases} \qquad (30)$$

If the factor for Laplace transformation is as p, the transformed forms of $U_1(t)$ and $i_1(t)$ are defined as $U_1(p)$ and $I_1(p)$. Firstly, four constants such as α, β, γ and λ are defined as

$$
\begin{cases}
\alpha = \dfrac{R_p L_{ls0} C_D + R_0 L_{lp} C_D}{L_{lp} L_{ls0} C_D}, \ \beta = \dfrac{R_0 R_p C_D + L_{lp} + L_{ls0}}{L_{lp} L_{ls0} C_D} \\[2mm]
\gamma = \dfrac{R_p + R_0}{L_{lp} L_{ls0} C_D}, \ \lambda = \dfrac{1}{L_{lp} L_{ls0} C_D}
\end{cases}
\tag{31}
$$

Define the amplitude and pulse duration of square voltage pulse source as U_s and T_0 respectively. $U_1(t)$ is as

$$
U_1(t) = \begin{cases} 0, & t < 0 \ or \ t \geq T_0 \\ U_s, & 0 \leq t < T_0 \end{cases}.
\tag{32}
$$

Equations (30) can be solved by Laplace transformation and convolution, and there are three states of solutions such as the over dumping state, the critical dumping state and the under dumping state. In the transformer circuit, the resistors are always small so that the under dumping state usually appears. Actually, the under dumping state is the most important state which corresponds to the practice. In this section, the centre topic focuses on the under dumping state of the circuit.

Define constants a, b, ω, ξ (a, $b<0$; $\omega>0$), A_1, A_2 and A_3 as (33).

$$
\begin{cases}
A_1 = \dfrac{1}{(a-b)^2 + \omega^2}, \ A_2 = \dfrac{-1}{(a-b)^2 + \omega^2}, \ A_3 = \dfrac{2b-a}{(a-b)^2 + \omega^2} \\[2mm]
a = \xi^{\frac{1}{3}} - \xi^{-\frac{1}{3}}(3\beta - \alpha^2)/9 - \alpha/3 \\[2mm]
b = \xi^{-\frac{1}{3}}(3\beta - \alpha^2)/18 - \alpha/3 - \xi^{\frac{1}{3}}/2 \\[2mm]
\omega = \sqrt{3}[(3\beta - \alpha^2)\xi^{-\frac{1}{3}}/9 + \xi^{\frac{1}{3}}]/2 \\[2mm]
\xi \triangleq \sqrt{\dfrac{\beta^3 + \alpha^3\gamma}{27} - \dfrac{\alpha^2\beta^2}{108} - \dfrac{\alpha\beta\gamma}{6} + \dfrac{\gamma^2}{4} - \dfrac{3\gamma - \alpha\beta}{6} - \dfrac{\alpha^3}{27}}
\end{cases}
\tag{33}
$$

The under dumping state solution of (30) is as

$$
U_2(t) = \begin{cases}
0, & t \leq 0 \\
\lambda R_0 U_s\{A_1 \exp(at) + [A_2 \cos(\omega t) + (A_2 b + A_3)\sin(\omega t)/\omega]\exp(bt)\}, & 0 < t \leq T_0 \\
\lambda U_s R_0\{A_1 \exp(at) + [A_2 \cos(\omega t) + (A_2 b + A_3)\sin(\omega t)/\omega]\exp(bt)\} - \lambda U_s R_0\{A_1 \\
\exp[a(t - T_0)] + [A_2 \cos\omega(t - T_0) + (A_2 b + A_3)\sin\omega(t - T_0)/\omega]\exp[b(t - T_0)]\}, & t > T_0
\end{cases}
\tag{34}
$$

The load current $i_1(t) = U_2(t)/R_0$. From (34), response voltage pulse $U_2(t)$ on load consists of an exponential damping term and a resonant damping term. The resonant damping term which has main effects on the front edge of pulse contributes to the high-frequency resonance at the front edge. Constant a defined in (33) is the damping factor of the pulse droop of square pulse $U_2(t)$, b is the damping factor of the resonant damping term, and ω is the resonant angular frequency. Substitute $\lambda R_0 U_s$ by U_0, and define two functions $f_1(t)$ and $f_2(t)$ as

$$f_1(t) = \begin{cases} 0, & t \leq 0 \\ U_0[A_2\cos(\omega t) + (A_2b + A_3)\sin(\omega t)/\omega]\exp(bt)\}, & 0 < t \leq T_0 \\ U_0[A_2\cos(\omega t) + (A_2b + A_3)\sin(\omega t)/\omega]\exp(bt)\} - U_0 & \\ \quad [A_2\cos\omega(t - T_0) + (A_2b + A_3)\sin\omega(t - T_0)/\omega]\exp[b(t - T_0)]\}, & t > T_0 \\ \quad f_2(t) = U_0[A_2\cos(\omega t) + (A_2b + A_3)\sin(\omega t)/\omega] & \end{cases} \qquad (35)$$

$f_1(t)$ is just the resonant damping term divided from (34), while $f_2(t)$ is the pure resonant signal divided from $f_1(t)$. If pulse width $T_0=5\mu s$, the three signals $U_2(t)$, $f_1(t)$ and $f_2(t)$ are plotted as Curve 1, Curve 2 and Curve 3 in Fig. 18 respectively. In the abscissa of Fig. 18, the section when $t < 0$ corresponds to the period before the time when the square pulse appears. Obviously, Curve 1~ Curve 3 all have high-frequency resonances with the same angular ω. The resonances of Curve 1 and Curve 2 at the front edge are in superposition. Under the under dumping state of the circuit, the rise time t_r of the response signal is about half of a resonant period as (36).

$$t_r = \pi / \omega. \qquad (36)$$

From (33) and (36), the rise time of the response signal $U_2(t)$ is determined by the parasitic inductance, leakage inductances (L_{lp} and L_{ls0}) and distributed capacitance C_D. The rise time t_r of the front edge can be minimized through increasing the resonant angular frequency ω. In the essence, the high-frequency "L-C-R" resonance is generated by the leakage inductances and distributed capacitance in the circuit.

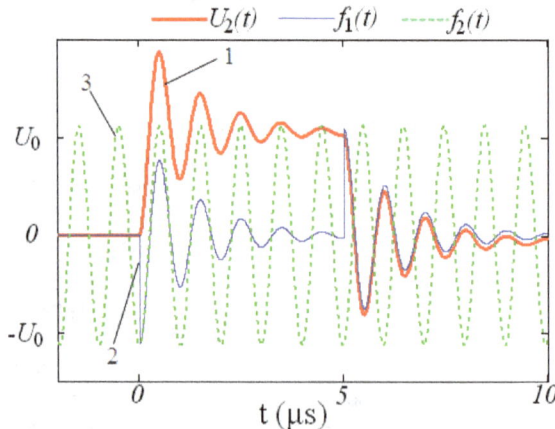

Figure 18. Typical pulse response waveforms of pulse transformer to the front edge of square pulse

The conclusion is that the rise time of the front edge of response pulse can be improved by minimizing the capacitance C_D and leakage inductance L_{lp} and L_{ls0} of the transformer. The waveform of the response voltage signal can be improved through increasing the damping resistor of the circuit in a proper range.

3.2.2. Pulse droop analysis of transformer response

In Fig.17, when the front edge of pulse is over, $U_c(t)$ of C_D and the currents flowing through L_{lp} and L_{ls0} all become stable. And these parameters have little effects on the response to the flat top of square pulse. During this period, load voltage signal $U_2(t)$ is mainly determined by L_μ. So, the simplified schematic from Fig.17 is shown as Fig.19 (a). The circuit equations are as

$$\begin{cases} U_1(t) = i(t)R_p + i_1(t)R_0 \\ U_2(t) = i_1(t)R_0 = L_\mu di_3(t)/dt. \\ \quad i(t) = i_1(t) + i_3(t) \end{cases} \tag{37}$$

The initial conditions are as $i_3(0)=0$ and $U_2(0)=R_0U_s/(R_0+R_p)$. The load voltage $U_2(t)$ is obtained as (38) through solving equations in (37).

$$U_2(t) = \frac{R_0 U_s}{R_p + R_0}\exp(-\frac{t}{\tau}), \quad \tau = \frac{L_\mu(R_p + R_0)}{R_p R_0}, \quad 0 < t < T_0. \tag{38}$$

In (38), τ is the constant time factor of the pulse droop. When L_μ increases which leads to an increment of τ, the pulse droop effect is weakened and the pulse top becomes flat. If $U_{20}=R_0U_s/(R_p+R_0)$, the response signal to the flat top of square pulse is shown in Fig. 19(b). When pulse duration T_0 is short at μs range, the pulse droop effect ($0<t<T_0$) of $U_2(t)$ is not obvious at all. However, when T_0 ranges from 0.1ms to several milliseconds, time factor τ has great effect on the flat top of $U_2(t)$, and the pulse droop effect of the response signal is so obvious that $U_2(t)$ becomes an triangular wave.

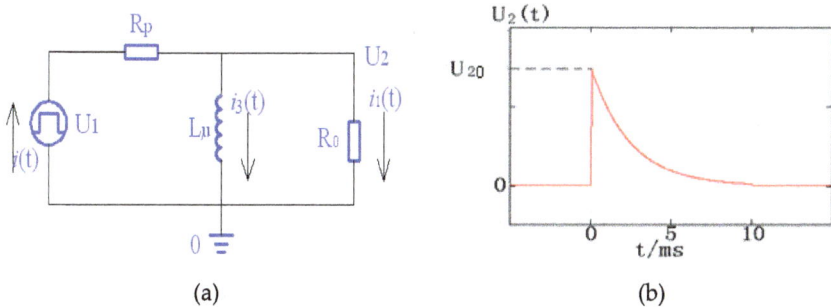

(a)   (b)

Figure 19. Schematic and response pulse of transformer to the flat-top of square pulse. (a) Equivalent schematic of transformer for flat top response of square pulse; (b) The pulse droop of the response pulse of transformer

3.2.3. Response to the back edge of square pulse

When the flat top of square pulse is over, all the reactive components in Fig. 17 have stored certain amount of electrical or magnetic energy. Though the main pulse of the response signal is over, the stored energy starts to deliver to the load through the circuit. As a result, high-frequency resonance is generated again which has a few differences from the resonance at the front edge of pulse. In Fig. 17, U_1 and R_P have no effects on the pulse tail response when the main pulse is over. C_D which was charged plays as the voltage source. Combine L_{lp} and L_{ls0} as L_l. The equivalent schematic for pulse tail response of transformer is shown in Fig. 20.

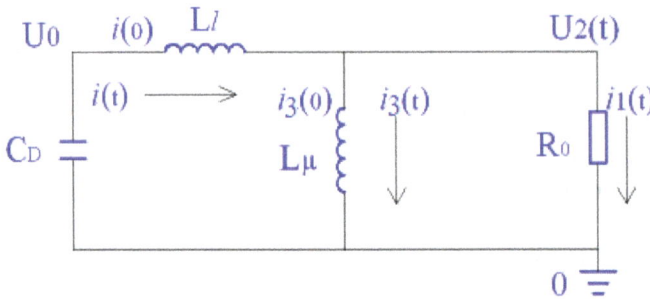

Figure 20. Equivalent schematic for back edge response of transformer to square pulse

The circuit equations are presented in (39) with initial condition as $U_C(0)=U_{c0}$.

$$\begin{cases} U_{c0} - \int i(t)dt / C_D = L_l di(t) / dt + i_1(t)R_0 \\ U_2(t) = i_1(t)R_0 = L_\mu di_3(t) / dt \\ i(t) = i_1(t) + i_3(t) \end{cases} \quad (39)$$

$i_1(0)$ and $i_3(0)$ are determined by the final state of the pulse droop period. There are also three kinds of solutions, however the under dumping solution usually corresponds to the real practices. So, this situation is analyzed as the centre topic in this section. Define six constants α_1, β_1, γ_1, α_2, β_2 and γ_2 as (40).

$$\begin{cases} \alpha_1 = \dfrac{R_0}{L_\mu} + \dfrac{R_0}{L_l}, \ \beta_1 = \dfrac{1}{L_l C_D}, \ \gamma_1 = \dfrac{R_0}{L_l L_\mu C_D} \\ \alpha_2 = R_0[i(0) - i_2(0)], \ \beta_2 = \dfrac{R_0 U_0}{L_l}, \ \gamma_2 = -\dfrac{R_0 i_2(0)}{L_l C_D} \end{cases} \quad (40)$$

The under dumping solution of (39) is calculated as

$$U_2(t) = B_1 \exp(a_1 t) + \exp(b_1 t)[B_2 \cos(\omega_s t) + \frac{B_2 b_1 + B_3}{\omega_s} \sin(\omega_s t)]. \tag{41}$$

In (41), B_1, B_2 and B_3 are three coefficients while a_1, b_1, ω_s and ξ_1 are another four constants as

$$\left\{\begin{array}{c} B_1 = \dfrac{\alpha_2 a_1^2 + \beta_2 a_1 + \gamma_2}{(a_1 - b_1)^2 + \omega_s^2}, \; B_2 = \alpha_2 - \dfrac{\alpha_2 a_1^2 + \beta_2 a_1 + \gamma_2}{(a_1 - b_1)^2 + \omega_s^2}, \; B_3 = \dfrac{(\alpha_2 a_1^2 + \beta_2 a_1 + \gamma_2)(b_1^2 + \omega_s^2)}{a_1[(a_1 - b_1)^2 + \omega_s^2]} - \dfrac{\gamma_2}{a_1} \\[2ex] a_1 = \xi_1^{\frac{1}{3}} - \xi_1^{-\frac{1}{3}}(3\beta_1 - \alpha_1^2)/9 - \alpha_1/3 \\[1ex] b_1 = \xi_1^{-\frac{1}{3}}(3\beta_1 - \alpha_1^2)/18 - \alpha_1/3 - \xi_1^{\frac{1}{3}}/2 \\[1ex] \omega_s = \sqrt{3}[(3\beta_1 - \alpha_1^2)\xi_1^{-\frac{1}{3}}/9 + \xi_1^{\frac{1}{3}}]/2 \\[1ex] \xi_1 \triangleq \sqrt{\dfrac{\beta_1^3 + \alpha_1^3 \gamma_1}{27} - \dfrac{\alpha_1^2 \beta_1^2}{108} - \dfrac{\alpha_1 \beta_1 \gamma_1}{6} + \dfrac{\gamma_1^2}{4} - \dfrac{3\gamma_1 - \alpha_1 \beta_1}{6} - \dfrac{\alpha_1^3}{27}} \end{array}\right. \tag{42}$$

The responses to front edge and back edge of square pulse have differences in essence, as the exciting sources are different. Define functions $f_3(t)$ and $f_4(t)$ as (43), according to (41).

$$\left\{\begin{array}{l} f_3(t) = B_1 \exp(a_1 t) \\[1ex] f_4(t) = \exp(b_1 t)[B_2 \cos(\omega_s t) + \dfrac{B_2 b_1 + B_3}{\omega_s} \sin(\omega_s t)] \end{array}\right. \tag{43}$$

The response signal $U_2(t)$ in (41) also consists of an exponential damping term $f_3(t)$ and a resonant damping term $f_4(t)$.

Define $B_1 + B_2$ as U_0'. In order to help to establish direct impressions, a batch of parameters are selected (C_D=2.14µF, L_μ=12.6µH and L_l=1.09µH) for plotting the response pulse curves. According to (41) and (43), signals $U_2(t)$, $f_3(t)$ and $f_4(t)$ are plotted as Curve 1, Curve 2 and Curve 3 respectively in Fig. 21(a) for example. Because the damping factor a_1 defined in (42) is large, the amplitude of $f_3(t)$ which corresponds to Curve 2 is very small with slow damping. The resonant damping term $f_4(t)$ which is damped faster determines the resonant angular frequency ω_s. The resonant parts of $U_2(t)$ and $f_4(t)$ are also in superposition at the back edge of pulse. When $f_4(t)$ is damped to 0, $U_2(t)$ becomes the same as $f_3(t)$. The half of the resonant period t_d is as

$$t_d = \pi / \omega_s. \tag{44}$$

According to (40) and (42), R_0 has effects on the damping factors of $f_3(t)$ and $f_4(t)$. The resonant frequency is mainly determined by leakage inductance, magnetizing inductance and distributed capacitance of transformer.

Fig. 21(b) shows an impression of the effect of L_μ on the tail of response signal. When L_μ changes from 0.1µH to 1mH while other parameters retain the same, the resonant waveforms with the same frequency do not have large changes. So, the conclusion is that, t_d and the resonant angular frequency ω_s are not mainly determined by L_μ. Fig. 21(c) shows the effect of leakage inductances of transformer on the pulse tail of response signal. When L_l is small at 10nH range · the back edge of pulse (Curve 2) is good as which of standard square pulse. When L_l increases from 0.01μH to 1μH range, the resonances become fierce with large amplitudes. If L_l increases to 10μH range, the previous under damping mode has a transition close to the critical damping mode (Curve 4). The fall time t_d of the pulse tail increases obviously. Fig. 21(d) shows the effect of distributed capacitances of transformer on the pulse tail of response signal. The effect of C_D obeys similar laws obtained from L_μ. So, the conclusion is that the pulse tail of the response signal can be improved by a large extent through minimizing the leakage inductances and distributed capacitances of transformer windings. Paper [24] demonstrated the analysis above in experiments.

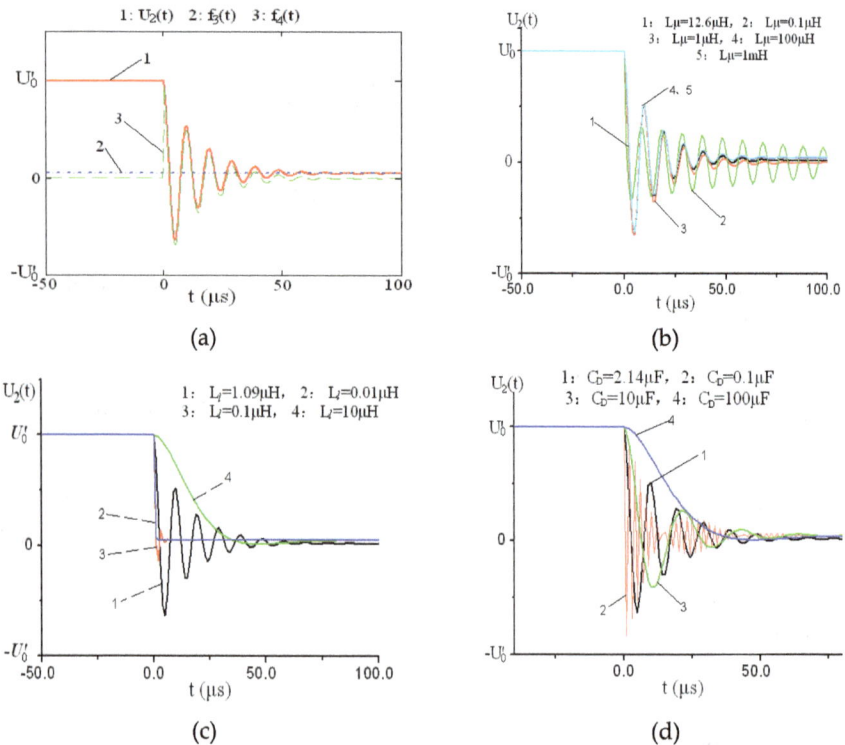

Figure 21. Typical back tail response signals of pulse transformer to the square pulse.
(a) The typical back edge response signals to square pulse in theory; (b) Effects of magnetizing inductance on the back edge response of transformer; (b) Effects of leakage inductance on the back edge response of transformer; (c) Effects of distributed capacitance on the back edge response of transformer;

4. Analysis of energy transferring in HES based on pulse transformer charging

As an important IES component, the pulse transformer is analyzed and the pulse response characteristics are also discussed in detail. The analytical theory aforementioned is the base for HES analysis based on pulse transformer charging in this section. In Fig. 2, the HES module based on capacitors and transformer operates in three courses, such as the CES course, the IES course and the CES course. Actually, the IES course and the latter CES course occur almost at the same time. The pulse transformer plays a role on energy transferring. There are many kinds of options for the controlling switch (S_1) of C_1, such as mechanical switch, vacuum trigger switch, spark gap switch, thyristor, IBGT, thyratron, photo-conductive switch, and so on. S_1 has double functions including opening and closing. S_1 ensures the single direction of HES energy transferring, from C_1 and transformer to C_2. In this section, the energy transferring characteristics of HES mode based on transformer charging is analyzed in detail.

The pulse signals in the HES module are resonant signals. According to the analyses from Fig. 15 and Fig.16, the common used pulse transformer shown in Fig. 9(a) has good frequency response capability in the band ranging from several hundred Hz to several MHz. Moreover, C_1 and C_2 in HES module are far larger than the distributed capacitances of pulse transformer. So, the distributed capacitances can be ignored in HES cell. In the practical HES module, many other parameters should be considered, such as the junction inductance, parasitic inductance of wires, parasitic inductance of switch, parasitic resistance of wires, parasitic resistance of switch, and so on. These parameters can be concluded into two types as the parasitic inductance and parasitic resistance. As a result, the equivalent schematic of the HES module is shown in Fig. 22(a).

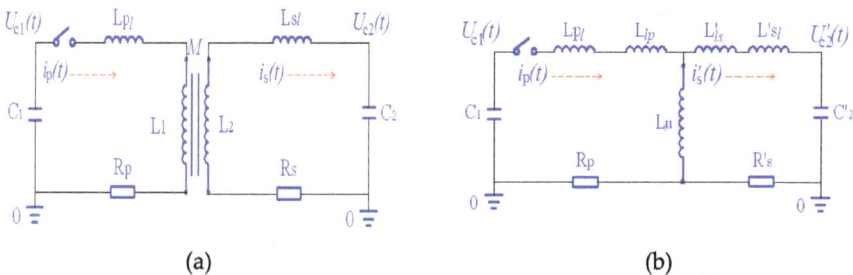

(a)   (b)

Figure 22. The basic hybrid energy storage (HES) system based on a source capacitor, a pulse transformer and a load capacitor. (a) Typical schematic of the transformer-based HES module; (b) Simplified schematic when the secondary circuit is equated into the primary circuit

In Fig. 22(a), C_1 and C_2 represent the primary energy-storage capacitor and load capacitor respectively. L_{pl} and L_{sl} represent the parasitic inductances in the primary circuit and secondary circuit, while R_p and R_s stand for the parasitic resistances in the primary circuit and secondary circuit respectively. L_1, L_2 and M of transformer are defined in (17) and (18). $i_p(t)$ and $i_s(t)$ represent the current in the primary and secondary circuit. The pulse transformer with closed magnetic core has the largest effective coupling coefficient (close to 1) in contrast to Tesla transformer and air-core transformer. Under the condition of large coupling coefficient, the transformer in Fig. 22(a) can be decomposed as Fig. 22(b) shows. L_μ, L_{lp} and L_{ls} are defined in (5), (11) and (14), respectively. Define the turns ratio of transformer as $n_s = (N_2/N_1)$. C_2, L_{ls}, L_{sl}, R_s and i_s in the secondary circuit also can be equated into the primary circuit as C_2', L_{ls}', L_{sl}', R_s' and i_s'. The equating law are as $C_2' = C_2 n_s^2$, $L_{ls}' = L_{ls}/n_s^2$, $L_{sl}' = L_{sl}/n_s^2$, $R_s' = R_s/n_s^2$, and $i_s' = i_s n_s$. The initial voltage of C_1 and C_2 are as U_0 and 0 respectively.

The voltages of C_1 and C_2 are $U_{c1}(t)$ and $U_{c2}(t)$, respectively. According to Fig. 22(a), the circuit equations of HES module are as

$$\begin{cases} U_0 - \dfrac{\int i_p(t)}{C_1} = R_p i_p(t) + (L_\mu + L_{lp} + L_{pl})\dfrac{di_p(t)}{dt} - M\dfrac{di_s(t)}{dt} \\ M\dfrac{di_p(t)}{dt} = (L_\mu n_s^2 + L_{ls} + L_{sl})\dfrac{di_s(t)}{dt} + R_s i_s(t) + \dfrac{\int i_s(t)}{C_2} \end{cases} \quad (45)$$

In view of Fig. 22(b), the circuit equations of HES module can also be established as

$$\begin{cases} (L_\mu + L_{pl} + L_{lp})\dfrac{d^2 i_p}{dt^2} + R_p\dfrac{di_p}{dt} + \dfrac{i_p}{C_1} = L_\mu\dfrac{d^2 i_s'}{dt^2} \\ (L_\mu + L_{sl}' + L_{ls}')\dfrac{d^2 i_s'}{dt^2} + R_s'\dfrac{di_s'}{dt} + \dfrac{i_s'}{C_2'} = L_\mu\dfrac{d^2 i_p}{dt^2} \end{cases} \quad (46)$$

The initial conditions are as $i_p(0)=0$, $i_s(0)=0$, $U_{c1}(0)=U_0$ and $U_{c2}(0)=0$. In view of that $i_p(t)=-C_1 dU_{c1}(t)/dt$ and $i_s(t)=-C_2 dU_{c2}(t)/dt$, Equations in (45) can be simplified as

$$\begin{cases} \dfrac{d^2 U_{c1}(t)}{dt^2} + 2\alpha_p\dfrac{dU_{c1}(t)}{dt} + \omega_p^2 U_{c1} - k_p\dfrac{d^2 U_{c2}(t)}{dt^2} = 0 \\ \dfrac{d^2 U_{c2}(t)}{dt^2} + 2\alpha_s\dfrac{dU_{c2}(t)}{dt} + \omega_s^2 U_{c2} - k_s\dfrac{d^2 U_{c1}(t)}{dt^2} = 0 \end{cases} \quad (47)$$

In (47), ω_p and ω_s are defined as the resonant angular frequencies in primary and secondary circuits, while k_p and k_s are defined as the coupling coefficients of the primary and secondary circuits respectively. These parameters are presented as

$$\begin{cases} \omega_p^2 = 1/[(L_1 + L_{pl})C_1] \;,\; \omega_s^2 = 1/[(L_2 + L_{sl})C_2] \\ \alpha_p = R_p/2(L_1 + L_{pl}) \;,\; \alpha_s = R_s/[2C_2(L_2 + L_{sl})] \\ k_p = MC_2/C_1(L_1 + L_{pl}) \;,\; k_s = MC_1/C_2(L_2 + L_{sl}) \end{cases} \quad (48)$$

Define the effective coupling coefficient of the HES module based on transformer charging as k, and the quality factors of the primary and secondary circuits as Q_1 and Q_2 respectively. k, Q_1 and Q_2 are presented as

$$\begin{cases} k^2 = \dfrac{M^2}{(L_1 + L_{pl})(L_2 + L_{sl})} = \dfrac{L_\mu^2}{(L_\mu + L_{lp} + L_{pl})n_s^2(L_\mu + L_{ls}' + L_{sl}')} = k_p k_s \\[3mm] Q_1 = \dfrac{\omega_p(L_1 + L_{pl})}{R_p}, \; Q_2 = \dfrac{\omega_s(L_2 + L_{sl})}{R_s} \end{cases} \tag{49}$$

Equations (47) have general forms of solution as $U_{c1}(t)=D_1 e^{xt}$ and $U_{c2}(t)=D_2 e^{xt}$. Through substituting the general solutions into (47), linear algebra equations of the coefficients D_1 and D_2 are obtained. The characteristic equation of the linear algebra equations obtained is calculated as

$$(1-k^2)x^4 + 2(\alpha_p + \alpha_s)x^3 + (\omega_p^2 + 4\alpha_p\alpha_s + \omega_s^2)x^2 + 2(\alpha_p\omega_p^2 + \alpha_s\omega_s^2)x + \omega_p^2\omega_s^2 = 0. \tag{50}$$

x in the characteristic equation (50) represents the characteristic solution. As a result, x, D_1 and D_2 should be calculated before the calculations of $U_{c1}(t)$ and $U_{c2}(t)$. Obviously, the characteristic solution x can be obtained through the solution formula of algebra equation (50), but x will be too complicated to provided any useful information. In order to reveal the characteristics of the HES module in a more informative way, two methods are introduced to solve the characteristic equation (50) in this section.

4.1. The lossless method

The first method employs lossless approximation. That's to say, the parasitic resistances in the HES module are so small that they can be ignored. So, the HES module has no loss. Actually in many practices, the "no loss" approximation is reasonable. As a result, equation (50) can be simplified as

$$(1-k^2)x^4 + (\omega_p^2 + \omega_s^2)x^2 + \omega_p^2\omega_s^2 = 0. \tag{51}$$

In (51), it is easy to get the two independent characteristic solutions defined as $x_\pm$. $U_{c1}(t)=D_1 e^{xt}$ and $U_{c2}(t)=D_2 e^{xt}$ can also be calculated combining with the initial circuit conditions. Finally, the most important four characteristic parameters such as $U_{c1}(t)$, $U_{c2}(t)$, $i_p(t)$ and $i_s(t)$, are all obtained as

$$\begin{cases} U_{c1}(t) = \dfrac{(1+T)L_\mu - L_\Sigma}{(1+T)^2 L_\mu - TL_\Sigma} U_0 [\dfrac{(1+T)L_\mu}{(1+T)L_\mu - L_\Sigma}\cos(\omega_+ t) + T\cos(\omega_- t)] \\[3mm] U_{c2}(t) = \dfrac{(1+T)L_\mu - L_\Sigma}{(1+T)^2 L_\mu - TL_\Sigma} \sqrt{\dfrac{L_1 + L_{pl}}{L_2 + L_{sl}}} \dfrac{C_1 U_0}{k C_2}[\cos(\omega_+ t) - \cos(\omega_- t)] \\[3mm] i_p(t) = \dfrac{(1+T)L_\mu - L_\Sigma}{(1+T)^2 L_\mu - TL_\Sigma} C_1 U_0 [\dfrac{(1+T)L_\mu \omega_+}{(1+T)L_\mu - L_\Sigma}\sin(\omega_+ t) + T\omega_-\sin(\omega_- t)] \\[3mm] i_s(t) = \dfrac{(1+T)L_\mu - L_\Sigma}{(1+T)^2 L_\mu - TL_\Sigma} \sqrt{\dfrac{L_1 + L_{pl}}{L_2 + L_{sl}}} \dfrac{C_1 U_0}{k}[\omega_+\sin(\omega_+ t) - \omega_-\sin(\omega_- t)] \end{cases} \tag{52}$$

In (52), L_Σ represents the sum of the parasitic inductances and leakage inductances, while ω_+ and ω_- stand for the two resonant angular frequencies existing in the HES module ($\omega_+ \!\gg\! \omega_-$). Parameters such as T, L_Σ, ω_+ and ω_- are as

$$\begin{cases} T \triangleq \omega_s^2 / \omega_p^2, \ L_\Sigma = L_{pl} + L_{lp} + (L_{ls} + L_{sl}) / n_s^2 \\[2mm] \omega_+^2 = \dfrac{1+T}{L_\Sigma C_1}, \quad \omega_-^2 = \dfrac{T}{(1+T)L_\mu C_1} \end{cases} \tag{53}$$

In (52), the voltages of energy storage capacitors have phase displacements in contrast to the currents. All of the voltage and current functions have two resonant angular frequencies as ω_+ and ω_- at the same time, which demonstrates that the HES module based on transformer with closed magnetic core is a kind of double resonant module. The input and output characteristics and the energy transferring are all determined by (52).

4.2. The "little disturbance" method

The "little disturbance" method was introduced to analyze the Tesla transformer with open core by S. D. Korovin in the Institute of High-Current Electronics (IHCE), Tomsk, Russia. Tesla transformer with open core has a different energy storage mode in contrast to the transformer with closed magnetic core. Tesla transformer mainly stores magnetic energy in the air gaps of the open core, while transformer with closed core stores magnetic energy in the magnetic core. So, the calculations for parameters of these two kinds of transformer are also different. However, the idea of "little disturbance" is still a useful reference for pulse transformer with closed core [24-25]. So, the "little disturbance" method is introduced to analyze the pulse transformer with closed magnetic core for HES module.

The "little disturbance" method employs two little disturbance functions $\Delta x_\pm$ to rectify the characteristic equation (50) or (51). That's to say, the previous characteristic solutions $x_\pm$ are substituted by $x_\pm + \Delta x_\pm$. In HES module, the parasitic resistances which cause the energy loss still exist, though they are very small. So, the parasitic resistances also should be considered. Define j as unit of imaginary number, and variable x_j as $-jx/\omega_s$. Equation (50) can be simplified as

$$(x_j^2 - \frac{j}{Q_1 \alpha^{\frac{1}{2}}} x_j - \frac{1}{\alpha})(x_j^2 - \frac{j}{Q_2} x_j - 1) = k^2 x_j^4. \tag{54}$$

Through substituting x_j by $x_\pm + \Delta x_\pm$ in (54), the characteristic equation of $\Delta x_\pm$ can be obtained. If the altitude variables are ignored, the solutions of the characteristic equation of $\Delta x_\pm$ are presented as

$$\begin{cases} x_+ = [\dfrac{(1+T)L_\mu}{TL_\Sigma}]^{\frac{1}{2}}, \quad x_- = (\dfrac{1}{1+T})^{\frac{1}{2}} \\[4mm] \dfrac{1}{\frac{1}{Q_2}}(x_\pm^2 - 1) + \dfrac{1}{Q_2}(x_\pm^2 - \dfrac{1}{\alpha}). \\[4mm] \Delta x_\pm = \dfrac{j}{2}\dfrac{\alpha^2 Q_1}{2x_\pm^2(1-k^2)-1-\dfrac{1}{\alpha}} \end{cases} \tag{55}$$

The solutions of (50) are as $x = jx|\omega_s = j(x_\pm + \Delta x_\pm)\omega_s$. $\Delta x_\pm$ shown in (55) describes the damping effects of the parasitic resistances in the circuit. The two resonant angular frequencies $\omega_\pm$ are rectified as

$$\omega_\pm = x_\pm \omega_s , \quad \Delta \omega_\pm = \Delta x_\pm \omega_s. \tag{56}$$

Define two effective quality factors of the double resonant circuit of HES module as

$$Q_{eff+} = \frac{\omega_+}{2|\Delta\omega_+|} = \frac{\rho_1}{R_1 + R_2/n_s^2} , \quad Q_{eff-} = \frac{\omega_-}{2|\Delta\omega_-|}. \tag{57}$$

$$\begin{cases} U_{c1}(t) = G_1 e^{-\beta_+ t}[\cos(\omega_+ t) + \dfrac{\sin(\omega_+ t)}{2Q_{eff+}}] + G_2 e^{-\beta_- t}[\cos(\omega_- t) + \dfrac{\sin(\omega_- t)}{2Q_{eff-}}] \\[4mm] U_{c2}(t) = G_3[e^{-\beta_+ t}(\cos(\omega_+ t) + \dfrac{\sin(\omega_+ t)}{2Q_{eff+}}) - e^{-\beta_- t}(\cos(\omega_- t) + \dfrac{\sin(\omega_- t)}{2Q_{eff-}})] \\[4mm] i_p(t) = -C_1\{G_1 e^{-\beta_+ t}[-\beta_+(\cos(\omega_+ t) + \dfrac{\sin(\omega_+ t)}{2Q_{eff+}}) + \omega_+(\dfrac{\cos(\omega_+ t)}{2Q_{eff+}} - \sin(\omega_+ t))] + \\[4mm] \qquad G_2 e^{-\beta_- t}[-\beta_-(\cos(\omega_- t) + \dfrac{\sin(\omega_- t)}{2Q_{eff-}}) + \omega_-(\dfrac{\cos(\omega_- t)}{2Q_{eff-}} - \sin(\omega_- t))]\} \\[4mm] i_s(t) = -C_2 G_3[-\beta_+ e^{-\beta_+ t}(\cos(\omega_+ t) + \dfrac{\sin(\omega_+ t)}{2Q_{eff+}}) + \omega_+ e^{-\beta_+ t}(\dfrac{\cos(\omega_+ t)}{2Q_{eff+}} - \sin(\omega_+ t)) + \\[4mm] \qquad \beta_- e^{-\beta_- t}(\cos(\omega_- t) + \dfrac{\sin(\omega_- t)}{2Q_{eff-}}) - \omega_- e^{-\beta_- t}(\dfrac{\cos(\omega_- t)}{2Q_{eff-}} - \sin(\omega_- t))] \end{cases} \tag{58}$$

In (57), ρ_1 represents the characteristic impedance of the resonant circuit, and $\rho_1 = [L_\Sigma(1+T)/C_1]^{1/2}$. According to (55), the general solutions of (49) ($U_{c1}(t) = D_1 e^{xt}$ and $U_{c2}(t) = D_2 e^{xt}$) are clarified. When the initial circuit conditions are considered, the important four characteristic parameters such as $U_{c1}(t)$, $U_{c2}(t)$, $i_p(t)$ and $i_s(t)$ are obtained as (58). In (58), $\beta_\pm = |\Delta\omega_\pm| = |\Delta x_\pm|\omega_s$, coefficients such as G_1, G_2 and G_3 are defined as

$$G_1 = \frac{x_-^2(x_+^2 - 1)}{x_+^2 - x_-^2}U_0 , \quad G_2 = \frac{x_+^2 x_-^2 T}{x_+^2 - x_-^2}U_0 , \quad G_3 = \frac{x_+^2 x_-^2}{x_+^2 - x_-^2}\sqrt{\frac{L_1 + L_{pl}}{L_2 + L_{sl}}}\frac{C_1}{kC_2}U_0. \tag{59}$$

From (58), all of the voltage and current functions have two resonant frequencies. In many situations of practice, the terms in (58) which include $\cos(\omega \cdot t)$ and $\sin(\omega \cdot t)$ can be ignored, as $\omega_+ >> \omega_-$ and $\beta_+ >> \beta_-$. The resonant currents $i_p(t)$ and $i_s(t)$ in primary and secondary circuit are almost in synchronization as shown in Fig. 23, and their resonant phases are almost the same. The first extremum point of $U_{c2}(t)$ defined as $(t_m$, $U_{c2}(t_m))$ corresponds to the maximum charge voltage and peak charge time of C_2. Of course, t_m also corresponds to the time of minimum voltage on C_1. That's to say, t_m is a critical time point which corresponds to maximum energy transferring. As $i_s(t) = -C_2 dU_{c2}(t)/dt$, $i_s(t)$ gets close to 0 when $t = t_m$. If $\omega_+ >> \omega_-$, the maximum charge voltage and peak charge time of C_2 are calculated as

$$
\begin{cases}
t_m = \dfrac{\pi}{\omega_+} = \pi (\dfrac{L_\Sigma C_1}{1+T})^{\frac{1}{2}} \\[3mm]
U_{c2}(t_m) = -G_3(1 + \exp(-\dfrac{\pi}{2Q_{eff+}}))
\end{cases}
\tag{60}
$$

Obviously, when the switch of C_1 in Fig.2 opens while the switch of C_2 closes both at t_m, the energy stored in C_2 reaches its maximum, and the energy delivered to the terminal load also reaches the maximum. This situation corresponds to the largest efficiency of energy transferring of the HES module. Of course, if the switch in Fig. 22(a) is closed all the time, the HES module acts in line with the law shown in (58). The energy stored in C_1 is transferred to transformer and capacitor C_2, then the energy is recycled from C_2 and transformer to C_1 excluding the loss, and then the aforementioned courses operate repetitively. Finally, all of the energy stored in C_1 becomes loss energy on the parasitic resistors, and the resonances in the HES module die down.

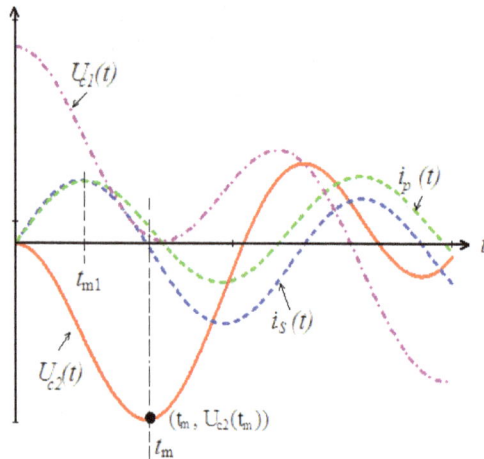

Figure 23. Typical theoretical waveforms of the output parameters of HES module based on pulse transformer charging, according to the "little disturbance" method

Under the condition $\omega_+ \gg \omega_-$, the peak time and the peak current of $i_P(t)$ are calculated as

$$
\begin{cases}
t_{m1} \approx \dfrac{\pi}{2\omega_+} = \dfrac{\pi}{2}(\dfrac{L_\Sigma C_1}{1+T})^{\frac{1}{2}} \\[3mm]
i_p(t_{m1}) = G_1 C_1 \omega_+ (1 + \dfrac{1}{4Q_{eff+}^2}) \exp(-\dfrac{\pi}{4Q_{eff+}})
\end{cases}
\tag{61}
$$

Usually, semiconductor switch such as thyristor or IGBT is used as the controlling switch of C_1. However, these switches are sensitive to the parameters of the circuit such as the peak current, peak voltage, and the raising ratios of current and voltage. The raising ratio of $U_{c1}(t)$ and $i_P(t)$ ($dU_{c1}(t)/dt$ and $i_P(t)/dt$) can also be calculated from (58), which provides theoretical instructions for option of semiconductor switch in the HES module.

Actually, the efficiency of energy transferring is also determined by the charge time of C_2 in practice. Define the charge time of C_2 as t_c, the maximum efficiency of energy transferring on C_2 as η_a, and the efficiency of energy transferring in practice as η_e. If the core loss of transformer is very small, the efficiencies of HES module based on pulse transformer charging are as

$$
\eta_a = \frac{\frac{1}{2}C_2 U_{c2}^2(t_m)}{\frac{1}{2}C_1 U_0^2} = \frac{C_1}{k^2 C_2} \frac{L_1 + L_{pl}}{L_2 + L_{sl}} (\frac{(1+T)L_\mu - L_\Sigma}{(1+T)^2 L_\mu - TL_\Sigma})^2 \ , \ \eta_e = \frac{\frac{1}{2}C_2 U_{c2}^2(t_c)}{\frac{1}{2}C_1 U_0^2} \leq \eta_a.
\tag{62}
$$

Actually, t_c corresponds to the time when S_2 closes in Fig. 2.

5. Magnetic saturation of pulse transformer and loss analysis of HES

5.1. Magnetic saturation of pulse transformer with closed magnetic core

Transformers with magnetic core share a communal problem of magnetic saturation of core. The pulse transformer with closed magnetic core consists of the primary windings (N_1 turns) and the secondary windings (N_2 turns), and it works in accordance with the hysteresis loop shown in Fig.24. Define the induced voltage of primary windings of transformer as $U_P(t)$, and the primary current as $i_P(t)$. If the input voltage $U_P(t)$ increases, the magnetizing current in primary windings also increases, leading to an increment of the magnetic induction intensity B generated by $i_P(t)$. When B increases to the level of the saturation magnetic induction intensity B_s, dB/dH at the working point (H_0, B_0) decreases to 0 and the relative permeability μ_r of magnetic core decreases to 1. Under this condition, magnetic characteristics of the core deteriorate and magnetic saturation occurs. Once the magnetic saturation occurs, the transformer is not able to transfer voltage and energy. So, it's an important issue for a stable transformer to improve the saturation characteristics of magnetic core and keep the input voltage $U_P(t)$ at a high level simultaneously.

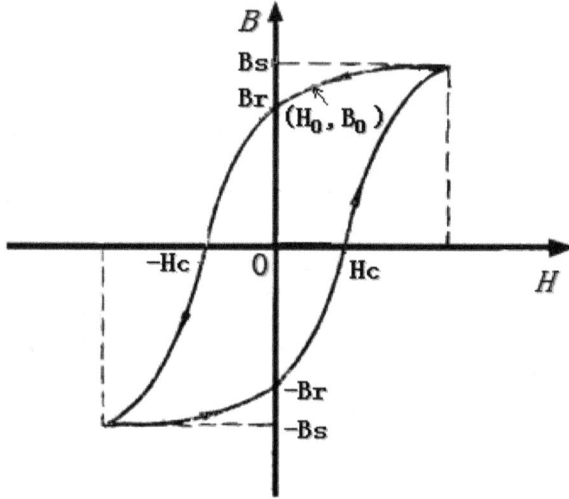

Figure 24. Typical hysteresis loop of magnetic core of pulse transformer

The total magnetic flux in the magnetic core is Φ_0 defined in (4). According to Faraday's law, $U_p(t)=d\Phi_0/dt$. Define the allowed maximum increment of B in the hysteresis loop as ΔB_{max}, and the corresponding maximum increment of Φ_0 as $\Delta\Phi$. Obviously, $\Delta B_{max}=B_s-(-B_r)$ and $\Delta\Phi=N_1\Delta B_{max}SK_T$, while parameters such as B_r, S and K_T are defined before (3). So, the relation between S and the voltage second product of core is presented as

$$S = \frac{\int_0^{t_c} U_p(t)dt}{N_1\Delta B_{max}K_T} = \frac{\int_0^{t_c} U_s(t)dt}{N_2\Delta B_{max}K_T}. \tag{63}$$

As parameters such as ΔB_{max}, N_1, N_2, S and K_T are unchangeable and definite in an already produced transformer, the charge time t_c defined in (62) can not be long at random. Otherwise, $\int_0^{t_c} U_p(t)dt > N_1\Delta B_{max}K_TS$, the core saturates and the transformer is not able to transfer energy. That's to say, (63) just corresponds to the allowed maximum charge time without saturation. If the allowed maximum charge time is defined as t_s, $\int_0^{t_s} U_p(t)dt = N_1\Delta B_{max}K_TS$.

According to (63), some methods are obtained to avoid saturation of core as follows. Firstly, ΔB_{max} and K_T of the magnetic material should be as large as possible. Secondly, the cross-section area of core should be large enough. Thirdly, the turn number of transformer windings (N_1) should be enhanced. Fourthly, the charge time t_c of transformer should be restricted effectively. Lastly, the input voltage $U_P(t)$ of transformer should decrease to a proper range.

Generally speaking, it is quite difficult to increase ΔB_{max} and K_T. The increment of N_1 leads to decrement of the step-up ratio of transformer. And the decrement of $U_P(t)$ leads to low voltage output from the secondary windings. As a result, the common used methods to avoid saturation of core include the increasing of S and decreasing the charge time t_c through proper circuit designing. Finally, the minimum cross-section design (S_{min}) of magnetic core in transformer should follow the instruction as shown in (64).

$$S \geq S_{min} = \frac{\int_0^{t_s} U_p(t)dt}{N_1 \Delta B_{max} K_T} \geq \frac{\int_0^{t_c} U_p(t)dt}{N_1 \Delta B_{max} K_T} = \frac{\int_0^{t_c} U_s(t)dt}{N_2 \Delta B_{max} K_T}. \tag{64}$$

In (64), $U_P(t)$ and $U_s(t)$ can be substituted by $U_{c1}(t)$ and $U_{c2}(t)$ calculated in (52) or (58). Moreover, small air gaps can be introduced in the cross section of magnetic core to improve the saturation characteristics, which has some common features with the Tesla transformer with opened magnetic core. Reference [40] explained the air-gap method which is at the costs of increasing leakage inductances and decreasing the coupling coefficient.

5.2. Loss analysis of HES

The loss is a very important issue to estimate the quality of the energy transferring module. In Fig. 22(a), the main losses in the HES module based on pulse transformer charging include the resistive loss and the loss of magnetic core of transformer. The resistive loss in HES module consists of loss of wire resistance, loss of parasitic resistance of components, loss of switch and loss of leakage conductance of capacitor. Energy of resistive loss corresponds to heat in the components. The wire resistance is estimated in (27), and the switch resistance and leakage conductance of capacitor are provided by the manufacturers. According to the currents calculated in (58), the total resistive loss defined as ΔW_R can be estimated conveniently. In this section, the centre topic focuses on the loss of magnetic core of transformer as follows.

5.2.1. Hysteresis loss analysis

In the microscope of the magnetic material, the electrons in the molecules and atoms spin themselves and revolve around the nucleuses at the same time. These two types of movements cause magnetic effects of the material. Every molecule corresponds to its own magnetic dipole, and the magnetic dipole equates to a dipole generated by a hypothetic molecule current. When no external magnetic field exists, large quantities of magnetic dipoles of molecule current are in random distribution. However, when external magnetic field exists, the external magnetic field has strong effect on these magnetic dipoles in random distribution, and the dipoles turn to the same direction along the direction of external magnetic field. The course is called as magnetizing, in which a macroscopical magnetic dipole of the material is formed. Obviously, magnetizing course of the core consumes energy which comes from capacitor C_1 in Fig. 2, and this part of energy corresponds to the hysteresis loss of core defined as W_{loss1}.

Define the electric field intensity, electric displacement vector, magnetic field intensity and magnetic induction intensity in the magnetic core as $\bar{E}$, $\bar{D}$, $\bar{H}$ and $\bar{B}$ respectively. The total energy density of electromagnetic field $W = \int (\bar{E} \cdot \partial \bar{D} / \partial t + \bar{H} \cdot \partial \bar{B} / \partial t) dt$. As the energy density of electric field is the same as which of magnetic field, the total energy density W in isotropic material can be simplified as

$$W = (\bar{E} \cdot \bar{D} + \bar{H} \cdot \bar{B}) / 2 = \bar{H} \cdot \bar{B} \quad or \quad dW = HdB. \tag{65}$$

The magnetizing current which corresponds to W_{loss1} is a small part of the total current $i_P(t)$ in primary windings. Define the magnetizing current as $I_m(t)$, the average length of magnetic pass as $<l_c>$, and the total volume of magnetic core as V_m. According to the Ampere's circuital law and Faraday's law,

$$H = N_1 I_m(t) / <l_c> \quad , \quad dB = -U_p(t)dt / N_1 SK_t. \tag{66}$$

According to (65) and (66), the hysteresis loss of magnetic core of transformer is obtained as

$$W_{loss1} = \int_0^{t_c} \frac{|U_p(t)I_m(t)| V_m}{<l_c> SK_T} dt \tag{67}$$

In some approximate calculations, the loss energy density is equivalent to the area enclosed by the hysteresis loop. If the coercive force of the loop is H_c, $W_{loss1} \approx 2H_c B_s V_m$.

5.2.2. Eddy current loss analysis

When transformer works under high-frequency conditions, the high-frequency current in transformer windings induces eddy current in the cross section of magnetic core. Define the eddy current vector as $\bar{j}'$, magnetic induction intensity of eddy current as $\bar{B}'$, magnetic field intensity of eddy current as $\bar{H}'$, magnetic induction intensity of $i_P(t)$ as $\bar{B}_0$, and magnetic field intensity of $i_P(t)$ as $\bar{H}_0$. As shown in Fig. 25, the direction of $\bar{j}'$ is just inverse to the direction of $i_P(t)$, so the eddy current field $\bar{B}'$ weakens the effect of $\bar{B}_0$. The eddy current heats the core and causes loss of transformer, and it should be eliminated by the largest extent when possible.

In order to avoid eddy current loss, the magnetic core is constructed by piled sheets in the cross section as Fig. 25 shows. Usually, the sheet is covered with a thin layer of insulation material to prevent eddy current. However, the high-frequency eddy current has "skin effect", and the depth of "skin effect" defined as δ is usually smaller than the thickness h of the sheet. As a result, the eddy current still exists in the cross section of core. Cartesian coordinates are established in the cross section of core as shown in Fig. 25, and the unit vectors are as $\vec{e}_x$, $\vec{e}_y$ and $\vec{e}_z$. To a thin sheet, its length ($\vec{e}_z$) and width ($\vec{e}_y$) are both much larger than the thickness h ($\vec{e}_x$). So, approximation of infinite large dimensions of sheet in $\vec{e}_y$ and $\vec{e}_z$ directions is reasonable. That's to say, $\partial/\partial y=0$ and $\partial/\partial z=0$. The "little disturbance" theory aforementioned before still can be employed to calculate the field $\vec{B}'$ generated by eddy current $\vec{j}'$.

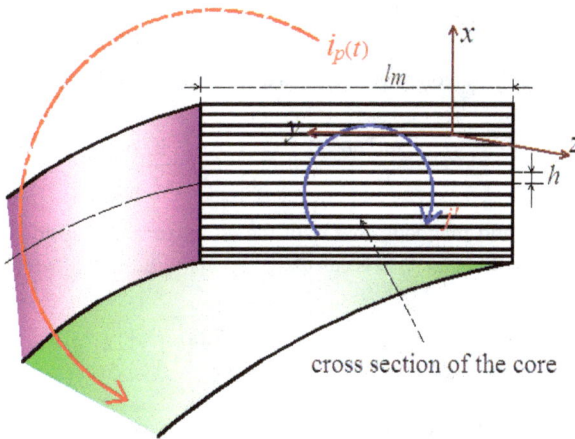

Figure 25. Distribution of eddy current in the cross section of toroidal magnetic core

The total magnetic induction intensity in the core is as $\vec{B}=\vec{B}_0+\vec{B}'=(B_0-B')\vec{e}_z$, and $B'\ll B_0$. $\vec{B}'$ generated by eddy current can be viewed as the variable of "little disturbance". According to Maxwell equations,

$$\begin{cases} \nabla \times \vec{E} = -\partial \vec{B} / \partial t \\ \nabla \times \vec{H} = \nabla \times (\vec{H}_0 + \vec{H}') = \vec{j}' + \partial \vec{D} / \partial t \end{cases} \tag{68}$$

From (68), it is easy to obtain the formula $\partial E_y / \partial x \approx -\partial B_0 / \partial t$, while (E_x, E_y, E_z) and (H_x, H_y, H_z) corresponds to vectors $\vec{E}$ and $\vec{H}$. Through integration,

$$\vec{E}(x) = -x(\partial B_0 / \partial t)\vec{e}_y \,, -h/2 \le x \le h/2. \tag{69}$$

Define the conductivity of the sheet in magnetic core as σ. From the second equation in (68), $(-\partial H_z' / \partial x)\vec{e}_y = \vec{j}' = \sigma \vec{E}$. It demonstrates that infinitesimal conductivity is the key factor to prevent eddy current. When working frequency is f, the depth of "skin effect" of the sheet is calculated as $\delta = (\pi f \mu \sigma)^{-1/2}$. According to (69), the "little disturbance" field of eddy current in isotropic magnetic material is presented as

$$\vec{H}'(x) = \frac{\sigma x^2}{2}\frac{\partial B_0}{\partial t}\vec{e}_z \,, \quad \vec{B}'(x) = \frac{\mu_0 \mu_r \sigma x^2}{2}\frac{\partial B_0}{\partial t} = \frac{1}{\omega}(\frac{x}{\delta})^2 \frac{\partial B_0}{\partial t}\vec{e}_z, \quad (\frac{-h}{2} \le x \le \frac{h}{2}). \tag{70}$$

Through averaging the field along the thickness direction ($\vec{e}_x$) of sheet,

$$\tilde{H}' = \frac{\sigma h^2}{24}\frac{\partial B_0}{\partial t} \,, \quad \tilde{B}' = \frac{1}{12\omega}(\frac{h}{\delta})^2 \frac{\partial B_0}{\partial t}. \tag{71}$$

As the electric energy and magnetic energy of the eddy current field are almost the same, the eddy current loss defined as W_{loss2} is calculated as

$$W_{loss2} = \int_0^{t_c} dt \iiint_V \tilde{B}'\tilde{H}'dV = \frac{\sigma h^2}{288\omega}(\frac{h}{\delta})^2 V_m \int_0^{t_c} (\frac{\partial B_0}{\partial t})^2 dt \tag{72}$$

From (72), W_{loss2} is proportional to the conductivity σ of the core, and it is also proportional to $(h/\sigma)^2$. As a result, W_{loss2} can be limited when $h \ll \delta$.

5.2.3. Energy efficiency of the HES module

As to the HES module based on transformer charging shown in Fig. 22(a), the energy loss mainly consists of ΔW_R, W_{loss1} and W_{loss2}. Total energy provided from C_1 is as $W_0 = \frac{1}{2}C_1 U_0^2$.

In practice, the energy stored in C_1 can not be transferred to C_2 completely, though the loss of the module is excluded. In other words, residue energy defined as W_{or} exists in C_1. Define the allowed maximum efficiency of energy transferring from C_1 to C_2 as η_{max}. So, η_{max} of the HES module is as

$$\eta_{max} = \frac{W_0 - (\Delta W_R + W_{loss1} + W_{loss2}) - W_{0r}}{W_0}. \tag{73}$$

From (62), η_a, η_e and η_{max} have relation as $\eta_e \leq \eta_a \leq \eta_{max}$.

Author details

Yu Zhang[*] and Jinliang Liu

College of Opto-Electronic Science and Engineering, National University of Defense Technology, Changsha, China

Acknowledgement

This work was supported by the National Science Foundation of China under Grant No.51177167. It's also supported by the Fund of Innovation, Graduate School of National University of Defense Technology under Grant No.B100702.

6. References

[1] Bialasiewicz J T (2008) Renewable energy systems with photovoltaic power generators: operation and modeling. IEEE Transactions on Industrial Electronics, 55(7): 2752-2758.

[2] Anderson M D and Carr D S (1993) Battery energy storage technologies. Proceedings of the IEEE, 81(3): 475-479.

[3] Schempp E and Jackson W D (1996) Systems considerations in capacitive energy storage. Energy Conversion Engineering Conference, 2: 666-671.

[4] Beverly R E and Campbell R N (2010) A 1MV, 10kJ photo-triggered Marx generator. IEEE Power Modulator and High-Voltage Conference: 560-563.

[5] Lehmann M (2010) High energy output Marx generator design. IEEE Power Modulator and High-Voltage Conference: 576-578.

[6] Simon E and Bronner G (1967) An inductive energy storage system using ignitron switching. IEEE Transactions on Nuclear Science, 14(5): 33-40.

[7] Gorbachev K V, Nesterov E V, Petrov V Yu, and Chernykh E V (2009) A helical-radial magnetic cumulation fast-growing current pulse generator. Instruments and Experimental Technology, 52(1): 58-64.

[8] Doinikov N I, Druzhinin A S, Krivchenkov Yu M (1992) 900MJ toroidal transformer-type inductive energy storage. IEEE Transactions on Magnetics, 28(1): 414-417.

---

[*] Corresponding Author

[9] Masugata K, Saitoh H, Maekawa H (1997) Development of high voltage step-up transformer as a substitute for a Marx generator. Review of Scientific Instruments, 68(5): 2214-2220.

[10] Bushlyakov A I, Lyubutin S K, Ponomarev A V (2006) Solid-state SOS-based generator providing a peak power of 4 GW. IEEE Transaction on Plasma Science, 34(5): 1873-1878.

[11] Rukin S, Lyubutin S, Ponomarev A (2007) Solid-state IGBT/SOS-based generator with 100-kHz pulse repetition frequency. IEEE Pulsed Power Conference: 861-864.

[12] Korovin S D, Kurkan I K, Loginov S V, (2003) Decimeter-band frequency-tunable sources of high-power microwave pulses. Laser and Particle Beams, 21: 175-185.

[13] Morton D, Weidenheimer D, Dasilva T, (2005) Performance of an advanced repetitively pulsed electron beam pumped KrF laser driver. IEEE Pulsed Power Conference: 1290-1293.

[14] Sethian J D, Myers M, Smith I D, (2000) Pulsed power for a rep-rate, electron beam pumped KrF laser. IEEE Transactions on Plasma Science, 28(5): 1333-1337.

[15] Sarkar P, Braidwoody SW, Smith I R, (2005) A compact battery-powered 500 kV pulse generator for UWB radiation. IEEE Transactions on Plasma Science, 33(5): 1306-1309.

[16] Mesyats G A, Korovin S D, Rostov V V, Shpak V G, and Yalandin M I, (2004) The RADAN series of compact pulsed power generators and their applications. Proceedings of the IEEE, 92(7): 1166–1178.

[17] Krasik Y E, Grinenko A, Sayapin A and Efimov S, (2008) Underwater electrical wire explosion and its applications. IEEE Transactions on Plasma Science, 36(2): 423-434.

[18] Boeuf J P, (2003) Plasma display panel: physics, recent developments and key issues. J. Physics D: Appl Phys, 36(6): 53-79.

[19] Choi Y W, Jeong I W, Rim G H, (2002) Development of a magnetic pulse compression modulator for flue gas treatment. IEEE Transactions on Plasma Science, 30(5): 1632-1636.

[20] Rossi J O and Ueda M, (2006) A 100kV/200A Blumlein pulser for high-energy plasma implantation. IEEE Transactions on Plasma Science, 34(5): 1766-1770.

[21] Kamase Y, Shimizu M, Nagahama T, (1993) Erosion of spark of square wave high-voltage source for ozone generation. IEEE Transactions on industry applications, 29(4): 793-797.

[22] Gaudreau M P J, Hawkey T, Petry J and Kempkes M A, (2001) A solid state pulsed power system for food processing. Digest of Technical Papers: Pulsed Power Plasma Science: 1174-1177.

[23] Gundersen M, Kuthi A, Behrend M and Vernier T, (2004) Bipolar nanosecond pulse generation using transmission lines for cell electro-manipulation. Power Modulator Symposium and High-Voltage Workshop Conference: 224-227.

[24] Zhang Y, Liu J L, Cheng X B, (2010) Output Characteristics of a Kind of High-Voltage Pulse Transformer with Closed Magnetic Core. IEEE Transactions on Plasma Science, 38(4): 1019-1027.

[25] Zhang Y, Liu J L, Cheng X B, (2010) A compact high voltage pulse generator based on pulse transformer with closed magnetic core. Review of Scientific instruments, 81(3): 033302.

[26] Rohwein G J, Lawson R N, Clark M C (1991) A compact 200kVpulse transformer system. Proceeding of the 8th IEEE Pulsed Power Conference: 968-970.

[27] Korovin S D, Gubanov V P, Gunin A V, Pegel I V and Stepchenko A S, (2001) Repetitive nanosecond high-voltage generator based on spiral forming line. The 28th IEEE international Conference on Plasma Science: 1249-1251.

[28] Hurley W G and Wilcox D J, (1994) Calculation of leakage inductance in transformer windings. IEEE Transactions on Power Electronics, 9(1): 121-126.

[29] Massarini A, Kazimierczuk M K, (1997) Self-capacitance of inductors. IEEE Transactions on Power Electronics, 12(4): 671-676.

[30] Blache F, Keradec J P and Cogitore B, (1994) Stray capacitances of two winding transformers: equivalent circuit, measurements, calculation and lowering. IEEE Pulsed Power Conference: 1211-1217.

[31] Grandi G, Kazimierczuk M K, Massarini A, Reggiani U, (1999) Stray capacitances of single-layer solenoid air-core inductors. IEEE Transactions on Industry Applications, 33(5): 1162-1168.

[32] Mostafa A E, Gohar M K, (1953) Determination of voltage, current, and magnetic field distributions together with the self-capacitance, inductance and HF resistance of single-layer coils. Proceedings of the I. R. E.: 537-547.

[33] Collins J A, (1990) an accurate method for modeling transformer winding capacitances. IEEE Pulsed Power Conference: 1094-1099.

[34] Kino G S and Paik S F, (1962) Circuit Theory of Coupled Transmission System. Journal of Applied Physics, 33(10): 3002-3008.

[35] Zhang Y, Liu J L, Fan X L, (2011) Characteristic Impedance and Capacitance Analysis of Blumlein Type Pulse forming Line Based on Tape Helix. Review of Scientific Instruments, 82(10): 104701.

[36] Lord H W, (1971) Pulse transformer. IEEE Transactions on Magnetics, 7(1): 17-28.

[37] Nishizuka N, Nakatsuyama M and Nagahashi H, (1989) Analysis of pulse transformer on distributed parameter theory. IEEE Transactions on Magnetics, 25(5): 3260-3262.

[38] Redondo L M, Silva J F and Margato E, (2007) Pulse shape improvement in core-type high-voltage pulse transformers with auxiliary windings. IEEE Transactions on Magnetics, 43(5): 1973-1982.

[39] Costa E M M, (2010) Resonance on coils excited by square waves: explaining Tesla transformer, IEEE Transactions on Magnetics, 46(5): 1186-1192.

[40] Zhang Y, Liu J L, Fan X L, Zhang H B, Feng J H, (2012) Saturation and Pulse Response Characteristics of the Fe-Based Amorphous Core with Air Gap. IEEE Transactions on Plasma Science, 40(1): 90-97.

Chapter 10

Dynamic Energy Storage Management for Dependable Renewable Electricity Generation

Ruddy Blonbou, Stéphanie Monjoly and Jean-Louis Bernard

Additional information is available at the end of the chapter

1. Introduction

The administrators of the distribution networks have to face the insertion of the decentralized electricity production of renewable origin. In particular, wind and photovoltaic electricity generation know a fast development supported by satisfying technical maturity, greater environmental concern and political will expressed by financial and statutory incentives.

However, one of the main handicaps of many renewable energies - and quite particularly wind energy and solar energy - is the temporal variation of the resource and the weak previsibility of its availability. Thus, connecting wind or photovoltaic farms to electrical networks is an important challenge for the administrators of distribution and transport networks of electricity. They impact on the planning, on the operation of distribution networks and on the safety of the electric systems.

For grid connected renewable generation, as long as the penetration rates of these productions are marginal, the network compensates, within few minutes, for the fast variations thanks to the power reserves, which is for example mobilized hydroelectric power reserve, or gas turbines in rotating reserve (which produce greenhouse gases). These power reserves have a cost that must be take into account in the economic analysis of the deployment of distributed renewable electricity generation.

On a regional scale, the presence of renewable generators often induces additional costs for network reinforcement to limit the risks of congestion. Indeed, the favorable conditions for wind energy exploitation are often found in remote area (windy coast, offshore) where the network infrastructures are weak or non-existent. Since a regional network is sized according to the maximum transit power, to prevent power congestion, it is necessary to size the network infrastructures to match the total installed capacity. As the load factor of wind energy is about 25 %, the ratio of the cost for network strengthening over the aggregated energy produced for wind (or solar) energy is higher to that of the other (non intermittent) sources of energy.

The renewable generations capacity will have to take into account the preservation of the reliability and the safety of the networks. The objective is that renewable generators must not entail the degradation of the supply security nor imply dramatic cost increase for the consumers.

Energy storage technologies are identified as key elements for the development of electricity generation exploiting renewable energy sources. They could contribute to remove the technical constraints that limit the contribution of renewables into electrical networks. As mentioned above, these technical limits are present both on the regional scale and on the scale of the whole network.

More generally, energy storage could propose valuable services by reducing the instantaneous variations of the injected power. A simple approach consists in storing a part of the random production that would be add up to the future production to decrease the amplitude of variation of the injected power. That approach, however, does not guarantee the availability of stored energy, nor the level at which the power will be injected to the network. Furthermore, a trade-off must be estimated carefully to ensure the benefits will surpass the cost associated with the deployment of energy storage facilities.

In this chapter, we present an advanced approach that uses power production forecasts to dynamically manage the power flow to and from the battery and the networks for grid connected wind or solar electricity production. The objective is to guarantee, some time in advance and with a predefined error margin, the level of power that will be sent to the network, allowing a more efficient management of these stochastic energy resources and the optimization of the sizing of the storage facility. We also propose, through an in-depth analysis of the wind to power transfer function, a discussion about the power limit setting and the sizing of storage capacity in the context of congestion management.

The chapter will be organized as follow. First, we review the available storage technologies through the lens of their compatibility with the proposed approaches including a short discussion on the envisaged power converter solution for coupling of renewable generations and storage. Then, we demonstrate the advantage of an in-depth analysis of the wind to power transfer function and the use of energy storage for the sake of the optimal sizing of transmission line capacity in the context of the transport of wind-originated electricity. The role of energy storage is emphasized further in the presentation of an advanced power flow and energy storage management scheme. We complete the chapter with the presentation of the results obtained by applying the proposed approach during a simulation using real wind energy production data. The interest of the proposed method is that he permits to guarantee, within a pre-set margin of error, the power that will be sent to the grid by automatically dispatching the power flows between the wind plants, the energy storage facility and the electrical network. We conclude the chapter with a short discussion on energy storage management dynamic strategies and the improvement perspective of such approach.

2. Energy storage technologies for renewable energy power smoothing

Energy-storage technologies are vital for the large-scale exploitation of renewable energies since they could ensure secure and continuous supply to the consumer from distributed and intermittent supply base.

Many techniques can be used to stored electrical energy [1]. First, it must be transformed into a storable form of energy that could be mechanical, chemical or thermal. Then, there must be a process that gets back the stored energy into a usable form. Within the scope of this chapter, we will focus on energy storage technologies for electrical applications.

The common belief that electricity cannot be stored at a realistic cost comes from the fact that electricity is mainly produced, transmitted and consumed in AC. Today, energy storage capacity roughly represents less than 3% of the total electricity production capacity. However, the emergence of cost effective power electronic solutions that can handle high power levels makes it possible to store electricity for grid applications.

The past decade was marked by strong evolution of the technological context in storage of energy [2,3]. At the same time, the static converters knew a strong development in the range of powers from the kW to about few MW, carried in particular by the development of the photovoltaic and wind productions. As shown in Figure 1, pumped hydroelectric storage represents more than 97 % of the total of 120 GW reported storage capacity, followed by the classic compressed air with 440 MW (250 times less). Other technologies adapted for a deployment in the electrical distribution networks comes then, with only some tenth of % of which a majority of NAS (sodium-sulfur) batteries.

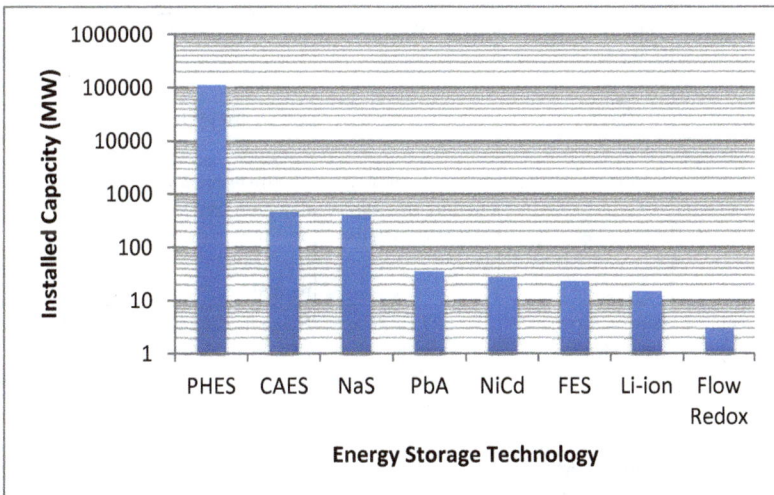

Figure 1. Installed capacity of various energy storage systems (from [2])

A summary of the various applications of energy storage aimed to support the electrical network expressly in the case of high rate of intermittent generations is reported in [3-7]. These articles review the characteristic of energy storage system in the scope of electrical networks with high renewable energies penetration rate.

The potential and opportunities of the storage of energy in the distribution networks is investigated in [2]. This study focuses on the technologies of storage susceptible to be installed on the levels of tension of distribution networks (unit power capacity in the range

Typology	Mechanical energy storage			Electrostatic energy storage
Technology	PHES	CAES	FES	SC
Rated Energy	500 MWh – 8000 MWh	500 MWh – 3000 MWh	5 kWh – 25 kWh	10 kWh
Rated Power	10 MW – 1 GW	Two plants in the world: 110 MW (USA) and 290 MW (Germany)	Few kW up to 10 MW	1W – 1 MW
Cycle Efficiency	65% – 80%	70%	85% - 95%	85% - 98%
Response time	Minutes	Seconds to minutes	< 1/50 sec.	<1/200 sec.
Cycling tolerance	50000	30000	> 10 million	1 million
Self discharge	Very low: water evaporation for long storage time	Very low: thermal loss in the storage tank. Pneumatic leaks.	Continuously; completely discharged within minutes	1%/day. Increase with temperature and SoC.
Power capital cost	500-1500 €/kW	<100 €/kW	400-25000 €/kW Application dependent	1000-20000 €/kW
Energy capital cost	10 – 20 €/kWh	5 – 70 €/kWh	400 – 800 €/kWh	6800 – 20000 €/kWh
Maturity status	Mature technology	Mature technology but only two plants in operation in the world	Mature technology; numerous units deployed in grids for frequency regulation.	Mature technology for embedded systems. Some stationary units reported.
Environmental or statutory constraints	Rely on favorable topology	Rely on favorable topology and availability of natural gas	No environmental risk. No emission.	Limited risk of toxicity, flammability of some used material.
Recycling ability	Dismantlement need to be planned	Dismantlement need to be planned	100%. No chemical compounds.	Dependent on material used

Table 1. Characteristics of energy storage systems

Typology	Electro-chemical energy storage					
	Conventional Batteries		High temperature Batteries		Redox flow Batteries	
Technology	PbA	Li-ion	ZEBRA	NaS	ZnBr	VRB
Rated Energy	1 kWh – 40 MWh	1 Wh – 50 MWh	Up to 100 KWh	400 kWh – 245 MWh	100 kWh – 2 MWh	2 MWh – 120 MWh
Rated Power	1W to 10 MW	Few W – 50 MW	5 kW - > 500 kW	50 kW - >10 MW	kW to MW	kW to MW
Cycle Efficiency	70% – 85%	80% - 90%	85% - 90%	85% - 90%	75-80 DC 65-70 AC/	80%-85% DC 65%-75% AC
Response time	1/200 sec.	1/200 sec.	1/200 sec.	1/200 sec.	1/200 sec.	1/200 sec.
Cycling tolerance	<1500 at 80% DoD	Up to 7000 at 80% DoD	3000 at 80 % DoD	4500 at 90 % DoD	1000-2000 at 80 % DoD	>13000 at 100% DoD
Self discharge	1% to 5% per month	2% to 10% per month	Due to thermal loss (up to 18%/days)	Due to thermal loss (up to 20%/days)	1 %/h due to diffusion of dibrome through the membrane	Up to 10% due to auxiliary consumption
Power capital cost	<500€/kW	500 – 2000 €/kW	- €/kW	1500-1800 €/kW	1000-2000 €/kW	1750€/kW
Energy capital cost	<50€/kWh (car batt.) to 250€/kWh	700 – 1500€/kWh	500€/kWh	200 – 250 €/kWh	600 – 800 €/kWh	215 €/kWh – 450
Maturity status	Mature technology for large number of applications	Mature technology for handheld electronic devices. Numerous demonstrators for electric cars and >1 MW stationary units.	Commercial units for embedded applications. Few stationary demonstrators.	Commercial units for stationary application (small market)	Prototypes and few industrial units	Prototypes and demonstration units. Few industrial units
Environmental or statutory constraints	Explosion risk if electrolyte gases leak.	Need electronic supervision of each cell for safe operation.	Low environmental impact if reactants are adequately confined	Low environmental impact if reactants are adequately confined	Limited environmental impact if reactants are adequately confined. Possible H2 emission to be accounted for.	Low environmental impact.
Recycling ability	90%	Recycling of the electrode metals	Up to 100%	Up to 98% with specific treatment of solid sodium		Recycling of the electrolyte

Table 2. Characteristics of electrochemical energy storage systems

of 10 to 20 MW in production and less than 40 MW in consumption). The author highlights the technologies that do not present major environmental or statutory constraints that could limit their deployment and that are susceptible to reach both technical and commercial maturity by 2015.

In this chapter, we consider energy storage technologies to tackle congestion relief and to smooth wind power variations on short time scales (up to several minutes). We are treating applications where energy storage systems are required to inject or absorb power during period of time in the order of minutes. Through these specific applications, we aimed to demonstrate the advantage of dynamic management of energy storage to raise the acceptance level of variable renewable energy sources for electricity generation.

Several criteria have to be analyzed to identify the storage technologies that are pertinent for the aforementioned applications. These applications require storage technologies with high power, short discharge period and good resilience to high number of charge-discharge cycles. Tables 1 and 2 report the main characteristics of a selection of energy storage technologies.

2.1. Tables nomenclatures

PHES: Pumped Hydro energy storage
CAES: Compressed Air Energy Storage
FES: Flywheel energy system
SC: Supercapacitors
PbA: Lead-Acid
Li-ion: Lithium-Ion
ZEBRA: Sodium Nickel Chlorides
VRB: Vanadium-Vanadium
ZnBr: Zinc – Bromine
NaS: Sodium – Sulphur

3. Pumped hydroelectric energy storage

Pumped hydroelectric storage (PHES) systems exploit gravitational potential energy. Energy is stored by pumping water from a lower reservoir to an upper reservoir. The amount of stored energy is proportional the volume of water in the upper reservoir. When needed, water flows from the upper reservoir to the lower reservoir to release the stored energy with round trip efficiency in the range of 70% to 80%. PHES is the major energy storage technology; it account for 97% of the world total storage capacity [2]. The energy can be stored several days and high power ramp can be achieved during both the charge and discharge phases (0–1800 MW in 16 s at the Dinorwig pumping station for example, [8]). The PHS technology suffers low modularity and can only be installed on site with particular topology. PHES is a key asset for wind energy as it enables the grid to operate securely while incorporating high wind penetrations. There may be additional benefits when using PHES that can charge and discharge at the same time (see Figure 2). This can be achieved in a single PHES facility by installing two penstocks as point out in [9]; a double penstock system enables the PHES to store excess wind energy while at the same time providing ancillary services to the grid. The results of the techno-economic studies [9] suggest that, the double penstock system could be economically credible while enable the wind energy penetration to increase above 40%. However, the economic value of PHES is sensitive to changes in fuel prices, interest rates, and total annual wind production.

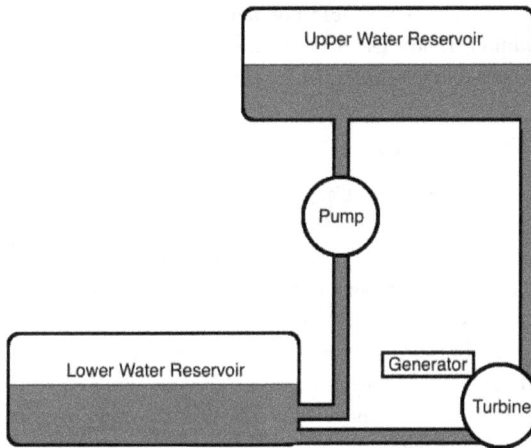

Figure 2. A double penstock PHES system

4. Batteries

The terminology "batteries" encompasses electrochemical storage cellular technologies that consist of an arrangement (in series or in parallel) of cell units. Each cell is made of two electrodes and an electrolyte secured into a sealed container. Batteries store chemical energy and generate electricity by a reduction-oxidation (redox) reaction. Batteries energy storage systems have been studied for almost 150 years, most research effort now aimed at cost reduction and high power application. The following section proposes a description of some promising batteries technologies. An overview of electrochemical energy storage systems is given in [10].

4.1. Lead-acid batteries

Lead-Acid batteries are the most used devices for low to medium scale energy storage application. Lead-acid batteries have a low-cost ($300–600/kW), high reliability, high power ramp capabilities and efficiency in the range (65%–80%). However, the performance of Lead-Acid battery will deteriorate quickly in the case of frequent charge-discharge cycles. The weak tolerance to high number of cycles limits the use of PbA batteries in application such as wind variations smoothing.

4.2. Lithium-ion batteries

Lithium-ion batteries are ideal for portable applications; they are widely use in mobile phone and in almost any other electronic portable device. They tolerate over 3000 cycles, have 95% efficiency at 80% depth of discharge and have high power ramp capability. Nowadays, the emergence of electric cars drives numerous researches on Li-Ion technology and materials to obtain reliable high power system [11]. Since their lifetime is related to the cycles' depth of discharge, Li-Ion should not be use in application where they may be fully discharged. In addition, Li-Ion technology must be operated with a protection circuit to ensure safe voltage and temperature operation ranges.

4.3. Sodium-sulphur batteries

NaS batteries are one of the most promising options for high power energy storage applications. The anode is made of sodium (Na), while the cathode is made of sulphur (S). The electrolyte enables the transfer of sodium ions to the cathode where they combine with sulphur anions and produce sodium polysulphide (NaSx). During the charge cycle, the opposite reaction occurs; sodium polysulphide is decomposed into sodium and sulphur. NaS batteries have good resilience to cycling (up to 4500 cycles), and can discharge quickly at high power [2-4]. NaS technology is modular; a single unit's rated power starts from 50 kW. Additionally, NaS batteries have low self-discharge and require low maintenance. However, the operating temperature must be kept at about 350°C.

5. Flow batteries

Flow batteries store at least one of its electrolytes in an external storage tank. During operation, the electrolytes need to be pumped into the electrochemical cell to produce electricity. Unlike conventional batteries, the power capacity of flow batteries is independent of the storage energy capacity and self-discharge is almost inexistent. The energy capacity depends on the stored volume of electrolyte and the power delivered depends only on the dimension of the electrodes and the number of cells. Additionally, flow batteries have a very short response time, can be fully discharged without consequences and are able to store energy over long period of time. Compared to conventional batteries, flow batteries have an unlimited life in theory and no memory effects. However, the necessity to control the electrolytic flows induces high operating cost.

5.1. Vanadium redox-flow batteries (VRB)

Among the various redox-flow batteries technology (Zinc Bromine, Polysulfide Bromide, Cerium-Zinc, ...), VRB exhibits the best potentiality, thanks to its competitive cost, its simplicity and since it contains no toxic materials [12]. Energy is stored in two reservoirs; a catholytic reservoir and an anolytic reservoir. VRB low specific energy, <35 Wh/kg, limits its use in non-stationary applications.

Figure 12 illustrates one major advantage of flow batteries. The maximum number of cycles tolerated during the lifetime of the batteries is plotted versus the depth of discharge for four technologies; PbA, Li-ion, NaS and VRB. The tolerable number of cycles decreases for PbA, Li-ion and NaS but remain constant at a relatively high value for VRB.

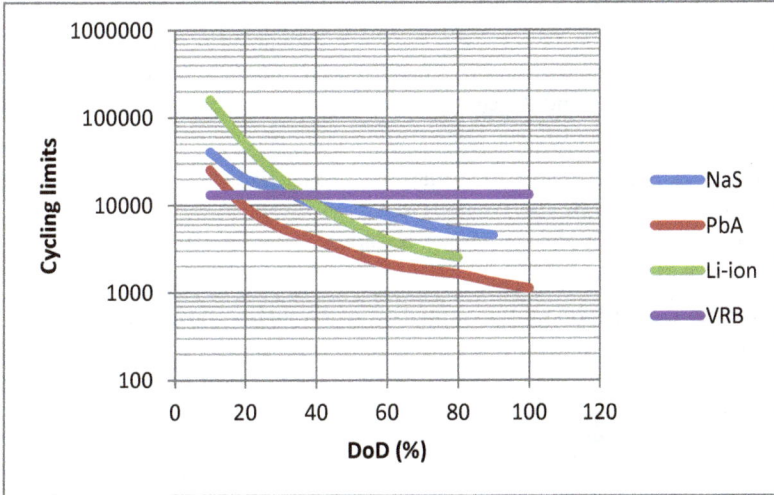

Figure 3. Cycling ability of various energy storage systems (from [2])

6. Super capacitors

Battery systems do not seems fully adequate for smoothing wind or solar power applications due to their limited tolerance to large number of charge - discharge cycles. Super capacitors (SC) or ultracapacitors, are electrochemical capacitor with remarkable high energy density, as compared to conventional capacitors, and high power density as compared to batteries. Moreover SC tolerate over a million charge – discharge cycles [13,14]. However, the voltage of an ultracapacitor tends to decrease during discharge. This affects the efficiency of the subsequent power converter and can undermine the energy utilization of the capacitor. In [15], parallel-series ultracapacitor shift circuits are employed to improve the energy utilization and minimize the voltage drop. The principal drawback of SC is its high cost (up to 20000€/kWh).

7. Power conversion solutions for coupling of renewable generations and storage

The present chapter deals with the combination of renewable electricity generation with energy storage system for the sake of renewable power smoothing. The preceding section focuses on the appropriate storage technologies. For wind or solar power smoothing, the storage technology should tolerate high number of cycles with partial DoD, be capable of high power ramp and short response time while keeping high efficiency. In addition, a chain of power conversion is necessary to pair the energy storage system with renewable energy sources and to adapt the voltage output of the ensemble to the network's voltage. The Figure 4 shows the structure of a DC-coupled hybrid power system with renewable sources and energy storage along with its control chain.

DC-coupled structures are flexible since they can accommodate with different type of energy sources and energy storage technology [12]. In a DC-coupled structure, the renewable energy sources and the energy storage devices are generally connected through static power converters to a DC bus. These power converters can be either:

- DC/DC buck-boost converters; to control the voltage variations of DC energy sources such as supercapacitors.
- AC/DC inverters; for storage devices requiring a mechanical training with variable speed, such as flywheel.

Power flows to the electrical grid from the DC-bus through a DC/AC inverter and a grid transformer.

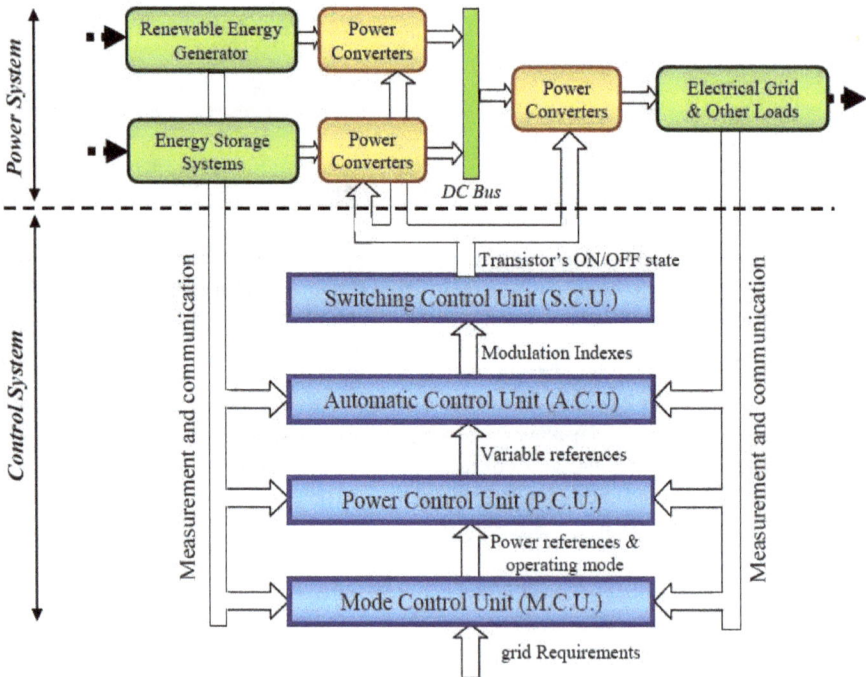

Figure 4. DC-coupled renewable and energy storage power conversion system (source [16])

The structure of the control chain involves four different levels explained below:

1. *The Switching Control Unit (SCU)* is the active interface between the power converters and the control units of higher level. The SCU opto-couplers and the modulation modules generate the power converters' transistors ON/OFF signal.

2. *The Automatic Control Unit (ACU).* The ACU's control algorithms calculate the modulation indexes of each power converter in accordance with the reference values set by the PCU.

3. *The Power Control Unit (PCU)* performs the instantaneous power balancing of the entire hybrid power system. The PCU's algorithm calculates the values of the parameters for the regulation of the voltages and the currents in accordance with the power reference values set by the MCU.

4. *The Mode Control Unit (MCU)* supervises the entire power system. The MCU sets the operating mode and the power references by taking into account the grid requirements from the network operator and the state of the power system. The state of the power system may include: the renewable energy generation capacity that is a function of the local climate, the SoC of the energy storage system and the grid operating condition at the injection point (voltage and frequency measurements).

The extent of the functions to be performed by the control chain and the level of complexity depend on the considered application and more specifically, on the typology of the storage system. Including for example, an imperative supervision at the level of every element in the case of Li-ion. In every case, this real time supervision of the storage unit is useful to the diagnosis in case of default or for the anticipation of needs in maintenance.

8. Sizing the storage capacity for the management of wind power induced congestion.

This sub-section discusses the sizing of transmission line capacity in the context of the transport of wind-originated electricity. A regional network is sized according to the maximal power that could transit through it. To prevent power congestion, it is necessary either, to size the network infrastructures to match the maximal expected power production or to limit the level of power that could transit through the transmission lines.

This last strategy calls for judicious arbitration between the loss of income due to the power limitation and the associated infrastructure cost reduction. A refine analysis of the production on a given site allows the developer to size sensibly the power level limit above which excess production will be rejected. To reduce energy waste, the excess of energy could be stored and re-injected later, during periods of low production. The aim here is mainly to avoid congestion while reducing the costs linked to infrastructures reinforcement and maximizing the energy output of the installation.

As the load factor of wind energy is about 25 %, the ratio of the cost for network strengthening over the aggregated energy produced for wind (or solar) energy is higher to that of the other (non intermittent) sources of energy.

In [17], the authors proposed an in-depth analysis of the wind speed variations and the related electrical power variations, based on a probabilistic approach that gives, for a specified wind speed range, the distribution of the expected wind farm power output. This method is used here to evaluate with more precision, the load factor of wind energy, in order to size the level of power curtailing and to estimate the required storage capacity to avoid energy waste.

Concerning the wind speed to electrical power conversion, many studies have investigated wind turbines response to wind variations. Figures 5 shows the plots of a two-months (61 days) sequence of wind speed and associated wind farm power output. Under the influence of meteorological conditions wind speed fluctuates over time. These variations occur on different time scales: from seconds to years. The response of a wind turbine, in term of power output variations, depends on the wind turbine technology [18,19]. Some smoothing effect can also be obtained due to the turbine inertia and size. For a group of turbines, further smoothing can be expected due to the spatial distribution of the turbine within the area. For large area, wind energy overall variability can be much lower than the variability of a single wind turbine since the meteorological fluctuations do not affect each wind cluster at the same time.

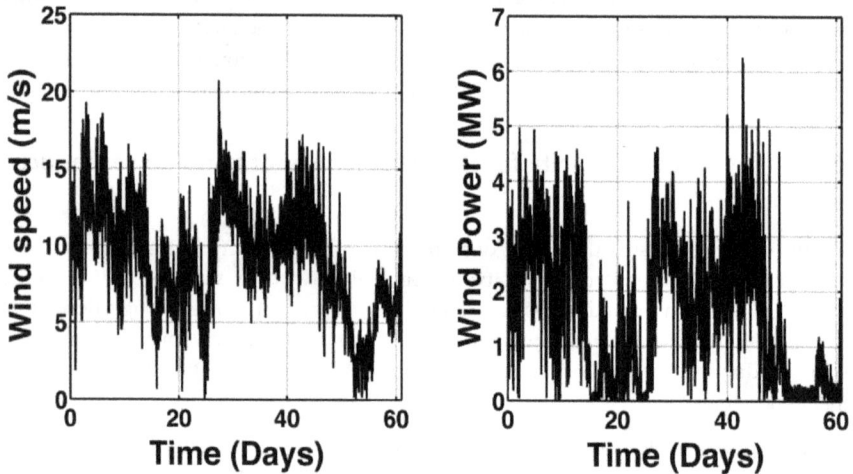

Figure 5. A two-month sample of wind speed and wind power.

To further investigate the wind speed – electrical power relationship, various tools from the Fourier analysis can be used. The plots in Figure 6 are the power spectral density of a sample of wind speed and electrical power produced by a cluster of wind turbines. The spectra are plotted for the frequency range between 1.3×10^{-4} Hz and 2×10^{-2} Hz using the periodogram method. Within this frequency range, no frequency peak can be observed from the two spectra. Moreover, we calculated the magnitude square coherence between the wind velocity V_{wind} and the power signal $P_{cluster}$. The magnitude square coherence is defined by:

$$C_{VP}(f) = \frac{|PowerSpect_{VP}(f)|^2}{PowerSpect_{VV}(f)PpowerSpect_{PP}(f)} \tag{1}$$

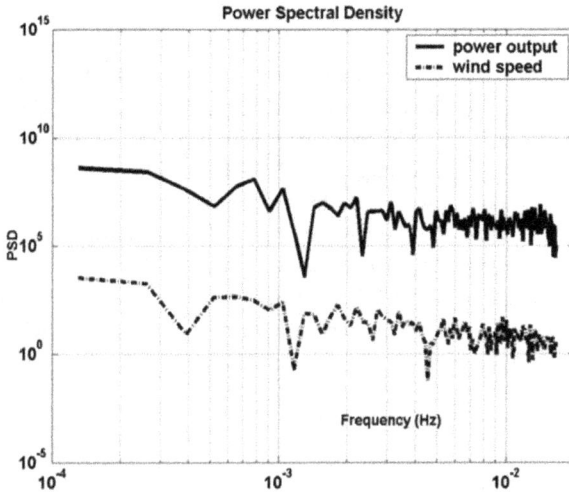

Figure 6. Wind power and wind speed power spectral densities

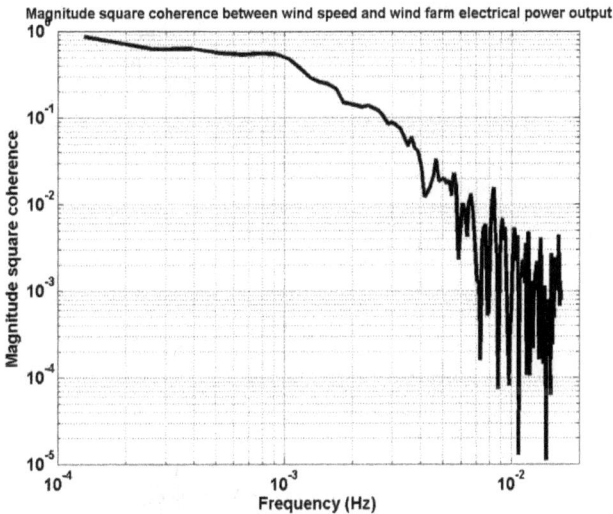

Figure 7. Magnitude square coherence between wind speed and electrical power

It is equal to the cross spectrum of V_{wind} and $P_{cluster}$ divided by the product of the power spectra of V_{wind} and $P_{cluster}$. This quotient is a real number between 0 and 1 that measures the correlation between V_{wind} and $P_{cluster}$ at the frequency f. The plot of Figure 7 shows the coherence C_{VP} dropping below 0.2 for frequency above $(2.10^{-3}$ Hz). This indicates a weak coherence between wind speed and power fluctuations for time scales lower than 8 minutes (frequencies larger than 2.10^{-3} Hz).

The plot of the gain of the wind speed– cluster's electrical power output transfer function completes the plot of the magnitude square coherence. The transfer function is the quotient of the cross spectrum of $P_{cluster}$ and the power spectrum of V_{wind}. The gain of this transfer function, plotted in Figure 8, drops by more than 10 dB below its maximum value (obtained for large time scales) for frequencies larger than 2.10^{-3} Hz. The phase of the transfer function, plotted in Figure 9 remains close to zero for frequencies below 2.10^{-3} Hz. From both the magnitude square coherence and the gain and phase of the transfer function, we deduce that, on large time scales, power variations are well correlated with wind speed variations. The large time scales wind variations affect the whole wind farm almost simultaneously. On the other hand, the short time scales power variations are not correlated with the variations of the wind speed.

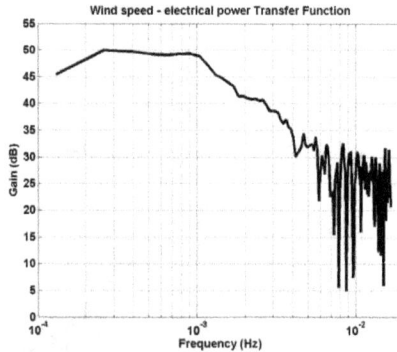

Figure 8. Gain of the wind speed - electrical power transfer function

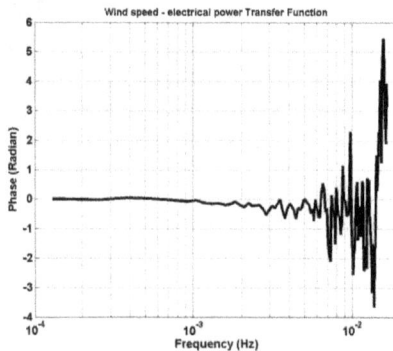

Figure 9. Phase of the wind speed - electrical power transfer function

Another way to represent the wind to power mapping is to consider the conditional probability that the observed mean electrical power $\overline{P}_{Cluster}$ takes on a value less than or equal to a given threshold P, given a mean wind speed $\overline{V}_{Wind}$. This conditional probability is the conditional cumulative distribution function:

$$F(P|\overline{V}_{Wind}) = prob(\overline{P}_{Cluster} \leq P|\overline{V}_{Wind}) \qquad (2)$$

The value of the threshold P is expressed as a fraction of the transmission line capacity. Initially, we choose to set the transmission line capacity to $PMAX_{2\,months} = 6.3\ MW$, the maximum observed wind farm power output over the two months test period. This value is also close to the wind farm installed capacity. In figure 10, the iso-percentages of the function $F(P|\overline{V}_{Wind})$ are plotted in the plane $(\overline{V}_{Wind}, \%P_{Line\ Capacity})$ for an averaging time equals to 5 minutes.

The conditional cumulative distribution function gives the probability that the power produced by a wind farm, exceeds a given power threshold. It is now possible to precisely set the level above which, the wind power will be stored or dumped rather than sent through the transmission line.

In the example illustrated in Figure 10, the probability that the instantaneous power exceed 55% of the initial transmission line capacity is 10%. Thus, setting the level of the curtailment to 55% of $PMAX_{2\,months}$, ensure that 90% of the production will not be rejected. The remaining 10% can be stored and re-injected whenever the transmission capacity is not fully used.

In the presented case, the excess electrical energy, produced when the wind farm power exceeds 55% of the transmission line capacity amounts to $Eexcess_{2\,months} = 26.8\ MWh$, which represents 1.2% of the total produced energy $ETOTAL_{2\,months} = 2.26\ GWh$ for the two months period.

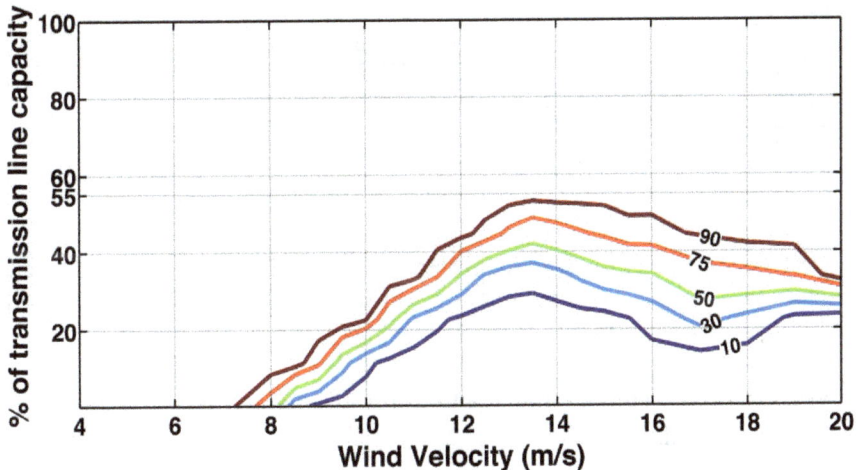

Figure 10. Iso-percentages of the conditional cumulative distribution function of the wind farm output power as a function of wind velocity. The power threshold levels are expressed as percentage of the transmission line capacity which, here, equals $PMAX_{2\,months}$.

The objective is to optimize the use of a transmission line by reducing its capacity while avoiding line saturation. Therefore, we set the transmission line capacity to 55% of $PMAX_{2\,months}$. A generic energy storage system is used to store all or part of the excess energy. We tested different level of storage capacity. For the tests, we set the storage system efficiency to 75% and limit the depth of discharge (DoD) to 80%.

If, during the test, the state of charge (SoC) of the storage system reaches 100%, the subsequent excess energy will be rejected as long as the SoC remains at 100%. To limit this event, as long as the SoC remains above 20%, the stored energy will be discharged whenever the instantaneous wind farm output power drops below 50% of $PMAX_{2\,months}$. This contributes to keep the storage SoC below 100% as often as possible, allowing more excess energy to be stored and redirected through the grid.

Table 3 gives the percentage of energy effectively sent through the transmission line in respect with $ETOTAL_{2\,months}$, as a function of the storage system capacity. These elements added with a cost analysis of power transmission lines, enable calculations to investigate the profitability of such a power curtailment scheme.

The plots of Figure 11 show the evolution of the storage system SoC and its temporal derivative (storage's charge and discharge power) during a test conducted with a capacity of storage equals 1.33 MWh, which represents 5% of $Eexcess_{2\,months}$. These results demonstrate the benefit of this approach. With a relatively limited storage capacity, it is possible to exploit 99.92% of the energy produced by the wind farm while limiting the transmission line capacity at 55% of the maximum wind power observed during the two-months test period.

Storage system capacity		Energy sent through the transmission line	
($\%Eexcess_{2\,months}$)	MWh	($\%ETOTAL_{2\,months}$)	GWh
0% (no storage)	0	98,81%	2,2353
1%	0,269	99,40%	2,2486
5%	1,34	99,92%	2,2603
10%	2,68	99,98%	2,2617

Table 3. Percentage of energy sent through the transmission line as a function of the storage system capacity

The ideal storage system characteristics could be deduced from this analysis. For the presented case, the adequate storage system should have a rated power of 2,5 MW, a rated energy of 1,5 MWh and efficiency around 75%.

9. Dynamic energy storage management for wind electricity injection into electrical grids

In the application presented above, the power variations induced by the wind's fluctuations are not anticipated for. We propose here to consider the potential benefit of wind energy production forecast to improve the reliability of wind-originated electricity.

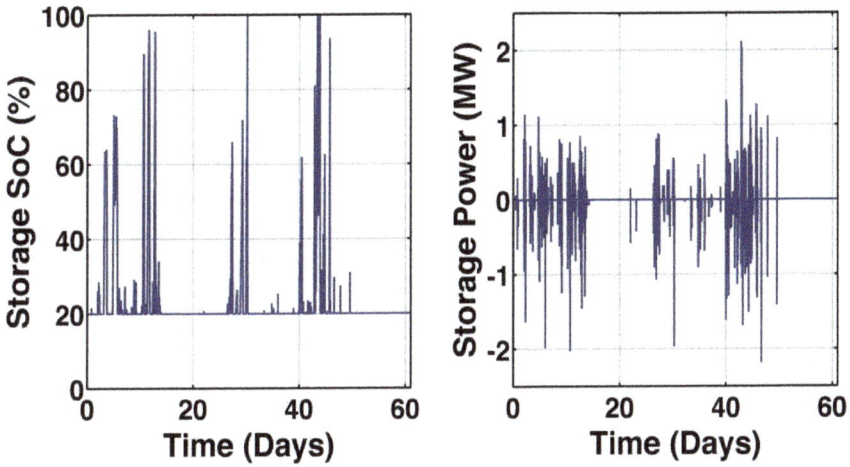

Figure 11. Time evolution of the storage system SoC and power during the test.

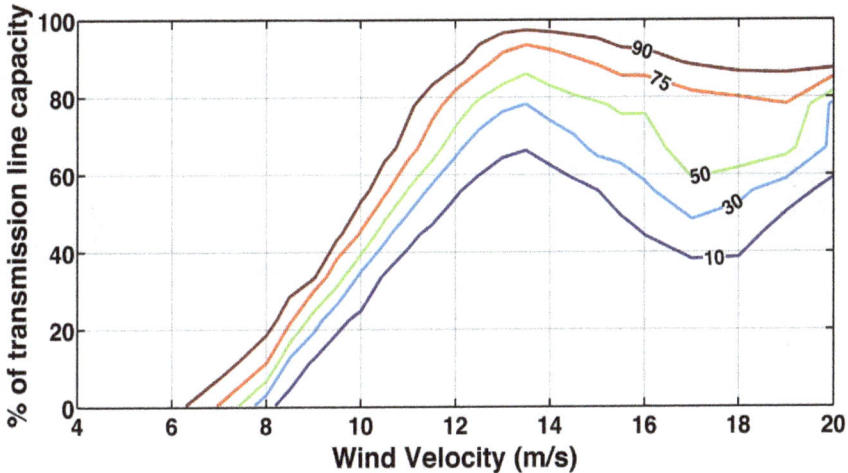

Figure 12. Iso-percentages of the conditional cumulative distribution function of the wind farm output power as a function of wind velocity. The power threshold levels are expressed as percentage of the curtailed transmission line capacity which, here, equals 55% $PMAX_{2\,months}$.

Efficient forecasting scheme that includes some information on the likelihood of the forecast and based on a better knowledge of the wind variations characteristics along with their influence on power output variation is of key importance for the optimal integration of wind energy in power system. In [20], the author has developed a short-term wind energy prediction scheme that uses artificial neural networks and adaptive learning procedures based on Bayesian approach and Gaussian approximation. We propose to illustrate how such a prediction tool combined with an energy storage facility could help to smooth the wind power variation and improve the consistency of wind electricity injected into utility network.

Energy storage technologies could be valuable to the development of wind and PV electricity generation. The main objectives of an energy storage management scheme for the sake of wind or photovoltaic electricity productions are:

- To guarantee energy on-demand application for stand-alone renewable generation
- To inform, in advance, about the dispersion of the incoming production.
- To guarantee the power level of the injected electricity by limiting the variability of the production with the help of energy storage systems.
- To guarantee optimized and safe exploitation of the energy storage system.

We seek, in this section, to illustrate the advantage of using energy or power prediction to develop an energy storage management scheme aimed at reducing the uncertainty on the incoming wind power that will be injected into the grid while maintaining a reasonable storage capacity.

10. The energy prediction scheme

The proposed method is based on the very short-term prediction scheme of the wind energy outlined in Figure 13. At any given time t, a neural network (referred hereafter as the predictor) gives an estimation of future wind energy values $E_{Wind}(t + \tau)$, with τ the horizon of prediction. The energy function $E_{Wind}(t)$ gives the energy produced by a group of wind turbine during the 30 minutes period preceding time t.

The input of the predictor is made of n_{inp} recent samples of wind power values, $[P_{Wind}(t - T_{short\_mem}), \cdots, P_{Wind}(t)]$. These n_{inp} components of the input are chosen from the values measured during the time interval preceding t; $[t - T_{short\_mem}, t]$.

In order to adjust the predictor's parameters continuously, the optimization of the parameters is performed every time a prediction is needed, during an adaptive training session. Throughout this training session, the predictor learns to emulate the desired input-output relationship using a collection of recorded inputs-outputs called the training set. Extensive description of neural network's training algorithm can be found in [21].

The predictor's training set is made of inputs $[P_{Wind}(t_i - T_{short\_mem}), \cdots, P_{Wind}(t_i)]$ along with their associated prediction targets $E_{Wind}(t_i + \tau)$, with $t_i \in [t - T_{long\_mem} - \tau, t]$. Once the neural network's optimization is judged successful, the trained predictor is used to calculate the prediction $E_{Wind}(t + \tau)$ using the most recent input vector available at time t.

The Figure 14 illustrates prediction scheme's performance. Three values of the horizon of prediction were tested; $\tau = 15\ minutes$, $\tau = 30\ minutes$ and $\tau = 45\ minutes$. The overall performance is good when compared with the benchmark persistent based prediction model. However, the level of improvement with regards to the persistent prediction model tends to evolve unfavorably with the horizon of prediction. An evaluation of this adaptive prediction scheme is reported in [22].

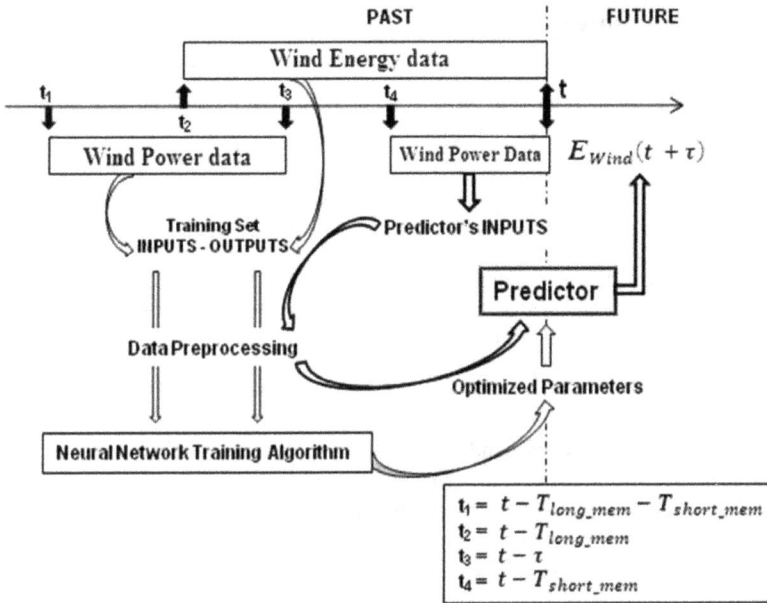

Figure 13. The adaptive wind energy prediction scheme

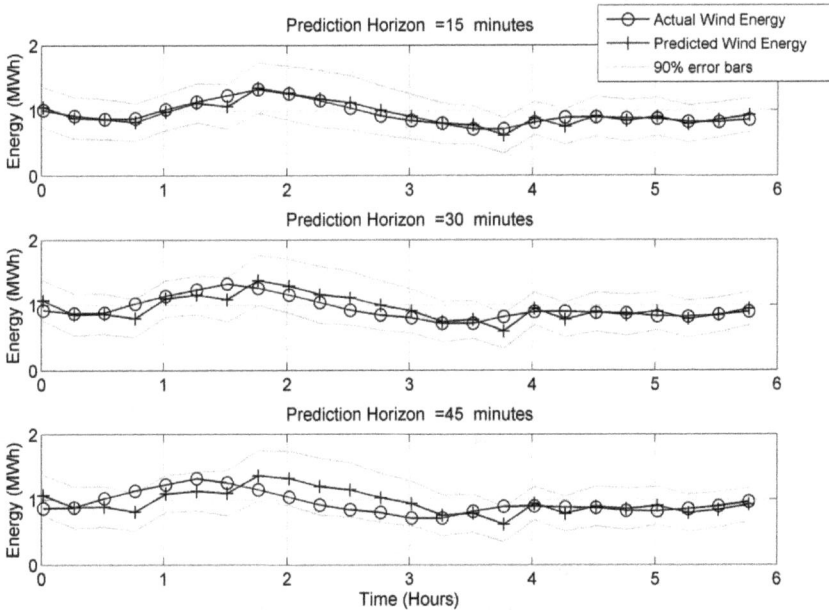

Figure 14. Actual and predicted wind energy

11. The dynamic power-scheduling algorithm

Once the predictor anticipated the energy production of the wind farms ahead in time. A scheduling algorithm is able to calculate P_{Sched}. In this chapter, we report the result of tests were P_{Sched} estimates the power level at which electrical energy will be delivered to the grid, throughout the future time interval $[t + 15 minutes, t + 45\ minutes]$. Therefore, the grid operator will have, 15 $minutes$ ahead in time, valuable information about the availability of wind energy. The Figure 15 gives a sketch of the power-scheduling plan.

To comply with the power assignment P_{Sched}, the electrical energy will come from the wind farm, supported if needed, by an energy storage system. The value of the scheduled power P_{Sched} is calculated under the following constraints:

1. Energy must be delivered to the grid at power levels that remain within the +/- 5% interval around P_{Sched}.
2. If the actual instantaneous wind power is above the top of this interval, the energy excess is sent to the energy storage. The algorithm takes into account the charging efficiency of the storage system. We set the charging efficiency to 85%.
3. If the actual instantaneous wind power falls below the bottom of this interval, energy discharged from the energy storage compensates the energy shortage, as long as the energy in the storage system is larger than a pre-set level to account for the lower DoD limit. The algorithm also takes into account the storage system's discharging efficiency. We set the discharging efficiency to 85%.

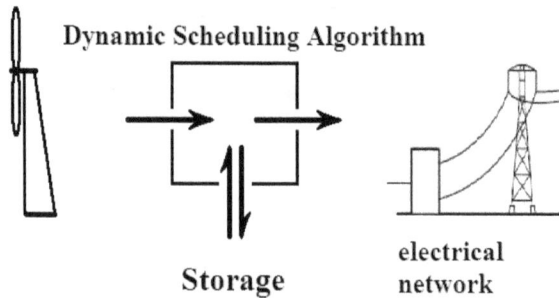

Figure 15. Dynamic power-scheduling

At any given time t, the scheduling algorithm evaluates P_{sched} from the predicted energy, the current storage reserve level and the previously observed deviation between the actual and predicted energy.

For the calculation of P_{sched}, the algorithm takes into account the stored energy only at the level of 50 % in case the previously observed deviation between the actual and predicted energy is positive and 0 % if the deviation is negative. The objective is to ensure there is some energy left in the storage system to compensate the power shortage and possible prediction errors at a later time.

The Figures 16 and 17 below, show the injected and scheduled power plots superimposed to the actual wind power plot. This nine-hours demonstration started with zero energy in the energy storage system. The dynamic scheduling algorithm manages to maintain the injected power within the +/- 5% of P_{Sched} interval while maintaining the energy reserve strictly above zero as shown by the plot of Figure 17. Notice that the required energy capacity remains relatively low, around 3% (less than 500 kWh) of the total energy supplied by the wind farms during the whole duration of the demonstration (about 15 MWh supplied in 9 h). The required charge or discharge power of the storage system is estimated at 500 KW.

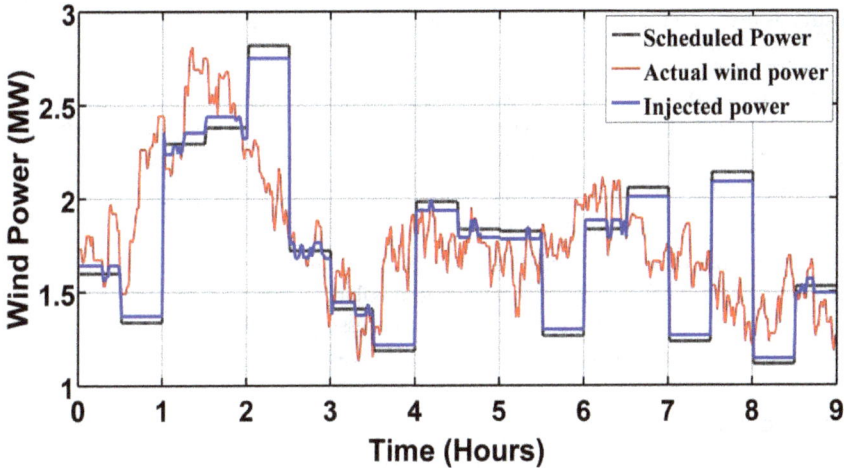

Figure 16. Scheduled and injected electrical power superimposed to the actual wind power

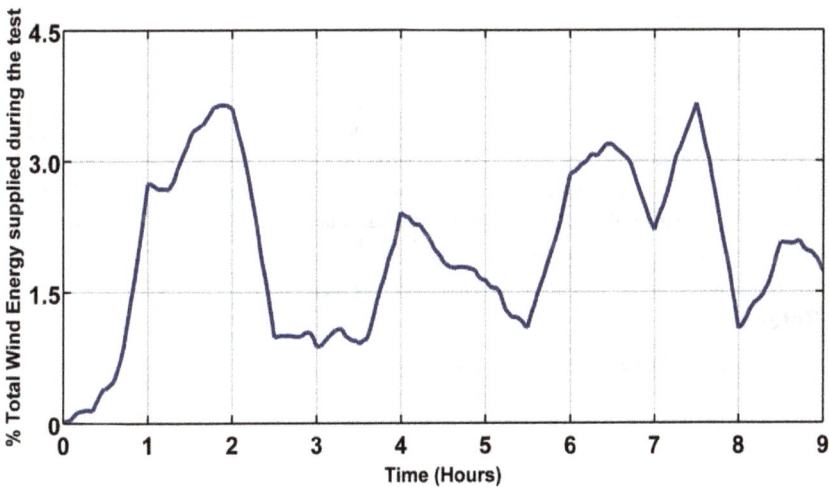

Figure 17. Evolution of the energy reserve of the storage system

The interest of the proposed method is that he permits to guarantee the power at which energy will be sent to the grid by automatically dispatching the power flows between the wind plants and the energy storage facility.

12. Conclusion

Energy storage technologies are identified as key elements for the development of electricity generation exploiting renewable energy sources. In this chapter, we have illustrated, through two simulations cases, how they could contribute to remove the technical constraints that limit the contribution of renewables energy sources into electrical networks.

The sector of energy storage technologies sees new solutions to emerge every day. The arrival on the market of electric vehicles contributes largely to this profusion of innovation. Our objective was to show that a dynamic approach of the management of the charge and the discharge at the level of the energy storage system insures a good quality of service (energy efficient power curtailment, power smoothing and uncertainty reduction) with a reduced storage capacity.

We illustrated our proposed approaches by dealing with the case of wind energy. However, the proposed methods are immediately transposable to the case of the photovoltaic electricity production given preliminary preprocessing of the solar insolation and photovoltaic-production data.

Author details

Ruddy Blonbou*, Stéphanie Monjoly and Jean-Louis Bernard
Geosciences and Energy Research Laboratory, Université des Antilles et de la Guyane, Guadeloupe, France

Acknowledgement

This work was performed, in part in the frame of the ERA/Alizeole Project (Contract No. 1/1.4/32315) and funded by the European Commission. Acknowledgments are due to, the Vergnet Caraïbes and Aerowatt companies, for providing technical help and EDF for providing the power data for this work. Special thanks to the Regional Council of Guadeloupe for its financial support.

13. References

[1] Hall P J, Bain E J (2008) Energy-storage technologies and electricity generation. Energy Policy 36: 4352–4355

---

* Corresponding Author

[2] Delille G, (2007) Contribution du stockage à la gestion avancée des systèmes électriques. PhD thesis, Université Lille Nord-de-France.

[3] Beaudin M, Zareipour H, Schellenberglabe A, Rosehart W (2010) Energy storage for mitigating the variability of renewable electricity sources: An updated review. Energy for Sustainable Development 14:302–314

[4] Díaz-González F, Sumper A, Gomis-Bellmunt O, Villafáfila-Robles R (2012) A review of energy storage technologies for wind power applications. Renewable and Sustainable Energy Reviews 16:2154–2171

[5] Rahman F, Rehman S, Abdul-Majeed M A (2012) Overview of energy storage systems for storing electricity from renewable energy sources in Saudi Arabia. Renewable and Sustainable Energy Reviews 16:274–283

[6] Ibrahim H, Llinca A, Perron J (2008) Energy storage systems - Characteristics and comparisons. Renewable and Sustainable Energy Reviews 12: 1221–1250

[7] Rena L, Tanga Y, Shia J, Doua J, Zhoub S, Jinb T (2012) Techno-economic evaluation of hybrid energy storage technologies for a solar–wind generation system. http://dx.doi.org/10.1016/j.physc.2012.02.048

[8] First Hydro Company website, http://www.fhc.co.uk/pumped storage.htm

[9] Connolly D, Lund H, Mathiesen B V, Pican E, Leahy M (2012) The technical and economic implications of integrating fluctuating renewable energy using energy storage. Renewable Energy (43): 47-60

[10] Antonucci P L, Antonucci V (2011) Electrochemical Energy Storage. In: Carbone R, editor. Energy Storage in the Emerging era of Smart Grids. Rijeka: InTech. pp. 3-20.

[11] Fergus J W (2010) Recent developments in cathode materials for lithium ion batteries. Journal of Power Sources 195: 939–954

[12] Blanc C, Rufer A, (2010) Understanding the Vanadium Redox Flow Batteries. In: Nathwani J, Ng A, editor. Paths to sustainable energy. Rijeka: InTech. pp. 333-358.

[13] Burke A. (2000) Ultracapacitors: why, how, and where is the technology. Journal of Power Sources; 91(1) : 37-50.

[14] Sharma P, Bhati T S (2010) A review on electrochemical double layer capacitors. Energy conversion and management 51: 2901-2912.

[15] Fang X, Kutkut N, Shen J, Batarseh A (2011) Analysis of generalized parallel-series ultracapacitor shift circuits for energy storage systems. Renewable Energy 36(10): 2599-2604.

[16] Zhou T, (2009) Control and energy management of a hybrid active wind generator including energy storage system. PhD Thesis, Ecole Centrale de Lille.

[17] Calif R, Blonbou R (2008) Analysis of the power output of a wind turbine cluster in the Guadeloupean archipelago. The International Scientific Journal for Alternative Energy and Ecology. ISJAEE" # 5(6).

[18] Baroudi J A, Dinavahi V, Knight A M, (2007) A review of power converter topologies for wind generators. Renewable Energy 32: 2369–2385.

[19] Joselin Herberta G M, Iniyanb S, Sreevalsan E, Rajapandian S, (2007) A review of wind energy technologies, Renewable and Sustainable Energy Reviews 11: 1117–1145.

[20] Blonbou R (2011) Very Short Term Wind Power Forecasting with Neural Networks and Adaptive Bayesian learning. Renewable Energy 36(3): 1118-1124.

[21] Bishop, C M (1995) Neural networks for pattern recognition. Oxford University Press

[22] Blonbou R, MONJOLY S, DORVILLE J F (2011) An adaptive short-term prediction scheme for wind power management. Energy Conversion and Management 52(6): 2412-2416.

Chapter 11

Dynamic Modelling of Advanced Battery Energy Storage System for Grid-Tied AC Microgrid Applications

Antonio Ernesto Sarasua,
Marcelo Gustavo Molina and Pedro Enrique Mercado

Additional information is available at the end of the chapter

1. Introduction

In the last decade, power generation technology innovations and a changing economic, financial, and regulatory environment of the power markets have resulted in a renewed interest in on-site small-scale electricity generation, also called distributed, dispersed or decentralized generation (DG) (Abdollahi Sofla & Gharehpetian, 2011). Other major factors that have contributed to this evolution are the constraints on the construction of new transmission lines, the increased customer demand for highly reliable electricity and concerns about climate change (Guerrero et al, 2010). Along with DG, local storage directly coupled to the grid (aka distributed energy storage or DES) is also assuming a major role for balancing supply and demand, as was done in the early days of the power industry. All these distributed energy resources (DERs), i.e. DG and DES, are presently increasing their penetration in developed countries as a means to produce in-situ highly reliable and good quality electrical power (Kroposki et al, 2008).

Incorporating advanced technologies, sophisticated control strategies and integrated digital communications into the existing electricity grid results in Smart Grids (SGs), which are presently seen as the energy infrastructure of the future intelligent cities (Wissner, 2011). Smart grids allow delivering electricity to consumers using two-way (full-duplex) digital technology that enable the efficient management of consumers and the efficient use of the grid to identify and correct supply-demand imbalances. Smartness in integrated energy systems (IESs) which are called microgrids (MG) refers to the ability to control and manage energy consumption and production in the distribution level. In such IES systems, the grid-interactive AC microgrid is a novel network structure that allows obtaining the better use of

DERs by operating a cluster of loads, DG and DES as a single controllable system with predictable generation and demand that provides both power and heat to its local area by using advanced equipments and control methods (Hatziargyriou et al, 2007). This grid, which usually operates connected to the main power network but can be autonomously isolated (island operation) during an unacceptable power quality condition, is a new concept developed to cope with the integration of renewable energy sources (RESs) (Katiraei et al, 2008).

Grid connection of RESs, such as wind and solar (photovoltaic and thermal), is becoming today an important form of DG (Mathiesen et al, 2011). The penetration of these DG units into microgrids is growing rapidly, enabling reaching high percentage of the installed generating capacity. However, the fluctuating and intermittent nature of this renewable generation causes variations of power flow that can significantly affect the operation of the electrical grid (Tiwaria et al, 2011; Kanekoa, 2011). This situation can lead to severe problems that dramatically jeopardize the microgrid security, such as system frequency oscillations, and/or violations of power lines capability margin, among others (Serban & Marinescu 2008). This condition is worsened by the low inertia present in the microgrid; thus requiring having available sufficient fast-acting spinning reserve, which is activated through the MG primary frequency control (Vachirasricirikul & Ngamroo, 2011).

To overcome these problems, DES systems based on emerging technologies, such as advanced battery energy storage systems (ABESSs), arise as a potential alternative in order to balance any instantaneous mismatch between generation and load in the microgrid (Molina, 2011). With proper controllers, these advanced DESs are capable of supplying the microgrid with both active and reactive power simultaneously and very fast, and thus are able to provide the required security level. The most important advantages of these advanced DESs devices include: high power and energy density with outstanding conversion efficiency, and fast and independent power response in four quadrants (Molina & Zobaa, 2011).

Much work has been done, especially over the last decades, to assess the overall benefits of incorporating energy storage systems into power systems (Hewitt, 2012; Schroeder, 2011; Maharjan et al, 2011; Qian, 2011). However, much less has been done particularly on advanced distributed energy storage and its utilization in emerging electrical microgrid, although major benefits apply (Molina, 2011; Vazquez et al, 2010). Moreover, no studies have been conducted regarding a comparative analysis of the modeling and controlling of these modern DES technologies and its dynamic response in promising grid-interactive AC microgrids applications.

In this chapter, a unique assessment of the dynamic performance of novel BESS technologies for the stabilization of the power flow of emerging grid-interactive AC microgrids with RESs is presented. Generally, electrochemical batteries include the classic and well-known lead-acid type as well as the modern advanced battery energy storage systems. ABESSs comprise new alkaline batteries, nickel chemistry (nickel-metal hydride–NiMH, and nickel-cadmium–NiCd), lithium chemistry (lithium-ion–Li-Ion, and lithium–polymer-Li-po), and sodium chemistry (sodium-sulfur–NaS, and sodium-salt–NaNiCl) (Molina & Zobaa, 2011; Molina & Mercado, 2006; Iba et al. 2008). In this work, of the various advanced BESSs nowadays existing, the foremost ones are evaluated. In this sense, the design and implementation of the proposed ABESSs systems are described, including the power conditioning system (PCS) used as interface with the grid. Moreover, the document provides a comprehensive analysis of both the dynamic modeling and the control design of the leading ABESSs aiming at enhancing the operation security of the AC microgrid in both grid-independent (autonomous island) and grid-interactive (connected) modes.

Section 2 details the general considerations for selecting batteries. Section 3 summarizes the key features of selected batteries. Section 4 defines the parameters to be considered in each type of battery and reviews some of the existing models for batteries. Finally, section 5 proposes a general model of batteries developed in MATLAB/Simulink and implemented in a test power system.

2. Selection criteria of BESS technologies

Unlike other commodities, there are not significant stocks or inventories of electricity to mitigate differences in supply and demand. Electricity must be produced at the level of demand at any given moment, and demand changes continually. Without stored electricity to call on, electric power system operators must increase or decrease generation to meet the changing demand in order to maintain acceptable levels of power quality (PQ) and reliability.

Electricity markets are structured around this reality. Presently, generating capacity is set aside as reserve capacity every hour of every day to provide a buffer against fluctuations in demand. In this way, if the reserve capacity is needed, it can be dispatched or sent to the grid without delay. There are costs, at times considerable, for requiring the availability of generating capacity to provide reserves and regulation of power quality. However, economic storage of electricity could decrease or even eliminate the need for generating capacity to fill that role.

For the selection of a specific energy storage technology in order to participate in the power reserve of a grid-tied AC microgrid, storage capacity must be defined in terms of the time that the nominal energy capacity is intended to cover the load at rated power. All storage technologies are designed to respond to changes in the demand for electricity, but on varying timescales. Thus, various types of existing storage technologies are adapted for different uses. Then, the power reserve range can be divided into two kinds:

- Power quality management (shorter timescales): Demand fluctuations on shorter timescales—sub-hourly, from a few minutes down to fractions of a second—require rapidly-responding technologies which are often of smaller capacity. Responding to these short-timescale fluctuations keeps the voltage and frequency characteristics of the grid electricity consistent within narrow bounds, providing an expected level of power quality. PQ is an important attribute of microgrid electricity, as poor quality electricity—momentary spikes, surges, sags (dips), or severe contingencies like outages—can harm electronic devices.

- Energy management (long timescales). Daily, weekly, and seasonal variations in electricity demand are fairly predictable. Higher-capacity technologies capable of outputting electricity for extended periods of time (up to some hours) moderate the extremes of demand over these longer timescales. These technologies aid in energy management, reducing the need for generating capacity as well as the ongoing expenses of operating that capacity. This is the case of serious failures of generation or disconnection of the MG from bulk power system. Variations in demand are accompanied by price changes, which lead to arbitrage opportunities, where storage operators can buy power when prices are low (hours of low consumption) and sell when prices are high (peak hours).

As a result, the permanent participation of the storage system in the AC microgrid is required for both situations: to control severe contingencies and to balance the demand and continuous changes of minor contingencies. In the first case the level of storage system performance is lower than in the second, but the power requirements and dynamic response are significantly higher. Conversely, in the case of energy management events it requires more energy, but less rapidly.

Based on the previous considerations and taking into account the considered grid-tied AC microgrid applications the following criteria for selection of battery energy storage systems are proposed:

- The possibility of build medium-scale units (MW), according to the size of the MG.
- Commercially available technology with applications in electric power systems. It is required that the technology has been proven by industry to ensure a real solution.
- High reliability. It is necessary that the equipment incorporated into the MG ensures high availability when required.
- Minimum requirements to allow the location of the storage systems next to the loads as a distributed energy storage device.
- Competitive costs (Installation and Operation-Maintenance). Storage devices should hold costs competitive with the benefit incorporated into the operation of the MGs.
- Long lifetime, exceeding 2000 cycles. Studies on the efficient use of new storage devices show that it takes more than 2000 cycles of charge/discharge to consider possible implementation in MGs.
- High electrical efficiency, defined as the ratio of the energy used to fully charge the storage device and the maximum extractable energy from it. This requirement requires maximum use of the device electrical storage, which will improve operating costs.
- Minimal environmental impact.
- Discharge time (bridging time) greater than a minute. According to the size and operation of the microgrid can be extended to several minutes or hours.
- Very short time response (less than a second), to improve the response of other alternatives.

- High discharge rate, which allows quickly cover large imbalances of power. This action will significantly improve regulation and reduce the impact of any disturbances in the main power system.
- High re-charge rate, to quickly restore the lost reserve from the BESS units and to allow quickly absorb large excesses of energy. For this particular case, it must maintain a state of optimal storage load to ensure a minimum level of storage when required by the control system.

These general guidelines serve as a basis for the selection of the storage device, but it should be taken into consideration that the final evaluation of the BESS device should be carried out conjointly with the power system with it is to interact and taking into account the policies of control-economy established.

3. Overview of BESS technologies

The term Battery contains the classic and well-known lead-acid (Pb-acid) type as well as the redox flow types batteries, and also include the so called advanced battery energy storage systems (ABESSs). ABESSs comprise new alkaline batteries, nickel chemistry (nickel metal hydride–NiMH, and nickel cadmium–NiCd), lithium chemistry (lithium polymer–Li-po, and lithium-ion–Li-Ion), and sodium chemistry (sodium sulfur–NaS, and sodium salt–NaNiCl). Based on the selection criteria previously described, the following batteries are studied:

- Lead-acid batteries
- Nickel cadmium and nickel metal hydride batteries
- Lithium ion and lithium polymer batteries
- Sodium sulfur batteries

3.1. Lead acid batteries

Each cell of a lead-acid battery comprises a positive electrode of lead dioxide and a negative electrode of sponge lead, separated by a micro-porous material and immersed in an aqueous sulphuric acid electrolyte. In flooded type batteries (with an aqueous sulphuric acid solution) during discharge, the lead dioxide on the positive electrode is reduced to lead oxide, which reacts with sulphuric acid to form lead sulphate; and the sponge lead on the negative electrode is oxidized to lead ions, that reacts with sulphuric acid to form lead sulphate. In this manner, electricity is generated and during charging this reaction is reversed. Valve regulated (VRLA) type uses the same basic electrochemical technology as flooded lead-acid batteries, except that these batteries are closed with a pressure regulating valve, so that they are sealed. In addition, the acid electrolyte is immobilized (Divya & Østergaard 2009).

Pb-acid batteries are the most commonly used batteries in various applications worldwide. They are within the category of less physical efficiency battery. They have also the lower energy densities and power per weight and volume (20 to 40 kWh/ton and 40 to 100 kWh/m³) (Nourai 2002). For this reason, Pb-acid batteries require more space and have greater weight than any other type of batteries. However, they have significant advantages that positions best suited for applications requiring high power and speed. The units are robust and secure, and allow extremely fast downloads, in periods of about 5 ms. The most important features are its low cost and high electrical efficiency. The cost of these batteries is in the order of $ 300 to $ 600 per kWh and performance can reach 90% (Chen 2009).

Another problem with these batteries is their relatively short lifetime measured in charge-discharge cycles, which reaches 500 cycles for the batteries most basic to 1000 cycles for the latest models (Chen 2009). The low amount of charge-discharge cycles is due to the high volumetric density of lead. Another major problem they have is the charging time of around three hours to the total load of batteries.

Despite these disadvantages, Pb-acid batteries have been used in many storage systems. Among them are the system built at the plant of 8.5MWh/1h BEWAG in Berlin, Germany, the system of 14 MWh/1.5h at the plant PREPA (Puerto Rico) and the greatest of all in Chinese (California, U.S.) of 10MWh/4h (Chen 2009). The earliest transportable battery system of lead-acid is located at the Phoenix distribution system is a multi-mode battery. The battery switches between power quality (2MW up to 15 s) and energy management (200 kW for 45min) mode (Divya & Østergaard 2009).

3.2. Nickel cadmium and nickel metal hydride batteries

Batteries of Ni-Cd type have a cadmium electrode (positive) and a nickel hydroxide (negative). The two electrodes are separated by nylon and potassium hydroxide. With sealed cells and half the weight of conventional lead acid batteries, these batteries have been used in a wide range of portable devices. Today, due to environmental problems and memory effect, Ni-Cd batteries are being replaced by Ni-MH or Li-Ion. Ni-Cd battery types are affected by the so-called memory effect. Memory effect, also known as battery effect or battery memory, is an effect that describes a specific situation in which Ni-Cd batteries gradually lose their maximum energy capacity if they are repeatedly recharged after being only partially discharged. The battery appears to remember the smaller capacity. The source of the effect is changes in the characteristics of the underused active materials of the cell.

Ni-Cd batteries have the advantage of a long life (up to 2000 charge-discharge cycles) and if they are charged and discharged properly maintain their properties to the end of its life. Each Ni-Cd cell can provide a voltage of 1.2 V and have a capacity between 0.5 and 2.3 Ah.

ABB and SAFT companies have developed a system based on Ni-Cd batteries for supporting the interconnected system of Alaska. The system is capable of delivering up to 40 MW during 15 minutes and is designed to act as a dumping reserve before activation of turbo-gas plants. So far, this battery system is the largest in the world.

Ni-HM batteries share several characteristics with Ni-Cd batteries. Each Ni-MH cell can also provide a voltage of 1.2 V and have a capacity between 0.8 and 2.7 Ah. Its energy density reaches 80 Wh/kg. They improve Ni-Cd batteries by changing the nickel hydroxide electrode and the other by a metal hydride alloy. Another advantage is that they have no memory effect. Their disadvantages are that they have less ability to release high peak power. They have also high self-discharge rate and are more susceptible to damage from overcharging.

3.3. Lithium ion, lithium polymer and lithium sulphur batteries

These batteries are built with alternating layers of electrodes, among which cyclically circulate lithium ions. The Li-Ion batteries have no memory effect and support recharge before being fully discharged. This is called the topping charge capacity. They have high energy density of the order of 115 Wh/kg.

The first lithium batteries were developed in 1979 and had a great attraction due to its high energy density, but low commercial development because of the risks of explosion. Subsequently, thanks to improvements developed by Sony with the Li-Ion batteries in 1990 were popularized in electronic equipment such as laptops or mobile phones. In addition, the flat design of the containers, the high energy density and the topping charge characteristic make them ideal for automotive applications.

This type of battery has a ratio of energy density three times greater than Pb-acid batteries. This difference is due to the characteristics of low atomic weight of lithium, about 30 times lighter than lead. In addition to having a higher voltage than lead-acid cells, this means fewer cells in series to achieve the desired voltage and lower manufacturing costs.

In addition to the strict selection of batteries with same voltage and internal resistance for connection in parallel or in series, it is also necessary that each battery cell should be charged to the same value as the other cells permanently. The voltage in the cell during discharge should not be less than 2.6V. The self-discharge of the lithium battery is approximately 5% per month. After a year unused, the capacity can be significantly reduced as well as the voltage level.

The big drawback with Lithium-ion type batteries is that they are not adaptable to permanent deep discharge duty cycles even in cases in which its nominal capacity is respected. Even more, this type of battery does not accept overloads.

The lithium polymer batteries are a variation of the Li-Ion. Their characteristics are very similar, but allow a higher energy density and a significantly higher discharge rate. The high initial costs are the main drawbacks. It is expected that once the mass production of Li-po is reached it will be priced lower than those of Li-Ion due to its simpler manufacturing.

Lithium sulphur batteries operate quite differently from Li-Ion batteries. The overall reaction between lithium and sulphur can be expressed as:

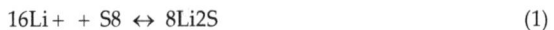

$$16Li+ \; + S8 \leftrightarrow 8Li2S \tag{1}$$

Based on the above complete reaction, sulphur cathode can offer a theoretical specific capacity of 1675 mAh/g and a theoretical energy density of 2600 Wh/Kg (Li et al 2010). Although investigated by many workers for several decades the practical development of the lithium/sulphur battery has been so far hindered by a series of shortcomings. A major issue is the high solubility in the liquid organic electrolyte of the polysulfides that forms as intermediates during both charge and discharge processes (Scrosati & Garche 2010)

The US the Department of Energy has sponsored a project by SAFT and SatCon Power Systems to design and construct two 100 kW/1 min Li-ion battery energy storage systems for use in providing power quality for grid connected micro-turbines (Naish 2008). Sanyo has developed a lithium-ion mega battery system with one of the world's largest capacities by installing approximately 1000 units of 1.6 kWh standard battery systems (a total of 1.5 MWh). This installation in the Kasai Green Energy Park, a massive testing site for large-scale, renewable power storage systems is located near Osaka (Japan). In the power storage building, economical late-night power is mainly used to charge batteries, which is then consumed during the day, while in the administration building, unconverted DC electricity from photovoltaic modules is the main source of power for charging batteries and direct consumption. The Standard Battery System for power storage is a storage battery unit with a capacity of approximately 1.6 kWh; containing 312 cylindrical lithium-ion battery cells often used in laptop PCs. Multiple systems can be connected to provide larger capacity. Batteries have a charge/discharge efficiency of 98% and are designed to last at least 10 years using the same rechargeable batteries (Panasonic).

3.4. Sodium sulphur batteries

Sodium sulphur batteries are one of the most favourable energy storage candidates for applications in electric power systems. They consist of an anode and a cathode of sodium and sulphur, respectively and a beta alumina ceramic material (beta-Al203) that is used as electrolyte and separator simultaneously. The tubular configuration of these batteries allows the change of state of the electrodes during charge and discharge cycles and minimizes the sealing area favouring the overall design of the cell (Wen 2008). Figure 1 shows the tubular design of each cell of sodium sulphur batteries.

The greatest advance in this type of battery has achieved very rapidly during the past two decades as a result of the collaboration between the Tokyo Power Company (TEPCO) and the NGK Insulators Company. TEPCO and NGK developed these batteries aiming at displacing the use of pumping stations.

Sodium sulphur batteries, usually work at temperatures between 300 and 350°C. At these temperatures, both sodium and sulphur and the reaction products are in liquid form, which facilitates the high reactivity of the electrodes. In this characteristic lies the high power density and energy of these batteries, nearly three times the density of lead acid batteries. They are environmentally safe because of the seal system with which they are constructed, thus not allowing any emissions during operation. Additionally over 99% of the battery materials can be recycled. They have a high efficiency in charge and discharge and a lifespan of approximately 15 years. The cells also have high efficiency (around 89%) and minimal degradation, which contributes to the life cycle, much larger than other cells (Baxter 2005). This type of battery has no self-discharge problems if they are kept at nominal operating temperature, which leads to having a high efficiency. For this purpose, the built containers have embedded heaters capable of maintaining the temperature with low energy consumption.

One of the most important characteristics of the sodium batteries is their ability to deliver power pulses of up to five times of its rated capacity over a period of time up to 30 seconds continuously. This is the fundamental reason because these batteries are considered economically viable for both power quality and energy managements applications. The pulse power capability is also available even if the unit is currently in the middle of a discharge process (Nourai 2002).The module of sodium batteries offered by TEPCO/NGK for power quality events have a nominal capacity of 50 kW, but the module can discharge up to 250 kW for 30 seconds or more, and comply with lower power levels for longer periods of time. Figure 2 shows the power vs. pulse duration of the discharge of a standard module with a capacity of 50 kW nominal power (Bito 2005).

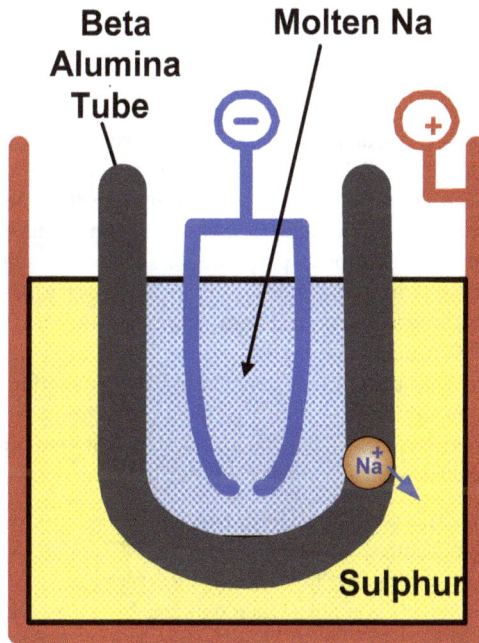

Figure 1. Schematic representation of a sodium-sulphur cell

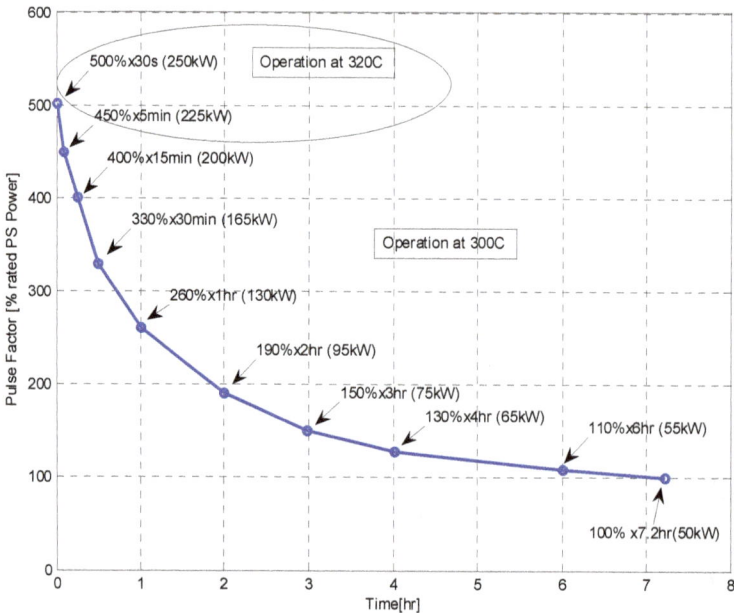

Figure 2. Pulsed power vs. discharge time of a NaS battery module

At 100% depth of discharge, sodium batteries last approximately 2500 cycles. Like other electrochemical batteries lower discharges extend its duration. At 90% depth of discharge, the cell has a lifespan of 4500 cycles, while 65% have a life of 6500 cycles and 20% a lifespan of 40 000 cycles. In practice, sodium battery discharge is limited to less than 100% of its theoretical capacity due to the corrosive properties of sodium polysulfide (Na2S3). This is the reason why the cells typically deliver 85-90% of its theoretical capacity. At 90% capacity of sodium polysulfide composition corresponds approximately to 1.82 V per cell. At this point, the main obstacles to large-scale applications of the sodium battery are its high cost of production which depends largely on the quantity of batteries produced. The approximate cost of these batteries, including the power electronic converters is $ 2500 to 3000 per kW (Iba at al 2006). According to (Gyuk 2003), the total system cost for a typical multifunctional NaS battery is $ 810 per kW, with 60% of this value attributable to the battery module.

Another obstacle in NaS batteries is given by the fact that the ceramic electrolyte is presently only commercially manufactured by one company, i.e. NGK. Moreover, the protection of intellectual property the company holds over the electrolyte difficult to study and implement appropriate models to simulate their dynamic behaviour (Hussien 2007).

The greatest sodium BESS installed is about 34 MW in Aomori, Japan, forming a hybrid system with a 51 MW wind farm. TEPCO/NGK commercializes sodium batteries under the trademark NaS in Japan and USA. So far the batteries TEPCO/NGK were the only ones available in the market for BESS, but POSCO, General Electric and Fiamm Sonick also develop sodium batteries. POSCO succeeded in developing a sodium sulfur battery for the first time in Korea, with the goal of commercializing by 2015 with RIST (a research institute wholly owned by POSCO). General Electric commercializes its Durathon battery which uses sodium metal halide chemistry and Fiamm Sonick battery is made up of salt (NaCl) and nickel (Ni). In China, research works began in the 70's and since 1980 the Chinese Institute SICCA has become the only institution outside of Japan with research in the area of sodium sulphur batteries.

4. Dynamic model of advanced BESS

The most important characteristics of a battery are determined by the voltage of their cells, the current capable of supplying over a given time (measured in Ah), the time constants and its internal resistance (Sorensen 2003). The two electrodes that supply or receive power are called positive electrodes (e_p) and negative (e_n), respectively. Inside the battery, the ions are transported between the negative and positive electrodes through an electrolyte. The electrolyte can be liquid, solid or gaseous. The electromotive force E_0 is the voltage difference between the electrode potentials for an open external circuit or at no load, defined as:

$$E_0 = Ee_p - Ee_n \tag{2}$$

The above description of the behaviour of the battery is in open circuit and the value of E_0 depends on the reduction potential of redox couple used (see Appendix for more details). During the initial discharge of the battery the battery voltage can be parameterized as:

$$V_0 = E_0 - \eta I R_0, \tag{3}$$

where I is the current consumption of the connected load, R_0 is the internal resistance of the cell and η the polarization factor. The polarization factor synthesizes or summarizes the contribution of complex chemical processes that can take part inside the cell between the electrodes through the electrolyte and are dependent of the battery type. Figure 3 shows an schematic with the potential difference across the cell with and without load.

Both the voltage V_0 and the resistance R_0 generally have a variable behaviour depending on the state of charge, the depth of discharge and also according to whether is charging or discharging the battery. That is why a more general and complete expression of equation (3) is the equation proposed in equation (4).

$$V_i = E_0 - \eta I R_i, \tag{4}$$

being $V_i = V_0 - K_v Q_d$ and $R_i = R_0 + K_R Q_d$. The voltage V_i in open circuit decreases linearly with the discharge Q_d in Ah, and the internal resistance R_i increases linearly with Q_d. That is, the open circuit voltage is lower and the internal resistance is higher in a state of partial discharge compared to the initial values V_0 and R_0 for fully charged battery. The constants K_v and K_R are constants that can be determined by battery testing and reflect the characteristics of the analyzed battery (Mukund 1999).

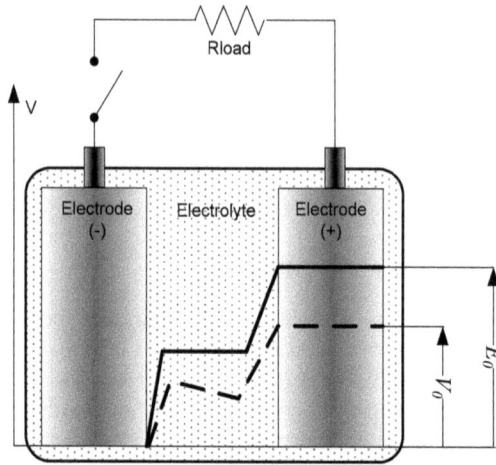

Figure 3. Potential distribution in an electrochemical cell. Solid line: unloaded cell, dashed line: loaded cell

The battery model taking into account equation (4) is very useful for steady-state studies, where the parameters K_v and K_R are constant. Figure 4 shows this battery model in schematic form.

In studies where it is necessary to study the dynamic behaviour of the battery system, possible variations of values of K_v and K_R should be taken into account. In these cases, the voltage and internal resistance of the battery does not have a linear behaviour as the one proposed in equation (4). The following sub-section briefly describes some characteristics of different types of batteries required for proposing a general model of the advanced BESSs.

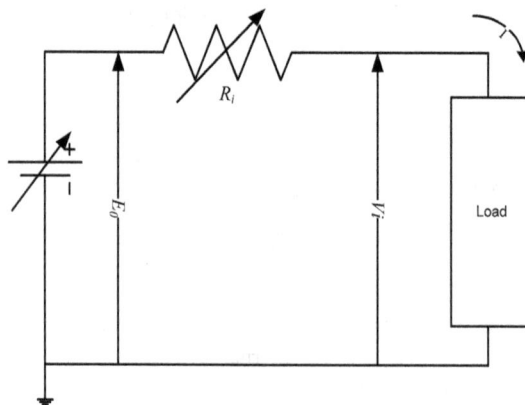

Figure 4. Equivalent electrical circuit of a steady-state battery

4.1. Analysis of performance characteristics

This sub-section discusses major performance characteristics curves of advanced BESS devices, obtained from the literature and by own experimental set-ups. These curves show indistinct of variations in voltage and/or internal resistance depending on the state of charge (SOC). In some of these curves instead of SOC they indicate the state of discharge (SOD). The relationship between these two states is given by equation (5).

$$SOC = 1 - SOD \qquad (5)$$

Batteries of Pb-acid type are characterized by an internal resistance which varies depending on the state of discharge. Figure 5 shows the variation of the internal resistance per cell vs. depth of discharge (CIEMAT 1992). This figure shows not only a nonlinear variation but also a hysteresis loop that clearly differentiates the broad difference that has the internal resistance in charging or discharging state.

In the case of Ni-MH batteries, Figure 6 shows the variation of open circuit voltage (Voc) and the internal resistance (Rseries) for different states of charge. This figure was constructed from testing a 750 mAh Ni-MH cell with discharge pulsed current from 75 mAh up to 750 mAh (Chen & Rincon-Mora 2006). As shown, the open circuit voltage varies with the SOC, but is almost independent of the depth of discharge. The internal resistance however, depends largely on the current drawn from the battery.

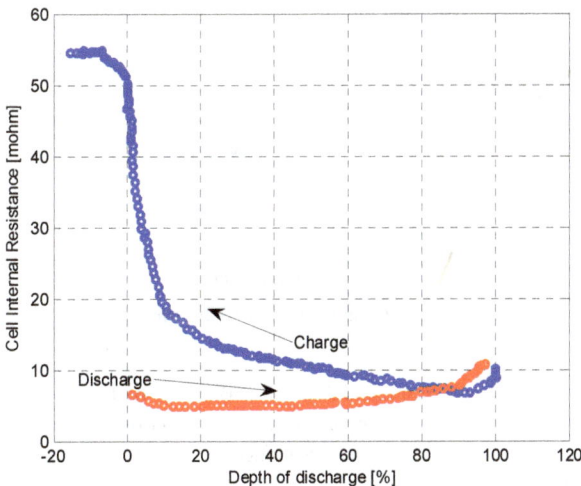

Figure 5. Internal resistance in charging or discharging state as a function of SOD for a Pb-acid battery at 25°C.

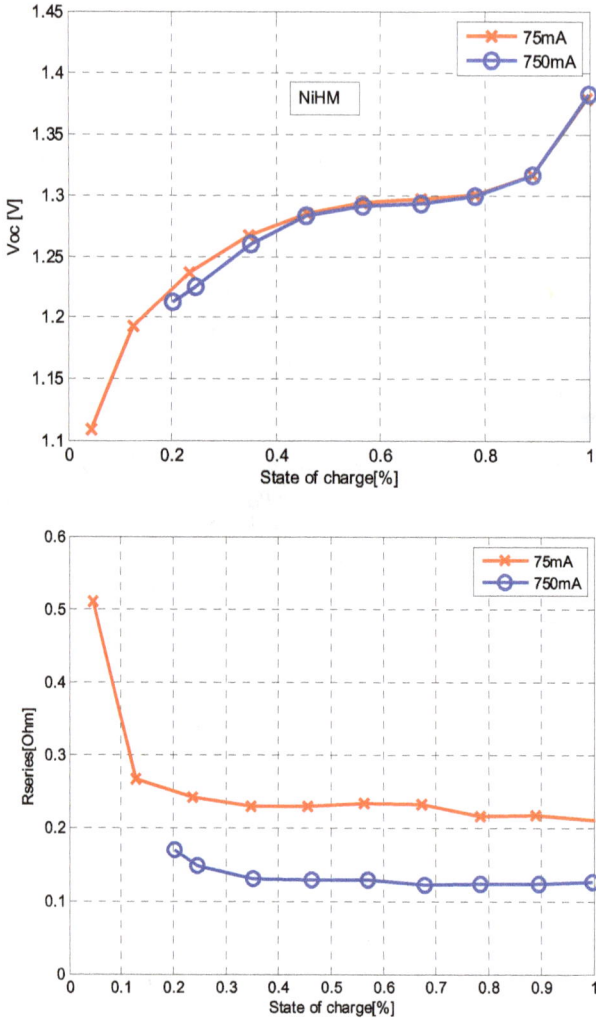

Figure 6. Variation of internal resistance (a) and voltage (b) depending on the state of charge for Ni-MH battery at room temperature.

Figure 7 was built from a test of a 850 mAh Li-Ion Polymer battery with discharge pulses from 80 mAh to 640 mAh (Chen & Rincon-Mora 2006). This figure shows the variation of open circuit voltage (Voc) and the internal resistance (Rseries) for different charge states. As shown, the open circuit voltage varies with SOC but is almost independent of the depth of discharge. On the other hand, in such batteries it can be seen that the internal resistance is not only independent of the state of charge, but also of the depth of discharge. The internal resistance remains almost constant from 20% SOC.

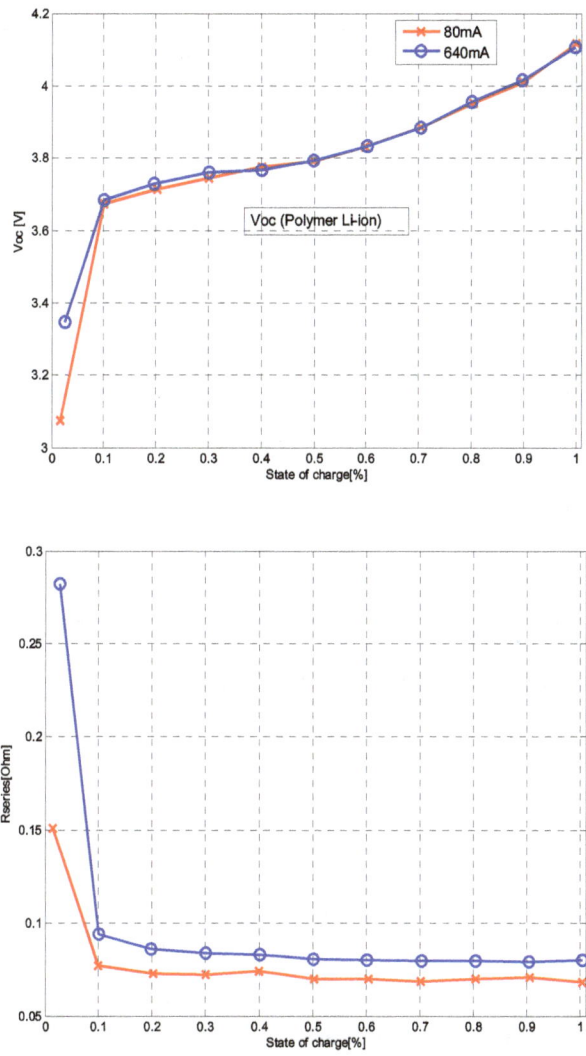

Figure 7. Variation of internal resistance (a) and voltage (b) depending on the state of charge for Li-ion polymer at room temperature

Figure 8 shows that for the case of NaS battery type, voltage changes with the depth of discharge of the battery (Hussien 2007). Due to their internal reactions, the electromotive force of the sodium battery is relatively constant, but decreases linearly after 60 to 75% depth of discharge (Van der Bosche 2006). Figure 9 also shows that depending on the state of charge, charge direction and the temperature at which the battery is operated, the internal resistance can vary up to four times its base value (Hussien 2007). It also clearly shows a hysteresis loop similar to that observed for lead acid batteries (Figure 5), in which the internal resistance value varies not only with temperature and SOD, but also whit the direction of current flow.

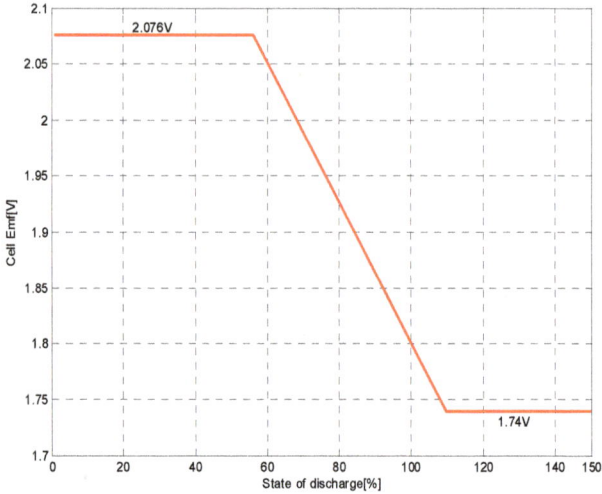

Figure 8. Voltage variation as a function of SOD for NAS-type battery cell

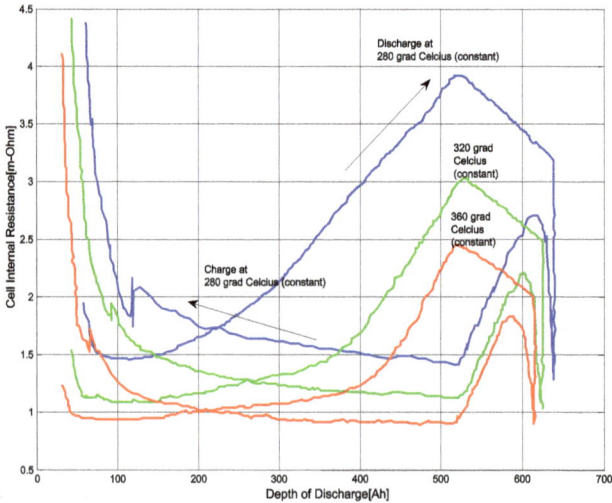

Figure 9. Internal resistance variation depending on the state of charge/discharge for various temperatures in a NAS-type battery cell

4.2. Proposed general model of BESS

Figures 5 through 9 shows a large nonlinearity in the behaviour of the most important batteries parameters. These features should be included in a model that wants to accurately represent the behaviour of batteries in power quality or energy management events.

Based on the analysis in the previous section, it can be seen that both, the battery voltage and the current capable of being delivered at any given time, generally depends on several factors. Among the most important ones are the following:

- The room temperature
- The amount of charge/discharge cycles the battery has been subjected to (cycles)
- The depth of charge/discharge
- The state of charge/discharge

If the aim of the battery model, for a given operating state, is to observe its behaviour in power quality or energy management events, this action sets a time within which the temperature can be considered constant for power quality-like events. While for energy management events lasting over an hour it should be considered a change of temperature; it may nevertheless be considered constant or studied to typical and/or extreme temperatures which would be the battery subjected to. In this way, for both cases (power quality or energy management events) the value of the parameters depends on the operating temperature of the battery.

It has been considered for the realization of the model that the battery is in a state of charge such that the characteristics curves are for the unit fully charged or discharged. This also sets the initial conditions of the model regarding the influence of the numbers of charge/discharge cycles that is capable of delivering the battery.

The depth of charge/discharge cycles influences not only the ability to the power or energy that the battery can deliver, but also its lifespan. In this sense, the depth of discharge must be taken into account in the maximum simulation time and the limitations recommended by the manufacturer.

The state of charge of the battery is the most important factor of all the above and should be taken into account directly in the model. This factor directly influences the value of power/energy that the battery can deliver in time of occurrence of events.

From the graphs shown above (Figures 5 through 9) it can be inferred a general model to simulate the battery considered. A model that includes all the batteries tested should consider that the open circuit voltage and internal resistance varies with the state and direction of the charge. The values of K_v and K_R can not be considered constant. The most convenient solution is to use directly the curves described in Figures 5 through 9 with the value of SOC.

Given the battery type, operating temperature and the depth of discharge, a model that takes into account these factors is show in Figure 10. This figure shows the outline of a general battery model, depicted as an example for a NaS-type battery.

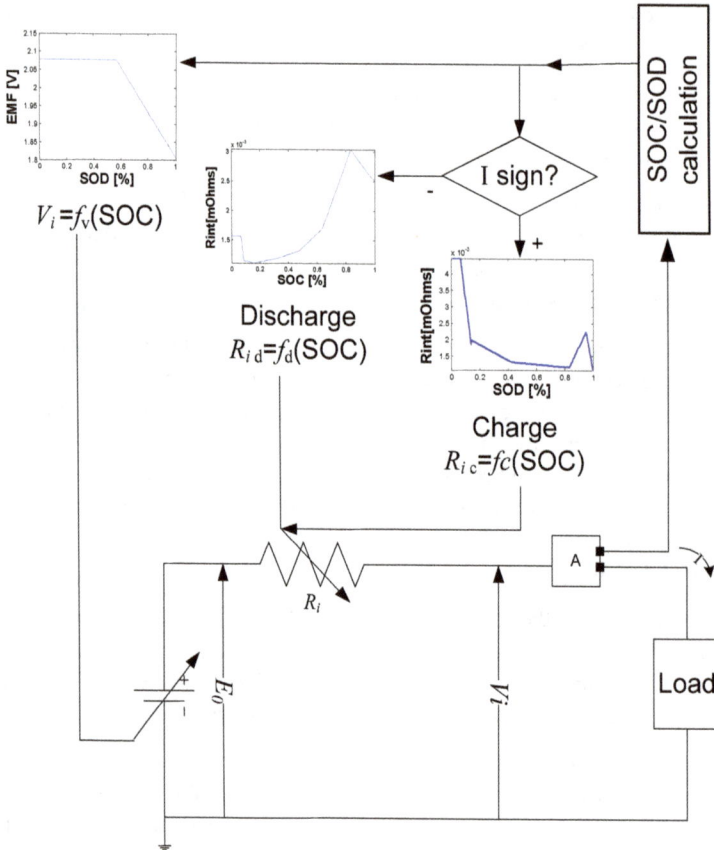

Figure 10. Proposed general model of the BESS

4.3. Test of proposed BESS model

The developed model of the BESS was tested using a single cell in order to validate the model. Figures 11 and 12 show the variation of internal resistance to changes of SOD for discharging and charging, respectively, in a NaS T5 type cell at 320°C. Figure 13 shows the electromotive force variation vs. SOD. Finally, Figure 14 shows a test system where a NaS PQ-G50 module (Hussien 2007) is connected through an IGBT DC/AC Inverter to an infinite bus. This module is a 50 kW pack consisting of 320 cells connected in series for obtaining a higher capacity storage device with higher voltage.

The module is simulated to perform a power quality event. From the simulation carried out up to 10s, the NAS battery delivers 50 kW. At 10s (around 10% SOD), the battery is commanded to deliver its maximum capacity (1 p.u. of 250 kW). Figure 15 shows the variation of the output active and reactive power, P and Q, on the AC side of the battery module throughout the duration of the simulation.

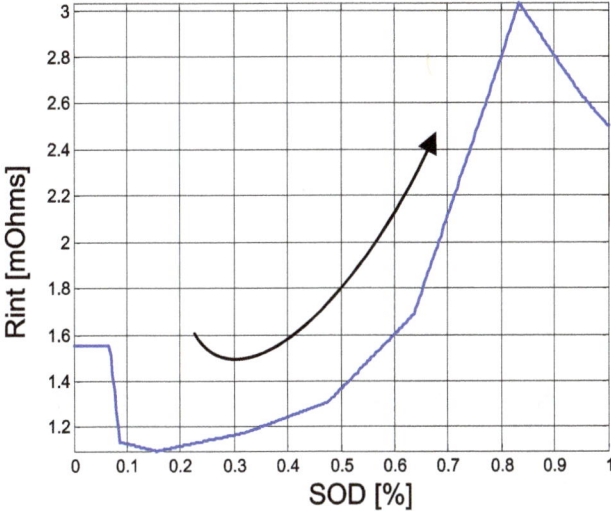

Figure 11. Simulation of cell resistance vs. SOD at 320°C for a discharge situation

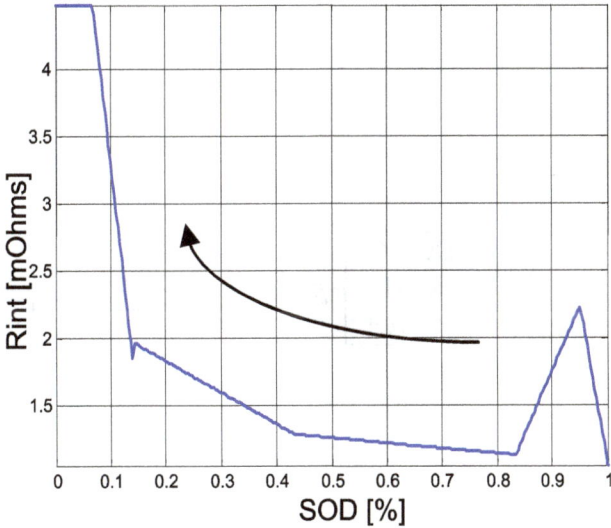

Figure 12. Simulation of cell resistance vs. SOD at 320°C for a charge situation

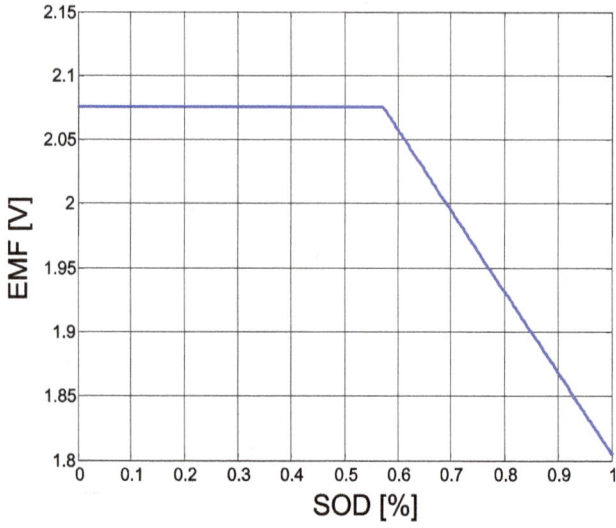

Figure 13. Simulation of cell electromotive force vs. SOD

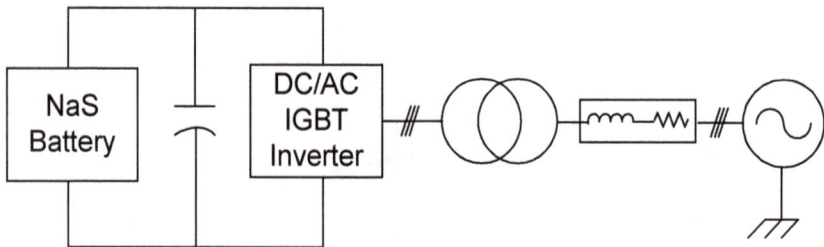

Figure 14. NAS module connected to a test power system

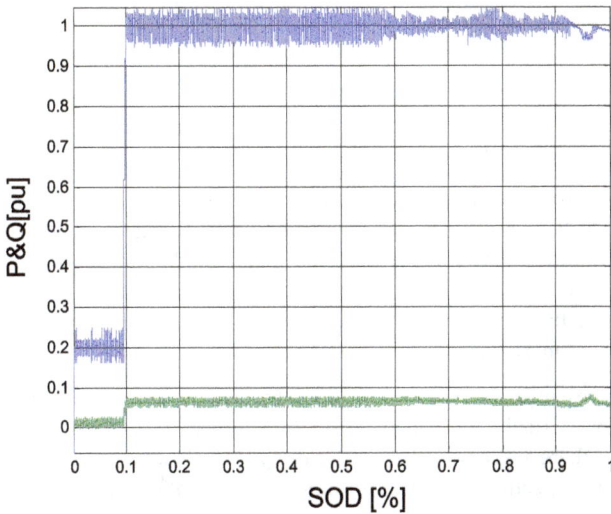

Figure 15. Variation of active and reactive power from the Battery module (AC side)

This model can be easily modified to operate also in energy management mode. Because of methodology of modelling used, the model can be easily modified to simulate the temperature variation.

5. Conclusion

With the exception of conventional lead-acid batteries, advanced batteries analyzed in this chapter represent the cutting edge technology in high power density BESS applications. Li-Ion batteries have the greatest potential for future development and optimization. In addition to small size and low weight of the Li-Ion, they offer higher energy density and high storage efficiency, making them ideal for portable devices and flexible grid-connected distributed generation applications in microgrids. However, some of the biggest drawbacks of Li-Ion technology are its high costs (due to the complexity arising from the manufacture of special circuits to protect the battery) and the detrimental effect of deep discharge in its lifespan (Divya & Østergaard 2009). Although the Ni-Cd and Pb-acid batteries can provide large peak power, they contain toxic heavy metals and suffer from high self-discharge.

Sodium sulfur-type BESS devices are best suited to the requirements set by modern microgrid applications. These batteries can act in contingencies where rapid action is required to maintain the adequate levels of the grid frequency, but also in the case of high penetration of renewable generation, such as wind or solar photovoltaic, since the NaS battery can operate as the perfect complement in valley hours. In this case, the excess energy can be stored for delivery in peak hours. They are environmentally safe and have low maintenance while operate at high temperatures; it does not represent a major drawback. The biggest drawbacks are the cost and the limited information about these type of batteries which difficult the development of experimental prototypes and computer models. It is expected however that the appearance of other vendors reduces costs and facilitate the modelling.

Appendix: Oxidation reduction (Redox) reaction

The oxidation-reduction reactions (also known as redox reaction) refer to all chemical reactions in which atoms have their oxidation state changed. Fundamentally, redox reactions are a family of reactions that are concerned with the transfer of electrons between species. Thus, in order to produce a redox reaction in the system, an element to yield electrons and one that will accept them must exist. This transfer occurs between a set of chemical elements, an oxidant and a reductant.

Oxidation involves an increase in oxidation number, while reduction involves a decrease in oxidation number. Usually the change in oxidation number is associated with a gain or loss of electrons, but there are some redox reactions (e.g., covalent bonding) that do not involve electron transfer. Depending on the chemical reaction, oxidation and reduction may involve any of the following for a given atom, ion, or molecule:

- Oxidation - involves the loss of electrons or hydrogen or gain of oxygen or increase in oxidation state
- Reduction - involves the gain of electrons or hydrogen or loss of oxygen or decrease in oxidation state

Oxidants are usually chemical substances with elements in high oxidation states (e.g., H_2O_2, CrO_3, OsO_4), or else highly electronegative elements (O_2, F_2, Cl_2, Br_2) that can gain extra electrons by oxidizing another substance. Reductants in chemistry are very diverse. Electropositive elemental metals, such as lithium, sodium, magnesium, iron, zinc, and aluminum, are good reducing agents. These metals donate or give away electrons readily.

In redox processes, the reductant transfers electrons to the oxidant. Thus, in the reaction, the reductant or reducing agent loses electrons and is oxidized, and the oxidant or oxidizing agent gains electrons and is reduced. The pair of an oxidizing and reducing agent that are involved in a particular reaction is called a redox pair or couple.

When a net reaction proceeds in an electrochemical cell, oxidation occurs at one electrode, the anode, and reduction takes place at the other electrode, the cathode. The cell consists of two half-cells joined together by an external circuit through which electrons flow and an internal pathway that allows ions to migrate between them. Since the oxidation potential of a half-reaction is the negative of the reduction potential in a redox reaction, it is sufficient to calculate either one of the potentials. Therefore, standard electrode potential is commonly written as standard reduction potential.

The sign of the potential depends on the direction in which the electrode reaction has elapsed. By convention, the electrode potentials refer to the semi-reduction reaction. The potential is then positive, when the reaction occurs in the electrode (facing the reference) is the reduction, and is negative when oxidation. The most common electrode as a reference electrode is called the reference or normal hydrogen, which has zero volts.

Finally, the voltage of a cell is determined by the reduction potential of redox couple used and is usually between 1 V and 4 V per cell. A complete table of the type of potential constituent of the electrode can be seen in (Linden & Reddy 2001).

Batteries in which the redox process is not reversible are called primary (non-rechargeable). For this work, are of interest only secondary batteries (rechargeable) which are based on some kind of reversible process and can be repetitively charged and discharged. In this way, only this type of batteries are considered here when batteries are referred.

Author details

Antonio Ernesto Sarasua
Instituto de Energía Eléctrica, Universidad Nacional de San Juan, Argentina

Marcelo Gustavo Molina and Pedro Enrique Mercado
CONICET, Instituto de Energía Eléctrica, Universidad Nacional de San Juan, Argentina

Acknowledgement

The authors wish to thank the CONICET (Argentinean National Council for Science and Technology Research), the UNSJ (National University of San Juan), and the ANPCyT (National Agency for Scientific and Technological Promotion) under grant FONCYT PICTO UNSJ 2009 – Cod. No. 0162, for the financial support of this work.

6. References

Abdollahi Sofla M. & Gharehpetian G.B. (2011). Dynamic performance enhancement of microgrids by advanced sliding mode controller. *Electrical Power and Energy Systems;* Vol. 33, No. 1, pp. 1–7.

Baxter, R. (2005). *Energy Storage: A Nontechnical Guide*, Issue 1, Pennwell Books. 978-1593700270. New York

Bito, A. (2005). Overview of the Sodium-Sulfur (NAS) Battery for the IEEE Stationary Battery Committee. Special Presentation from Sodium-Sulfur Battery Division, NGK Insulators, Ltd. Nagoya, Japan. June 15

Chen, H.; . Cong, T.N.; Yang, W.; Tan, Ch.; Li, Y. and Ding, Y. (2009). Progress in electrical energy storage system: A critical review. *Progress in Natural Science*, Vol. 19, pp. 291-312.

Chen, M. & Rincón-Mora, G.A. (2006). Accurate Electrical Battery Model Capable of Predicting Runtime and I-V Performance. *IEEE Transactions on Energy Conversion*, Vol 21, No. 2, pp. 504-511.

CIEMAT (1992). *Fundamentos, dimensionado, y aplicaciones de la energía solar fotovoltaica.* CIEMAT: Centro de Investigaciones Energéticas, medioambientales y tecnológicas. Universidad Politécnica de Madrid. 8478341684. Madrid

Divya, K.C. & Østergaard, J. (2009). Battery energy storage technology for power systems - An overview. *Electric Power Systems Research,* Vol. 79, pp.511-520.

Guerrero, J.M.; Blaabjerg, F.; Zhelev, T.; Hemmes, K.; Monmasson, E.; Jemei, S.; Comech, M.P.; Granadino, R. & Frau, J.I. (2010) Distributed generation: toward a new energy paradigm. *IEEE Industrial Electronics Magazine;* Vol. 4, No. 1, pp. 52–64.

Gyuk, I. (2003). EPRI-DOE Handbook of Energy Storage for Transmission and Distribution Applications. *Electric Power Research Institute (EPRI) and U.S. Department of Energy Inc.,* Final Report: 1001834, December 2003.

Hatziargyriou, N.; Asano, H.; Iravani, R. & Marnay, C. (2007). Microgrids. *IEEE Power & Energy Magazine;* Vol. 5, No. 4, pp. 78–94.

Hewitt, N.J. (2012). Heat pumps and energy storage – The challenges of implementation. *Applied Energy;* Vol. 89, No. 1, pp. 37-44.

Hussien, Z.F.; Cheung, L. W.; Siam, M.F. M. & Ismail, A. B. (2007). Modelling of Sodium Sulphur Battery for Power System Applications. *Elektrika,* Vol. 9, No. 2, pp. 66-72.

Iba, K.; Ideta, R. & Suzuki, K. (2006). Analysis and Operational records of a NAS battery. *Proceedings of the 41st International Universities Power Engineering Conference,* UPEC 2006, Vol. 2, pp. 491-495.

Iba, K.; Tanaka, K. & Yabe, K. (2008). Operation and Control of NaS Batteries on a University Campus. *16th PSCC,* Glasgow, Scotland, July 14-18, 2008. ISBN of Conference Proceedings: 978-0-947649-28-9

Kaneko, T.; Uehara, A.; Senjyua, T.; Yona, A. & Urasaki, N. (2011). An integrated control method for a wind farm to reduce frequency deviations in a small power system. *Applied Energy;* Vol. 88, No. 4, pp. 1049–1058.

Katiraei, F.; Iravani, R.; Hatziargyriou, N. & Dimeas, A. (2008). Microgrids management: Controls and operation aspects of Microgrids, *IEEE Power & Energy Magazine;* Vol. 6. No. 3, pp. 54–65.

Kroposki, B.; Lasseter, R.; Ise, T.; Morozumi, .S; Papatlianassiou, S. & Hatziargyriou, N. (2008) Making microgrids work. *IEEE Power & Energy Magazine;* Vol. 6, No. 3, pp. 40–53.

Li, K.; Wang, B.; Su, D.; Park, J.; Ahn, H. & Wang, G. (2010). Enhance electrochemical performance of lithium sulphur battery through a solution-based processing technique. *Journal of Power Sources*

Linden, D. & Reddy, T.B. (Ed(s).) (2002). *Handbook of Batteries.* Mac Graw Hill. 0071359788. New York

Maharjan, L.; Tsukasa, Y. & Akagi, H. (2010). Active-power control of individual converter cells for a battery energy storage system based on a multilevel cascade PWM converter. *IEEE Trans. Power Electronics;* Vol. PP, No. 99,. ISSN: 0885-8993

Mathiesen, B.V.; Lunda, H. & Karlsson, K. (2011). 100% Renewable energy systems, climate mitigation and economic growth. *Applied Energy;* Vol. 88, No. 2, pp. 488–501.

Molina, M. G. & Mercado, P. E. (2006). New Energy storage devices for applications on frequency control of the power system using FACTS controllers. *Proceedings of X Latin-American Regional Meeting of CIGRÉ (ERLAC), CIGRÉ*, pp. 222-226, May 2006 Iguazú, Argentina.

Molina, M. G. & Zobaa, A.F. (2011). Improving the performance of grid-tied AC microgrids including renewable generation by distributed energy storage. *3ʳᵈ Latin American Conference on Hydrogen and Sustainable Energy Sources (HYFUSEN 2011)*, June 6–9, Mar del Plata, Argentina.

Molina, M. G. (2011). Enhancement of power system security level using FACTS controllers and emerging energy storage technologies. In: Acosta, M. J., editor. Advances in Energy Research – Vol. 6. New York: Nova Science Publishers Inc; pp. 1–65.

Mukund, R. P. (1999). Wind and Solar Power Systems. *CRC Press*.

Naish, C.; McCubbin, I.; Edberg, O. & Harfoot, M. (2008). Outlook of Energy Storage Technologies. *Study of the European Parliament - Policy Department of Economic and Scientific Policy*, Info: IP/A/ITRE/FWC/2006-087/Lot 4/C1/SC2.

Nourai, A. (2002). Large-Scale Electricity Storage Technologies for Energy Management. *Proc. 2002 IEEE Power Engineering Society Summer Meeting*, Vol. 1, pp. 310-315.

Panasonic (n.d.). 1.5MWh Litium-ion Mega Battery System. Available from: *http://panasonic.net/sanyo/gep/guide/04.html*

Qian, H.; Zhang, J.; Lai, J. & Yu, W. (2011). A high-efficiency grid-tie battery energy storage system. *IEEE Trans. Power Electronics*; Vol. 26, No. 3, pp. 886–896.

Schroeder, A. (2011). Modeling storage and demand management in power distribution grids. *Applied Energy*; Vol. 88, pp. 4700-4712..

Scrosati, B. & Garche, J. (2010) Lithium batteries: Status, prospects and future. *Journal of Power Sources*. Vol. 195, pp 2419–2430.

Serban, I. & Marinescu, C. (2008). Power Quality Issues in a Stand-Alone Microgrid Based on Renewable Energy. *Revue Roumaine des Sciences Techniques, Série Électrotechnique et Énergétique*; Vol. 53, No. 3, pp. 285–293.

Sørensen, B. (2003) "Renewable Energy".Elsevier Science. Third Edition.

Tiwari, G.N.; Mishra, R.K. & Solanki, S.C. (2011). Photovoltaic modules and their applications: A review on thermal modeling. *Applied Energy*. Vol. 88, No. 7, pp. 2287–2304.

Vachirasricirikul, S. & Ngamroo, I. (2011). Robust controller design of heat pump and plug-in hybrid electric vehicle for frequency control in a smart microgrid based on specified-structure mixed H_2/H_∞ control technique. *Applied Energy*. Vol. 88, No. 11, pp. 3860–3868.

Van den Bossche, P.; Vergels, F.; Van Mierlo, J.; Matheys, J. & Van Autenboer, W. (2006). SUBAT: An assessment of Sustainable battery technology. *Journal of Power Sources*, Vol. 162, No. 2, pp. 913-919.

Vazquez, S.; Lukic, S.M.; Galvan, E.; Franquelo, L.G. & Carrasco, J.M. (2010). Energy storage systems for transport and grid applications. *IEEE Trans. Power Electronics*; Vol. 57, No. 12, pp 3881–3895.

Wen, Z.; Cao, J.; Gu, Z.; Xu, X.; Zhang, F. & Lin, Z. (2008). Research on sodium sulfur battery for energy storage. *Solid State Ionics*. Vol. 179, pp.1697-1701.

Wissner M. (2011). The Smart Grid – A saucerful of secrets? *Applied Energy*; Vol. 88, No. 7, pp. 2509–2518.

Chapter 12

In-Situ Dynamic Characterization of Energy Storage and Conversion Systems

Ying Zhu, Wenhua H. Zhu and Bruce J. Tatarchuk

Additional information is available at the end of the chapter

1. Introduction

The combustion of fossil fuels predominates in the commercial implementation of energy conversion and power generation. However, it brings severe problems to the environment due to inevitably incomplete combustion. Meanwhile, the price of fossil fuels keeps increasing due to the depletion of natural resources. The growing concerns of global warming, as well as the reducing availability of fossil fuels, require replacements of gasoline and diesel fuels, such as Fischer-Tropsch synthetic fuels [1-3], biofuels [4, 5], and hydrogen fuel [6, 7]. The thermal conversion efficiency of a traditional automobile engine is between 17% and 23% [8], limited by the intrinsic characteristics of Carnot cycle. Energy storage and conversion systems with low/zero emissions, high efficiency, and great durability are required by the development of sustainable energy and power economy.

The development of commercial applications of electrical vehicles (EVs) or hybrid electrical vehicles (HEVs) was propelled by increasing demands of an on-board rechargeable energy storage system. The advanced vehicle systems or processes with higher energy efficiency are preferred to use for saving fuels and improving the mileage of per unit fuel consumed. Hybrid power trains reduce undesirable emissions and also have their potential to improve fuel economy significantly. A highly efficient engine can charge the battery pack and propel the vehicle at the same time. The battery power assists the engine acceleration or propels the vehicle efficiently on its own at low speeds [9]. The battery pack is also returned with some energy from the electric motor, which is served as another generator in the regenerative braking or coasting mode. Therefore, the battery burst charge acceptance during frequent braking and power output capability during heavy acceleration are significantly important for the HEV fuel efficiency and engine emissions.

In addition to the on-board energy storage system, the development of energy conversion system explored another solution instead of combustion engines. A highly efficient approach of energy conversion enhances the application of hydrogen storage in many fields. Fuel cells can convert the energy stored in hydrogen to electricity without combustion at a higher level of energy efficiency. Furthermore, hydrogen fueled cells can produce electricity with almost zero emissions comparing to other energy conversion technologies. This is also beneficial to reduce the CO_2 emissions.

It is significantly important to obtain deep understanding of energy storage and conversion systems to approach both technical and commercial breakthrough in sustainable energy development. The purpose of system characterization is to find out how and to what degree the properties, kinetics, and other effects of a system influence its performance. The understanding of system performance also provides basis for system diagnosis to distinguish good ones from degraded ones. For a rechargeable battery, main performance limiting factors include actual capacity, rate performance, state-of-charge (SoC) and state-of-health (SoH). While for a fuel cell, attentions are mainly paid to electrode structures, electrolyte fabrications, conductivity mechanisms, reaction limitations, catalytic poisoning, and cell degradations.

Electrochemical Impedance Spectroscopy (EIS) is a sensitive, powerful, and non-destructive analytical technique capable of assessing the dynamic response of an electrochemical system. It is generally conducted by superimposing an *ac* signal on the *dc* output of the working system and measuring its resulting *ac* signal over a spectrum of frequencies. The obtained impedance is simulated by an equivalent circuit (EC) diagram, which produces a similar load response to the working system. Different types of circuit elements along with different values are able to simulate various processes occurring in the measuring system. A well validated EC diagram can be employed to mechanistically discriminate the kinetic and mass transfer processes that limit system outputs.

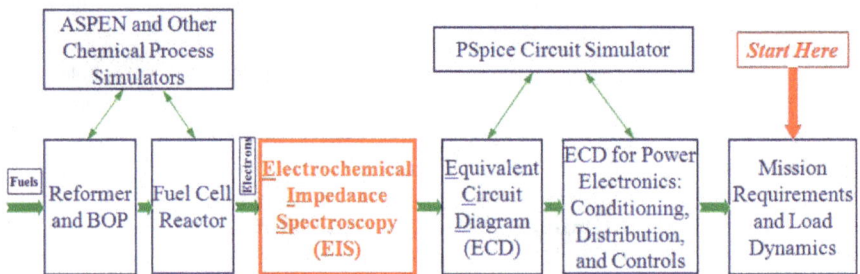

Figure 1. Methodology for integrated system design, operation, and control.

The great prospects of impedance analysis and EC simulation of energy storage and conversion systems lie in characterizing chemical reactors in terms of electronics. Chemical, electrochemical, and physical processes occurring in energy storage and conversion systems, such as batteries and fuel cells, are simulated in chemical process simulators (ASPEN for example), studying mass balance, thermodynamics, and kinetics of systems. Meanwhile, electrical circuits are simulated in analog circuit simulators (PSpice for example) to predict circuit behaviors and provide industrial standard solutions including the non-linear transient analysis for voltage and current versus time. It is EIS that establishes the connection between chemical processes (ASPEN) and power electronics (PSpice). The *in-situ* real-time measuring technique as well as EC simulation distinguishes the real processes occurring in batteries and fuel cells and interprets them into electronic circuit elements for system simulation. In this way, EIS systematically provides a competent methodology (Figure 1) for integrated system design, operation, and control.

This chapter highlights the competence of EIS, together with EC simulation, to dynamically characterize rechargeable batteries and fuel cells. Section 2 deals with the basic techniques of impedance measurement and EC simulation. Several key factors during the measurement are discussed. The emphasis is mostly addressed on EC element models and their physical interpretations. The impedance analysis and EC simulation of lead (Pb)-acid batteries and nickel metal-hydride (Ni-MH) batteries are presented in Section 3. And section 4 shows a research case of the EIS application to a proton exchange membrane (PEM) fuel cell stack system. This section also reviews the recent progress in impedance study of novel high temperature PEM fuel cells.

2. Electrochemical impedance spectroscopy

The history of impedance spectroscopy (IS) can be dated back to 1880s, when Heaviside initially introduced the concept of "impedance" [10, 11] to his research on electromagnetic induction. Later in 1883, Kennelly [12] extended the concept of "impedance" to generalized conductors, and mathematically defined the total impedance of a system, in a complex plane, as the vector sum of its resistance, its inductance-speed[1], and the reciprocal of its capacity-speed[2]. However, the technique of IS itself did not come out until Nernst [12] employed Wheatstone bridge to measure dielectric constants in 1894.

During the past century, impedance measurement contributed to the characterization of materials and devices, study of electrochemical reaction systems, corrosion of materials, and investigations of power sources. According to different materials and systems it applied to, IS can be classified into two branches [13]. The one following Nernst's initial achievement is called non-electrochemical IS. It applies to dielectric materials, electronically conducting materials, and other complicated materials with combining features [13]. The other branch, newly coming out based on the development of non-electrochemical IS, is named Electrochemical Impedance Spectroscopy (EIS). It focuses on IS applications to ionically conducting materials and electrochemical power sources [13]. The popularity of IS keeps improving especially after the prevailing of electronics and computers. Not only is it capable of dealing with complex processes, reactions, and variables through simple electrical elements, it is also a valid technique for power source diagnosis and system quality controls. This chapter focuses on the applications of EIS to rechargeable batteries and fuel cells.

---

[1] Inductance-speed: The product of angular frequency and inductance, ωL [12].
[2] Capacity-speed: The product of angular frequency and capacitance, ωC [12].

2.1. Measuring techniques

The EIS measurement is conducted by superimposing an electrical stimulus on the output of the tested electrochemical system and measuring the resulting signal. The impedance of the measuring system is then calculated from the stimulus and its resulting signal by transform functions and Ohm's Law. In this way, the performance of the system under measurement can be studied as a black box, which is described as "feeling an elephant that we cannot see" [14] in Mark Orazem and Tribollet's book. Thus, for an electrochemical system, it is possible to study the properties of its interfaces and materials without taking the system apart. The word "*in-situ*" generally refers to this type of techniques that characterizes electrochemical systems under operation with the help of voltage, current, and time, distinguishing from "*ex-situ*" techniques which studies individual components departed from electrochemical systems in a non-assembled and non-functional form [15].

Different kinds of electrical stimulus can be used in EIS measurements, including a step function of voltage, a random noise, a single frequency signal, and any other types of stimuli generated by combining the foregoing three ones [16]. With the increasing commercial availability of measuring instruments, the characterization of electrochemical systems generally employs an *ac* signal (either a small *ac* voltage or current). In most cases, it is also called "*ac* impedance".

Several impedance measuring instruments are commercially available, including products from EG&G Inc., Gamry Instruments, Scribner Associate Inc., and AMETEK (Solartron Analytical and Princeton Applied Research). An instrument set basically consists of a potentiostat, or known as an electrochemical interface, connecting to a frequency response analyzer (FRA). Besides, a complete connection circuit for measurement requires an electronic load. Specifications of different instruments are designed for different scales of measurements. The feasibility and accuracy of measurements are determined by the frequency resolution, frequency accuracy, and bandwidth of instruments and electronic loads. For example, batteries and fuel cells are low impedance systems, usually much lower than 1 Ω (sometimes even down to milliohms) [16]. Their impedance measurement requires a high current and a low frequency bandwidth.

It is worth to noting that the choice of *ac* signal value have great effect on impedance measurement. The reliability of EIS analysis is based on the assumption that the measured system is linear. It is also important to keep the system in a relatively steady state throughout the measurement to ensure the accuracy of the measured data. The measured impedance data may appear scattered and unregulated if the signal is too weak to excite a measurable perturbation. On the other hand, the signal has to be small enough to keep the measuring system within the range of pseudo-linearity. A larger signal also brings more extra heat that breaks the steady state of the system. The data measured at low frequencies behave more sensitive to the strength of the applied *ac* signal due to larger impedance at lower frequency [17]. Correspondingly, an appropriate value of the signal should be well picked to ensure the accuracy of the impedance data measured. It is not necessary to determine an exact value for the input signal, however, it has to be controlled in a certain range which is neither so small that the output signals are too weak to be measured, nor so large that a great distortion is introduced into measurements. Different signal values can be trailed before measurements to find out a suitable one.

There are two traditional modes for impedance measurements according to different regulating variables: potentiostatic mode and galvanostatic mode. Potentiostatic mode employs a small *ac* voltage signal with fixed amplitude and measures the resulting *ac* current. In this mode, the *dc* voltage is controlled at a certain value that facilitate the control of system linearity. Reversely, a small *ac* current signal with fixed amplitude is superimposed on the regulated *dc* current in galvanostatic mode. The resulting *ac* voltage is measured. Galvanostatic mode provides higher accuracy than potentiostatic mode when measuring low impedance systems because the voltage can be measured more accurately than controlled [18]. This mode is more welcomed where a steady *dc* current is required during the entire measurement. However, it is difficult to control the resulting *ac* voltage strictly within the linear range especially at low frequencies.

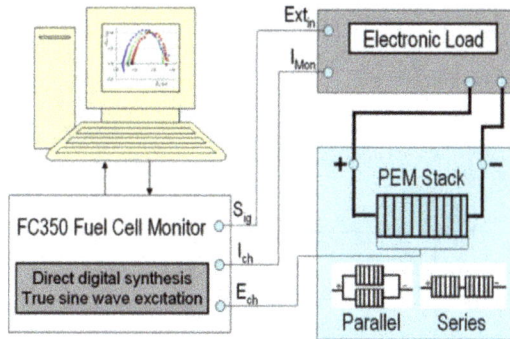

Figure 2. A typical connection diagram for *ac* impedance measurement conducted on PEM fuel cell stacks: single stack, two stacks in parallel arrangement, and two stacks in series arrangement [19].

A typical connection diagram for *ac* impedance measurement of a PEM fuel cell stack is presented in Figure 2 [19]. The commercial instrument connected in this measuring system is Gamry FC350™ Fuel Cell Monitor. It is featured by its capability to measure impedance at high current levels. The four-terminal connection is commonly used for low impedance measurement (such as batteries and fuel cells) to avoid measurement errors led by the impedance of cables and connections [16]. A novel measuring mode, hybrid mode, is employed instead of galvanostatic mode in order to overcome its drawback. In this mode, the values of a desired voltage perturbation and an estimated system impedance are set before measurements and the *dc* current is fixed during measurements. The current perturbation at each frequency is adjusted by the desired voltage perturbation and the estimated impedance in order to ensure that the *ac* voltage does not extend beyond the linear range [17]. The value of ac current signal is calculated from equation (1) [17]. In this equation, ΔI_n is the amplitude of the *ac* current signal employed to measured the nth frequency point (at a certain frequency). $\Delta V_{desired}$ is the desired voltage perturbation set before measurements. $|Z_{n-1}|_{measured}$ is the magnitude of the system impedance measured at the $(n-1)^{th}$ frequency point. The impedance value used to obtain the ac current signal for the first frequency point is the estimated impedance value set before measurements [17].

$$\Delta I_n = \frac{\Delta V_{desired}}{|Z_{n-1}|_{measured}} \qquad (1)$$

2.2. Data analysis and interpretation

2.2.1. Data presentation

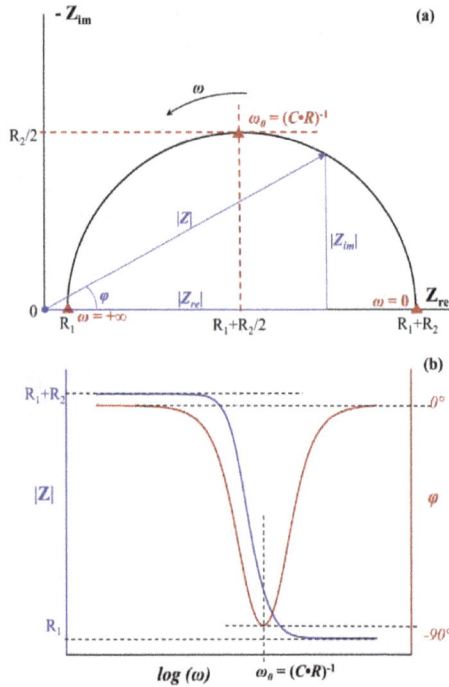

Figure 3. Examples of (a) Nyquist plot and (b) Bode Plot generated by the diagram $[R_1(CR_2)]$.

The two most preferred methods to present impedance data are the Nyquist plot and the Bode plot. Nyquist plot, named after Harry Nyquist, is a polar plot of the transfer function [20]. In some publications, it is also called the complex-plane plot or the Argand plot [16]. An example of Nyquist plot is shown in Figure 3a (plotting method adapted from [20]). This impedance spectrum is a set of continuous complex impedance points produced by the circuit diagram $[R_1(CR_2)]$ over a certain range of frequency (from zero to positive infinite in Figure 3a). The x-axis presents the real component of the impedance points and the y-axis presents their negative imaginary component. In this way, the capacitive impedance arcs are flipped to the first quadrant in the plot. As the frequency goes from zero to positive infinite, the impedance points move toward the origin of the plane. The impedance value and phase angle of the points can be calculated from the Nyquist plot:

$$|Z| = \sqrt{|Z_{re}|^2 + |Z_{im}|^2} \tag{2}$$

$$\tan\phi = \frac{|Z_{im}|}{Z_{re}} \tag{3}$$

Nyquist plot can provide plenty information of impedance variables but failed to present their frequency dependences. Bode plot, named after Hendrik W. Bode, makes up this inadequacy. It displays the frequency dependence of a linear, time-invariant system, usually shown in logarithmic axis. Figure 3b is an example of Bode plot generated by the diagram $[R_1(CR_2)]$. It includes a magnitude plot presenting the magnitude of impedance ($|Z|$) versus the logarithmic angular frequency (log ω), and a phase plot presenting the phase shifts of ac current to ac voltage (ϕ) against the logarithmic angular frequency (log ω).

2.2.2. Simulation and interpretation

There are mainly two methods to acquire models for impedance data interpretation, visually summarized and classified in a flow chart by Macdonald [13]. One mathematically establishes models based on the theory, which puts forward a hypothesis for physical and chemical processes contributed to the impedance. The other one utilizes empirical models, called equivalent circuits (EC). Some researchers also presented a combining method with both of them [22]. The values of certain variables, such as ohmic resistance, were acquired directly from the empirical EC simulation and used as known variables to establish the mathematical model. Whichever employed, the validations of data themselves are essential before simulation. The relations originally published by Kramers (1929) and Kronig (1926) (K-K Transforms) became a simple but effective method for data validation from 1980s, in order to ensure the causality, linearity, and stability of the measured systems [23, 24]. Both the mathematical models and the empirical ECs also have to be validated before data interpretation and system characterization. The fitting programs, generally following the procedures of complex nonlinear least squares (CNLS) fit algorithm (such as LEVM [25] and EQUIVCRT [26]), are employed to validate the derived models by estimating the parameter standard deviation and the goodness of fit [25].

Comparing to mathematical models, deriving an EC model is easier, faster, and more intuitive. An EC diagram is a physical electrical circuit which produces a similar load response to the measured system, derived based on both experiences and theories. The overpotential losses of the testing cell (electrochemical systems) are introduced by the impedances contributed by different physical and chemical processes occurring in the cell (electrochemical system). The impedances of different processes predominate different frequency regions. Thus, they can be identified and mechanistically discriminated by EC simulation according to different process relaxation times. However, the physical interpretation of circuit elements is not straightforward due to the uncertainty of EC diagrams. Different arrangements or combinations of EC elements can produce the same dynamic response when three or more elements are employed in one EC diagram. The only solution to overcome this difficulty is acquiring sets of impedance data with different variables and changing conditions.

Three basic EC elements along with their mathematical expressions and physical meanings [27] are summarized in Table 1. Z is the impedance of the elements. Y is the reciprocal of Z, called admittance. Generally, resistor R, capacitor C, and inductor L are three fundamental elements reflecting ideal processes. Figure 4a (plotting method adapted from [27]) sketches out the impedance behavior of the single elements calculated at $R = 10$ Ω, $C = 0.01$ F (from 0.04 Hz to 100 kHz), and $L = 0.01$ H (from 0.1 mHz to 7 kHz). The impedance of the ideal resistor does not change with frequency. The magnitude of the ideal capacitor's impedance decreases with increasing frequency. Whereas the magnitude of the ideal inductor's impedance increases with increasing frequency.

Element Symbol	Element Name	Impedance Expression	Physical Interpretation
R	Resistor	$Z_R = R$	Contributed by energy losses, dissipation of energy, and potential barrier
C	Capacitor	$Z_C = \dfrac{1}{j\omega C}$	Contributed by accumulations of electrostatic energy or charge carriers
L	Inductor	$Z_L = j\omega L$	Contributed by accumulations of magnetic energy, self-inductance of current flow, or charge carrier's movement

Table 1. The mathematical expressions and physical meanings [27] of ideal EC elements: R, C, and L.

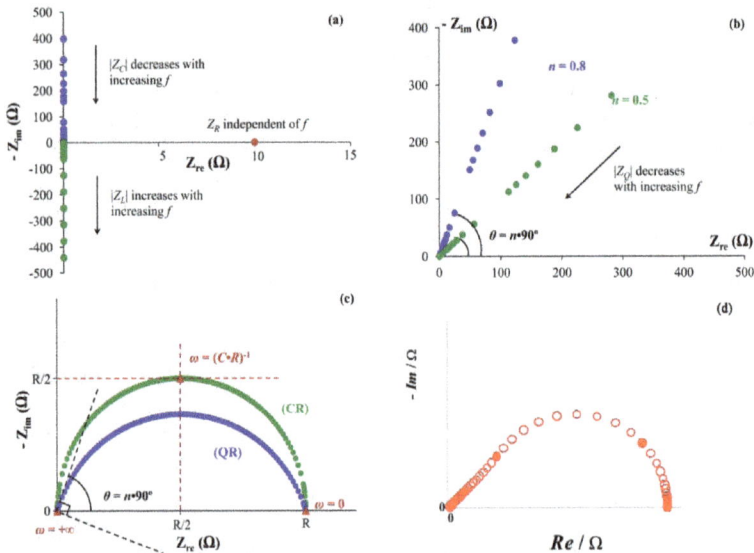

Figure 4. Nyquist plots of (a) R, C, L, (b) Q, (c) (CR) and (QR), and (d) O [27].

However, the ideal EC elements only are not able to reflect non-ideal factors in practical cases. A generalized element, constant phase element (CPE) Q, was developed to simulate non-ideal processes. Q reflects the exponential distribution of time constants. This non-ideal distribution may be caused by the surface roughness and vary thickness of electrodes, unevenly distributed current, and non-homogeneous reaction rate. In the impedance expression of Q [16] (Table 2), the exponent n reflects the degree of non-ideality. When the value of n in the expression of Q equals to 1, 0, and -1, it can be found that the expression becomes the same as capacitor C, resistor R, and inductor L, respectively. Figure 4b (plotting method adapted from [27]) shows two sets of impedance data produced by the expression of Q with $A = 100$ F^{-1}. They behaves as straight lines across the origin in Nyquist plot. When the value of n decreases from 0.8 to 0.5, the slope decreases. The angle of incline equals to the time of n and 90º.

The (CR) circuit is one of the basic combinations commonly used in EC simulations. The elements enclosed in parentheses are connected parallel to each other. And the brackets indicate that the elements enclosed are connected in series. For example, (CR) means a pure capacitor C and a pure resistor R are parallel in the connection; while $[CR]$ means a pure capacitor and a pure resistor are connected in series. Figure 4c presents the Nyquist plot of (CR). It is a semi-circle centering at $(R/2, 0)$ with a radius of $R/2$. As the angular frequency ω increases from 0 to ∞, the summit of the semi-circle is reached when:

$$\omega = \omega_0 = \frac{1}{CR} \tag{4}$$

The (CR) circuit can be employed to simulate an ideal double-layer process. The non-ideal one requires the circuit of (QR). The element Q replaces C to reflect the non-ideality of an double-layer process. (QR) behaves as a depressed semi-circle having its center dropped down below the Z_{real} axis (Figure 4c). The degree of the depression is presented by the exponent n of the Q element. The EC diagram $[R_1(CR_2)]$ mentioned previously in Figure 3 is another fundamental combination. Based on the spectrum of (CR), $[R_1(CR_2)]$ shifts horizontally along the positive Z_{real} axis in Nyquist plot (Figure 3a). The smaller intercept with the Z_{real} axis equals to the resistor R_1. And the larger intercept equals to the sum of R_1 and R_2. The center of the semi-circle is $(R_1+R_2/2, 0)$ and its radius is $R_2/2$.

Element Symbol	Element Name	Impedance Expression
Q	Constant phase element (CPE)	$Z_Q = \dfrac{A}{\sqrt[n]{j\omega}}$
W	Warburg element	$Z_W = \dfrac{\sigma}{\sqrt{\omega}} - j\dfrac{\sigma}{\sqrt{\omega}}$ $\|Z_W\| = \sqrt{2}\dfrac{\sigma}{\sqrt{\omega}}$
O	Finite diffusion element (FDE)	$Z_O = \dfrac{1}{Y_0\sqrt{j\omega}}\tanh(B\sqrt{j\omega})$ $Y_0 = \dfrac{(nF)^2 A}{R_{ig}T}\left(\sum \dfrac{1}{C_i\sqrt{D_i}}\right)^{-1}$ $B = \delta/\sqrt{D_i}$

Table 2. The mathematical expressions of non-ideal EC elements: Q [16], W [29], and O [30].

Based on the second Fick's law, Emil Warburg developed expressions for the impedance response of diffusion processes in 1899 [14]. This is now called Warburg diffusion element (W). It is used to simulate an one-dimensional unrestricted diffusion process to a large planar electrode [28]. The mathematical equations for Z_W [28] are listed in Table 2. It behaves as a line with unit slope in Nyquist plot, exactly the same as a Q with the exponent n of 0.5 (Figure 4b). Finite diffusion element (FDE) O, or sometimes called porous bounded Warburg , was established based on W. Its application is extended to a rotating disk electrode (RDE), where diffusion occurs over the Nernst Diffusion Layer (NDL), that is a diffusion layer with finite thickness. The expressions for Z_O [30] are listed in Table 2. Figure 4d [27] shows a typical Nyquist plot of an O element. Its impedance spectrum presents the same behavior as W at higher frequency region and changes to an arc similar to (CR) at lower frequency region.

Randles circuit $[R_\Omega(C_d[R_{ct}W])]$ considers the effect of a diffusion process. It consists of one ohmic resistance R_Ω, one parallel (C_dR_{ct}) sub-circuit behaving as a semi-circle in Nyquist plot, and one infinite diffusion element W behaving as a unit slope line at the lowest frequency region. Figure 5 [21] shows a sketch for the impedance spectrum of a Randles circuit. The dash lines illustrate the overlap region of (C_dR_{ct}) and W.

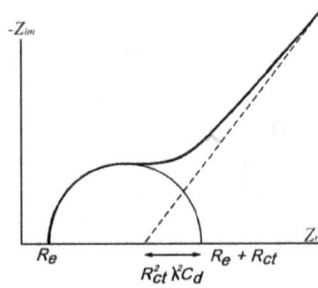

Figure 5. Nyquist plot of Randle's circuit ([21] Courtesy of Solartron Analytical).

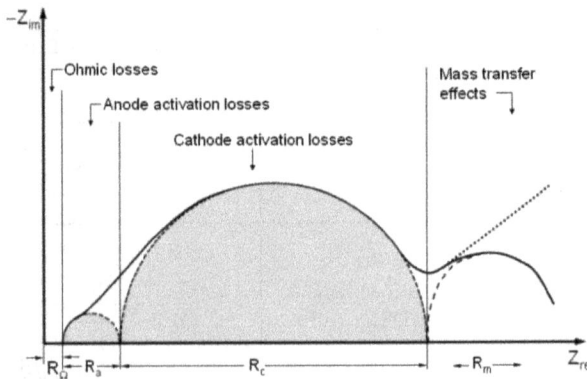

Figure 6. Nyquist plot of a typical EC model for batteries and fuel cells [31].

Figure 6 [31] presents the impedance spectra of a typical EC diagram commonly used to simulate batteries and fuel cells. This diagram consists of three time constants. Its total impedance is the sum of the ohmic losses Z_Ω, the anode polarization losses Z_a, the cathode polarization losses Z_c, and the mass transfer losses Z_m. The ohmic losses are simulated by one pure resistor R_Ω. Its value is represented as the smaller intercept on the real axis. The anode and cathode polarization losses are simulated by a parallel (CR) sub-circuit respectively. Their impedance spectra look like two overlapping semi-circle in Nyquist plot. Some electrochemical systems, like Ni-MH batteries, have strong Warburg behavior in the low frequency region for the mass transfer processes. The diagram can be expressed as $[R_\Omega(C_aR_a)(C_c[R_cW])]$. However, some do not always show this behavior obviously, such as PEM fuel cells. The low frequency arc can be simulated by a parallel (QR) sub-circuit instead of a pure Warburg element. The diagram is expressed as $[R_\Omega(C_aR_a)(C_c[R_c(Q_mR_m)])]$. The Warburg element and sub-circuit (Q_mR_m) are connected in series with R_c because the mass transfer processes are considered as processes on the cathode side.

3. EIS applications to rechargeable batteries

The phrase "electric battery" was innovatively re-defined in late 1740's by Benjamin Franklin [32] when he described a series of his experiments with electricity. However, Franklin's "electric battery" referred to the pile of glass plate capacitors set up in 1748. The production and storage of energy were not achieved until Alessandro Volta designed and built his "crown of cups" and the columnar pile in late 1790s [32]. The Volta pile published in 1800 is generally acknowledged as the first battery because it fulfilled one of the most important functions of batteries: It stored energy in chemicals and converted it into electrical energy by chemical reactions. Although the first reversible system was successfully developed by Johann Ritter in 1802 right after the birth of Volta's pile [33], real rechargeable batteries (also known as secondary batteries) did not come out until Gaston Planté invented the first lead (Pb)-acid battery in 1859 [8]. The original concerns to the emergence and development of rechargeable batteries focused on the environmental problems caused by the toxic substances used in batteries. The pollutant substance, generally mercury, used in primary batteries was greatly reduced by decreasing the use of mercury batteries. And the reuse and recycle of rechargeable batteries also moderated the pollutions introduced by itself.

The greater contribution of rechargeable batteries to human beings was explored and well developed with the advancing requirements of grid energy storage. Since rechargeable batteries are feathered by reversible electrochemical reactions, they are able to adapt the energy supplement to the energy demand, that is to store energy in chemicals during valley period and convert energy into electricity during peak period. Having developed for over 150 years, rechargeable batteries have been matured in the commercial applications to automotive starters, electronic products, and all other portable electrical devices daily used. Even EVs requiring light power consumption, such as bicycles and wheelchairs, have been well commercialized. The development and commercialization of pure electric automobiles powered by a build-in rechargeable battery makes it possible to save human lives out from energy crisis and environmental pollutions caused by fossil fuels and combustion engines.

The development of electric rail network began from the 19th century and prevailed in UK in most of the 20th century [8]. In the past few decades, electric bicycles also predominated in city transportations in China. One of the most challenge issues is the range limitation that an EV can travel with unit battery charge time.

Nelson [34] extensively analyzed and reviewed power requirements for batteries in the HEVs. He recommended for HEV applications that the hybrid driving duty cycle was able to be realistically performed at an operating window between 30% SoC and 70% SoC hybrid limits. He also proposed a nominal hybrid operation range of 40-60% SoC. There is an inefficient charging process above the hybrid limit (>70~80%), and the power output capability is insufficient for start-up, acceleration, or other low-speed driving needs below the hybrid limit (<20~30% SoC).

In general, batteries are expensive and necessary for operating maintenance, and also have limited cycle life. The impedance technique can be applied to dynamic modeling of battery behavior and *in-situ* diagnosis of battery systems including actual capacity, SoC and SoH [35]. Battery simulation is an important issue for design of different battery applications. It is desired to develop a reliable and easy-to-parameterized battery model. The following two sections present the impedance applications to Pb-acid batteries and Ni-MH batteries separately.

3.1. Pb-acid batteries

3.1.1. Characteristics of Pb-acid battery

As the first type of rechargeable battery, the Pb-acid battery system has features of high-rate discharge ability, suitable specific energy, highly acceptable reliability and toughness. Also, it is commercially beneficial in production and recycle processes because the Pb-acid battery is made primarily from the same low-cost initial material [36]. Recently, the advanced Pb-acid battery systems for on-board energy storage have been developed and applied to the HEVs [37-39]. A Pb-acid ultrabattery has been developed within a single battery container by Furkawa *et al.* [37]. This battery combines features of Pb-acid battery and supercapacitor and is functioned as an integrated hybrid energy storage system. This type of valve-regulated Pb-acid battery is designed specifically for high-rate HEV applications at partial state-of-charge (SoC). EnerSys Energy [40] also produced an improved Pb-acid battery, which is called as "Odyssey". The Pb-acid battery delivers much higher cranking power and also retains deep-cycle potentials. Because of the strong needs for development of low-cost energy-saving systems and low-emission vehicles, a few breakthroughs in the lead-acid battery have been made especially in new electrode structure design, rate and deep-cycle performance.

The concentrated H_2SO_4 solution is applied in the Pb-acid battery as electrolyte, which is potentially dried out due to the high-temperature working environment. The ionic resistance of the normal operating battery is negligible and makes a small contribution to the series resistance of the battery which depends on the SoC. The unique feature for the battery is that the electrode processes involve a dissolution–precipitation mechanism. The electrolyte concentration varies with the change of the battery SoC level due to generation of SO_4^{2-} ions during charge reaction and consumption of SO_4^{2-} ions during discharge process, i.e. both electrodes converted to lead sulfate - double-sulfate reactions. Many challenges for traditional Pb-acid batteries are described as relatively low cycle life (50-500 cycles), especially in electrical vehicle applications, limited energy density, sulfation caused poor charge retention, sulfation during long term storage, and health hazard in designs with antimony and arsenic in grid alloys.

3.1.2. Pb-acid battery test and impedance data analysis

The Gamry FC350™ system with an electronic load was applied to the battery impedance tests. The measuring circuit was connected as shown in Figure 2 [19], simply replacing the PEM stacks with the batteries. The sinusoidal current signal from the Gamry system, working in the hybrid-impedance mode, modulated the current from the Pb-acid batteries. The hybrid EIS mode was applied for the experiments in order to obtain the impedance data of the rechargeable batteries at various frequencies. The non-linear least squares (NLLS) fitting algorithm was utilized for the impedance spectrum to find the model parameters for best agreement between the simulated impedance spectra and the measured impedance data. The tests were conducted on the Pb-acid batteries by a unique impedance method, *i.e.* test the batteries at different loads. *In-situ* test, data analysis, and suitable impedance simulation were performed to develop a wide range design tool or an appropriate model for dynamic applications of Pb-acid batteries for different power reserve demands.

Impedance tests were conducted on three ordinary types of Pb-acid batteries in the experimental work [41]. One (AGN-8) was malfunctioning and taken out from a Toyota car after eight-years' SLI use. The second one (BDU-S) was at rest for *ca.* eight years, infrequently charged for maintenance requirements. The third battery (CDU-N) was ordered from a local store to substitute the AGN-8 Pb-acid battery. The hybrid EIS mode was utilized in the experiments to examine the EIS behavior at different frequencies. After each measurement, the consumed energy was approximately estimated and charged back in the Pb-acid battery. The battery was considered to be fully charged after overnight charging at a current of less than 1.00 A with a voltage no more than 13.5 V.

Figure 7 [41] shows the Nyquist plots of three Pb-acid batteries obtained at the load of 5 A. The interesting curve is the one collected from the AGN-8 battery after eight-year's SLI service in vehicle. The battery has been running for eight years and approximately close to its cycle life for the SLI work. The loop curve is clearly different from other Pb-acid batteries near the high frequency side. However, the curves simulated from the circuit diagram $[R_\Omega(R_1Q_1)(R_2Q_2)L]$ are well agreed with the collected impedance data during the battery test, particularly at 1 A load of current. At the high frequency side, the change related to the anode surface (Pb/PbSO$_4$) appears to be the major reason for the shape change of the impedance loop. A carbon plate functioned as a capacitor was attached to the negative plate (Pb/PbSO$_4$) [39] in order to improve the anode performance and slow down the electrode degradation rate. Not only the burden at the negative electrode is significantly lowered, the rate performance and pulse capability are also considerably improved because of the capacitance behavior from the attached carbon electrode.

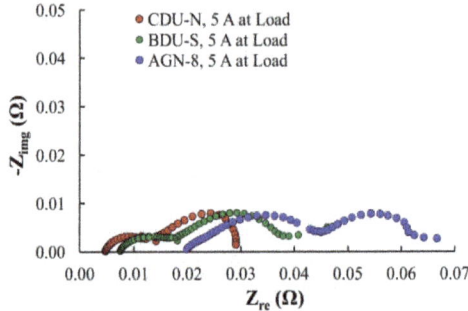

Figure 7. Nyquist plots of three batteries tested at 5A load [41].

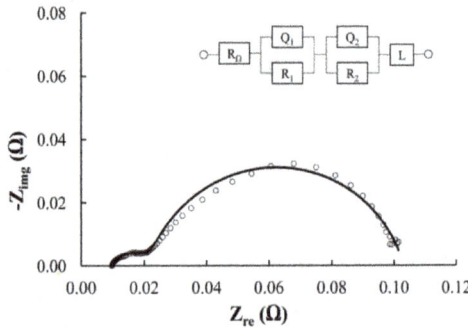

Figure 8. Non-destructive tests of the defective BDU-S battery at 1 A load [41].

The BDU-S battery was also analyzed based on the EC simulation with the circuit $[R_\Omega(R_1Q_1)(R_2Q_2)L]$ (Figure 8 [41]). The simulated curve is well fitted to the measured impedance spectra that presents the impedance behavior of the BDU-S battery when it was operated at a load of 1 A and produced an output voltage of $ca.$ 12.17 V [41]. However, it was no longer capable of operation at the output level of 12 V due to battery failure. Further diagnostics [41] shown that one cell in the BDU-S battery was suffering degradations. The negative ($Pb/PbSO_4$) electrode was in failure with inadequate charge retention because of sulfation on the anode surface.

3.1.3. Future performance improvement of rechargeable batteries

The simulation of the complexity of modern power electronics is very difficult and impedance-based battery models potentially provide useful physical elements and suitable parameters for system dynamic evaluations. The battery performance is considerably nonlinear and the dynamic performance is determined by numerous parameters such as battery life-time, operating temperature, SoC, and depth-of-discharge (DoD). EIS is a useful tool to get suitable chemical and physical parameters for simulation of battery power systems. The development of high-rate long-life HEV batteries and deep-cycle long-life EV batteries is significantly important for future batteries in vehicle applications. Flat thin-plate structures have been designed and made further improvement for cranking power needs and deep-cycle requirement. The amount of lead used in the battery has been greatly reduced using this related technology. It is necessary for future work to examine battery chemistry and advanced electrode processes, estimation of mass transfer limitations, and exploration of the failure mechanisms. Numerous important factors including active material, utilization of active material, current collector, support configuration, electrolyte, separator, and system safeguarding (thermal management and gas recombination) are related to battery performance and operating life-time. For SLI applications nearly more than 100 years, Pb-acid batteries have been utilized to provide automobile reserve power requirements. This is because the Pb-acid battery provides the greatest cost/performance ratio among all batteries. The battery designs using advanced concepts and structure enhancements likely produce novel battery power systems with required performance. These efforts potentially stimulate the further technical developments for the deep-cycle and high-rate power needs in energy storage and power reserve applications.

3.2. Ni-MH batteries

Unlike the traditional Ni-Cd and Pb-acid batteries based on the dissolution-precipitation mechanism with dendrite formation possibility during charge and discharge, the mechanism for a Ni-MH battery is the movement of hydroxide ions between a metal hydride (MH) electrode and nickel hydroxide electrode. This simple mechanism produces a long battery cycle life of more than 1000 cycles, higher power capability, and a dense electrode structure [42]. The battery has natural protection against overcharging and over-discharging with oxygen and hydrogen recombination inside the cell to form water. The overcharge process, over-discharge process, capacity retention, and the *in-situ* SoC parameter are quite important to the battery cycle life and calendar life time. The improvement of the battery energy storage and conversion efficiencies will significantly increase the overall energy efficiency and prolong the HEV's traveling miles per unit fuel use.

3.2.1. Ni-MH chemistry and hybrid operating range

Enhanced energy storage efficiency means less production of CO_2 greenhouse gases and lower emissions of NO_x and SO_x for improving acid rain environment. Battery energy storage and fuel efficiency are significantly important for the HEV fuel economics. The charge acceptance, power output capability, and battery cycle-life are key factors for its application in the energy rapid storage considerations. The basic principles and electrochemical reactions occurring in the Ni-MH cell are described as follows [43, 44],

At the positive electrode,

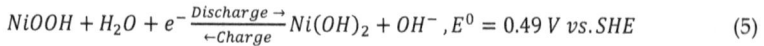

$$NiOOH + H_2O + e^- \xrightarrow[\leftarrow Charge]{Discharge \rightarrow} Ni(OH)_2 + OH^- , E^0 = 0.49\ V\ vs.SHE \tag{5}$$

At the negative electrode,

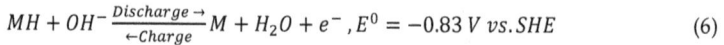

$$MH + OH^- \xrightarrow[\leftarrow Charge]{Discharge \rightarrow} M + H_2O + e^- , E^0 = -0.83\ V\ vs.SHE \tag{6}$$

The overall cell reaction is written as

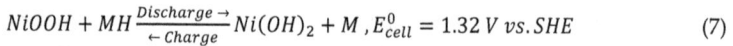

$$NiOOH + MH \xrightarrow[\leftarrow Charge]{Discharge \rightarrow} Ni(OH)_2 + M , E_{cell}^0 = 1.32\ V\ vs.SHE \tag{7}$$

The nickel electrode is thermodynamically unstable in the sealed cell and oxygen-evolution occurs at the electrode as a parallel and competing reaction. The parasitic reaction during charge and overcharge is expressed as

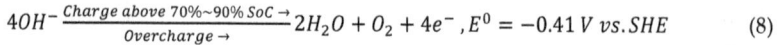

$$4OH^- \xrightarrow[Overcharge \rightarrow]{Charge\ above\ 70\%\sim90\%\ SoC \rightarrow} 2H_2O + O_2 + 4e^- , E^0 = -0.41\ V\ vs.SHE \tag{8}$$

This reaction happens during the battery processes of charge and overcharge. Reaction (8) starts as a parallel side-reaction, competing with the primary charging Reaction (5) at a certain state-of-recharge (SoR, i.e. the actual charge input as percent of the battery-rated capacity). At a higher charging rate, the difference between SoR and SoC may even start earlier due to higher potential and mass transfer limitation of the electrolyte. Hence, the HEV storage application preferably uses the 70%~80% SoC level as the higher hybrid operation limit. However, the nickel-based battery is normally designed that the cell capacity is limited by the positive electrode. The negative to positive capacity ratio varies from 1.5 to 2.0. The evolved oxygen from the positive electrode diffuses to the MH electrode and recombines to form water. Typically, the discharge reserve is approximately 20% of the positive capacity [43]. The range from 0 to 20% SoC level is called as deep discharge region. In order to ensure proper power output capability, the HEV energy storage considers the 20~30% SoC level as the lower hybrid operating limit [34].

3.2.2. Impedance test and equivalent circuit simulation

The impedance tests were conducted by the same equipment and measurement connection used for the tests of the Pb-acid batteries. Small amount of the consumed capacity of the Ni-MH battery was estimated and charged back after each test of impedance measurement. After the Ni-MH battery was prepared for impedance measurement at another SoR, the Gamry impedance system together with an electronic load, operating in hybrid impedance mode, modulated the current information from the working battery at load. At the same time, the current signal at the electronic load was sent back for the computer data management. The Gamry system collects these measured data and creates the data files for further impedance data processing and circuit simulation.

The Ni-MH battery was charged to a certain level of SoR at a 0.2 C rate and then conducted impedance tests with an electronic load. The battery capacity was measured as 3.7 Ahr at 0.2 C rate and impedance data were collected in condition of 0.37 mA load and 5 mV ac voltage at ca. 50% SoR. The ac impedance data were analyzed and the Nyquist plot of the Ni-MH battery is shown in Figure 9. The simulated curve via an EC model of $[R(Q_aR_a)(Q_c[R_cQ_m])]$ is well agreed with impedance data in the above Nyquist plot. This EC model is a modification based on the one illustrated in Figure 6. CPEs are used to replacing the pure capacitors and the Warburg element in order to take in account the non-ideal effects. In considering of energy efficiency, inefficient charge, capacity retention rate, power output needs, battery cycle-life, as well as Nelson's valuable work, the Ni-MH battery for on-board energy storage is preferred to work at 50±10% SoC with an operating limitation of 50±20% SoC. It is easy to accurately measure various battery parameters during the battery off-board break, but the challenge is to dynamically determine the current SoC level for on-board energy storage and in-time energy release when the battery pack is being operated inside a moving vehicle. Energy efficiency is important for HEV fuel economics and further improvement of system efficiency means less greenhouse gases and lower emissions. Although the ultra-capacitor is an alternative solution to the burst energy storage, the battery with a suitable capacitance behavior is still an ideal solution in considering of the total cost of the energy storage and conversion system.

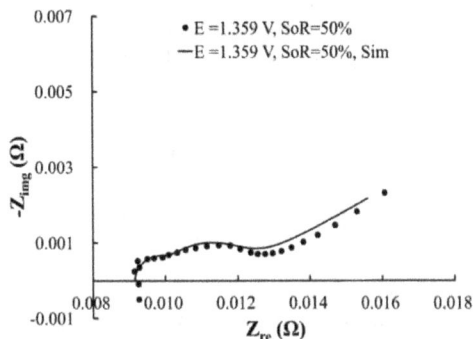

Figure 9. Simulated curve via an EC model of $[R(Q_aR_a)(Q_c[R_cQ_m])]$ well agreed with impedance data in the Nyquist plot of the Ni-MH battery.

4. EIS applications to proton exchange membrane fuel cells

Both the energy quick consumption and alternative fuel development require further improvement of energy efficiency for lowering costs and reducing emissions. The zero exhaust is the most attractive advantage of fuel cells over other energy conversion technologies. In comparison with batteries, fuel cells require a continuous fuel supplement as long as they convert chemical energy to electricity. The first fuel cell can be traced back to William Grove's "gaseous voltaic battery" developed in 1839 [45]. This prototype successfully proved that the reaction of hydrogen and oxygen could produce electricity.

After many attempts at improving the "gas battery" by several investigators, Ludwig Mond and Carl Langer significantly achieved the practical one. Their "new gas battery" published in 1889 [46] was considered as the prototype of current fuel cells. However, it was not until 1960s when the commercial application of a fuel cell was realized for the first time in NASA's Gemini program [47]. General Motors produced their hydrogen powered fuel cell vehicle in 1967 [47], inspiring the research and development of fuel cells to be commercially applied to automotives and replace combustion engines.

Generally, fuel cells are classified into five main types based on different electrolyte [16]. Alkaline fuel cell (AFC) utilizes aqueous alkaline solutions as electrolyte. It is now able to operate below 100°C. Proton exchange membrane (PEM) fuel cell is featured by the solid polymer electrolyte. Its low temperature operation and high energy conversion efficient make it become one of the most promising solution to combustion engines. Phosphoric acid fuel cell (PAFC) uses concentrated or liquid phosphoric acid (H_3PO_4, abbreviated to PA) as electrolyte and operates at around 200°C. Molten carbonate fuel cell (MCFC) and solid oxide fuel cell (SOFC), operating at extremely high temperature, overcome the poisoning issues of the other three types and reduce operation costs. An electrical efficiency of 60% were reported to be achieved in 2009 by a natural gas powered SOFC device (Ceramic Fuel Cells Limited).

Fuel cell systems produce much higher efficiency than combustion engines; however, two major challenges, high cost and low reliability, have to be overcome to implement its successful commercialization. Properly designed fuel cell systems can be a reliable and durable method to provide high efficient and environmentally friendly power sources for many applications, including global transportations, portable devices, and residential backups. Integrated systems, consisting with the subsystems of fuel processor, fuel cell, power electronics, and thermal management, can successfully fulfill the production of both electricity and heat simultaneously from the same power source, called combined heat and power (CHP) [15].

Comparing to battery investigations, EIS dedicated more contributions to the development and diagnostics of fuel cells. Impedance measurement, analysis and EC simulation can be conducted to investigate component fabrications, interfacial processes, transfer mechanisms and cell degradations of several types of fuel cells. Current fuel cell investigations are mainly focus on cell degradation, recoverable poisoning, state of MEA health, and stability of long-term operation. The EC models, plausible explanations, and remaining problems of conventional PEM fuel cells and high temperature PEM fuel cells will be reviewed and presented in this section.

4.1. Conventional proton exchange membrane fuel cells

4.1.1. Impedance measurement

The Ballard Nexa™ fuel cell system [19, 30, 48] was connected to the measuring circuit (Figure 2 [19]). The power module contains one PEM stack consisting of 47 single planar fuel cells, each with an active area of *ca.* 122 cm^2. It provides an unregulated *dc* power output of 1200 W at nominal output voltage of 26 Vdc. The rated output current of the stack reaches 44 A and its open circuit voltage normally rises up to 41 V. A controller board is embedded in the module to facilitate the automated stack operation. Measurement of the stack without the embedded controller board were accomplished by electrically isolating the PEM stack from the controller subsystem.

The tests were conducted on three PEM stacks numbered #308, #515, and #881, identical to each other. There were totally five sets of impedance tests of the stacks published with different scales listed below [19, 30, 48]. The impedance data of the stacks were collected at different current levels without embedded system controller in most of the tests. However, several sets of impedance data were collected from the stacks together with the embedded controller board. The Gamry FC350™ fuel cell monitor connecting to a TDI-Dynaload® RBL488 programmable load was employed to obtain impedance data. Different small *ac* signals were chosen according to different scales of the tests, ensuring the accuracy of the data without scattering.

1. Single cell tests

The single planar fuel cells in the stack #308 were tested separately at the temperature of 26ºC under varying current loads of 0.2 Adc, 0.5 Adc, and 1.0 Adc, without the embedded system controller [30]. The single cells are numbered from the anode side to the cathode side. An *ac* voltage signal changed from 10 mV to 20 mV was set as the desired voltage perturbation in hybrid mode, according to different system current loads. The impedance data were collected from the single cells numbered #10, #15, #25, #31, and #47. Three sets of them are shown in Figure 10 [30]. The impedance behavior of the 47th cell in the stack obviously differs from the impedance behaviors of others. It has the lowest ohmic resistance because the humidified air inlet increases the cell humidity. However, the observable mass transport losses are caused by impurity build-up. For each single cell, no significant difference in ohmic resistance is observed when the current changes. However, cells closer to the anode side bear higher resistance due to less humidity.

2. Group cell tests

Four fuel cell groups consisting with different number of single cells (12, 24, 36, and 47 cells respectively) in the stack #308 were tested at the temperature of 26ºC and a current load of 0.2 Adc without the embedded system controller [30]. A desired *ac* voltage signal of 10 mV was set for the cell groups in hybrid mode. There are no significant changes between the ohmic resistances of each group; however, the polarization resistance, which dominates the cell impedance, increases proportional with the increase of cell numbers.

3. Single stack test

The single stack tests were conducted individually on the stack #308 [30], #515, and #881 [19, 48]. Impedance data of the stack #308 were measured with and without the embedded system controller at relatively steady state under each current load using an *ac* current signal of 500 mA in galvanostatic mode over a frequency range from 10 µHz to 20 kHz. The #515 and #881 PEM stacks were measured at varying *dc* current loads with a desired *ac* voltage signal of 150 mV in hybrid mode over a frequency range from 10 mHz to 10 kHz, both with (Figure 11a [19]) and without (Figure 11b [19]) the embedded system. Though the impedance collected with the embedded system is larger than the one without, both sets of impedance spectra have the same behavior and change tendency.

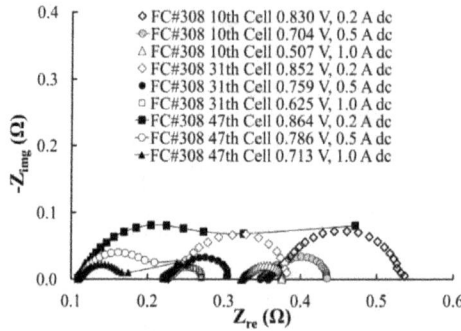

Figure 10. Nyquist plots of single cells numbered #10, #31, and #47 of the PEM fuel cell stack #308 in the Nexa™ PEM system [30].

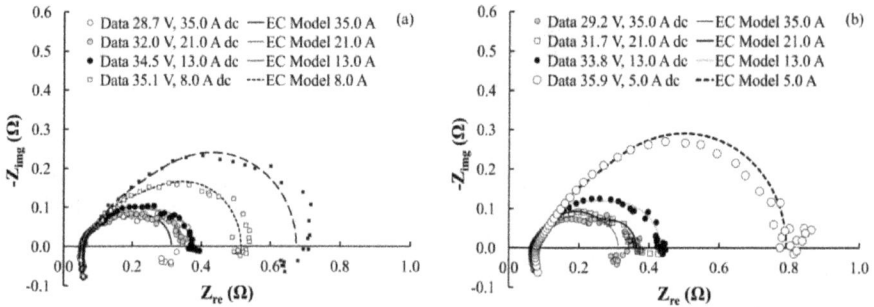

Figure 11. Nyquist plots of the PEM fuel cell stack #515 in Nexa™ PEM system. (a) The PEM fuel cell stack is equipped with embedded controller, compressor, and other electronic devices; (b) The PEM fuel cell stack is running while its controller board and other electronic devices uses an external power source [19].

4. Group stack tests – two stacks in series

The PEM stack #515 and #881 were operated in series as one power source [19]. The impedance of the stack group with the embedded system controllers was measured with a desired voltage perturbation of 150 mV in hybrid mode. The current of 5 Adc, 10 Adc, 15 Adc, and 30 Adc were loaded to the whole power system. The impedance spectra present the similar behavior as the single stack (Figure 12a [19]).

5. Group stack tests – two stacks in parallel

The PEM stack #515 and #881 were operated in parallel as one power source [19, 48]. The impedance of the stack group was measured following the same procedures as the measurement of the series stack group. A desired *ac* voltage of 150 mV was employed in hybrid mode. The current of 10 Adc, 21 Adc, 30 Adc, 40 Adc, and 60 Adc were loaded to the whole power system. The impedance behavior and variation tendency are consistent with the data set collected from the single stacks and the series stack group (Figure 12b [19]). However, the impedance of the parallel stack group is reduced to about one quarter of the series stack group. Thus, the parallel connection is preferred for commercial applications.

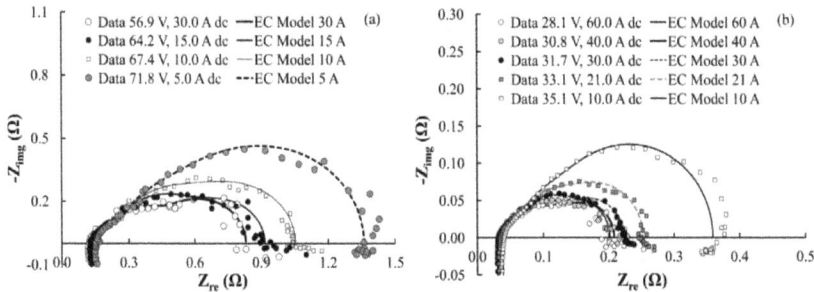

Figure 12. Nyquist plots of the PEM fuel cell stacks #515 and #881 (a) in series and (b) in parallel operation embedded with system controller, compressor, and other electronic devices [19].

4.1.2. EC simulation

A three time constant diagram, $[R_\Omega(R_aC_a)(R_cC_c)(R_mC_m)L]$, was applied to the sets of impedance data measured from the single PEM fuel cell stacks (#515 and #881 separately) and the two stack groups (#515 and #881 in parallel and in series). The Nyquist plots of measured impedance data and the simulation curves calculated from the EC model are shown in Figures 9, 10, and 11. It is possible to determine the bipolar plate contact and membrane electrolyte resistance from ohmic resistance R_Ω by switching the embedded control system to an external power supply and subtracting the wiring and contacting resistance. The following three parallel (RC) sub-circuits break down the polarization impedance into different parts contributed by different processes, including the anode activation process (R_aC_a), the cathode activation process (R_cC_c), and the gas diffusion process (R_mC_m). The catalyst loading level at the cathode side is two to three times higher than that at the anode side [49]. Thus, the cathode activation impedance is determined to be the large arc at middle frequency region, the resistance calculated from the EC model of which is about two to three times larger than the one ascribed to the anode activation impedance dominating the high frequency region. The stack impedance is mainly dominated by the cathode activation impedance, but with the increase of current loads, its contribution significantly decreases. Meanwhile, the contribution of the gas diffusion impedance greatly increases with the increasing current load, even up to almost half of the total stack impedance.

It can be observed that at high frequency region (the smaller semi-circles closer to the origin of the coordination) the simulated curves are well fitted to the measured data. On the other hand, the simulated curves at lower frequency region (the large semi-circle dominating the stack impedance) derivate from the measured data. The (R_mC_m) sub-circuit may not be accurate enough to simulate the impedance arc at low frequency region. Since it is determined to be a finite gas diffusion process, an O element should be much more suitable for the simulation. From the aspect of its physical meaning, an O element also facilitates data interpretation. The (R_mC_m) sub-circuit was kept other than being substituted by an O element to facilitate the following PSpice simulation. Although the impedance spectra simulated by the EC model slightly deviated from the impedance of the real PEM system, the pulse load test simulated in PSpice by this EC model presented acceptable results [19]. This preliminary test provided a realistic method for further design, operation, and control of an integrated system.

4.2. High temperature proton exchange membrane fuel cells

4.2.1. High temperature membranes

The performance of a conventional PEM fuel cell is limited by its operating temperature. The traditional perfluorosulfonic acid (PFSA) membrane exhibits great conductivity, excellent thermal and chemical stability, and considerable economic efficiency, but suffers from severe degradations at the temperature higher than 100°C. Thus, it is desired to develop alternatives to PFSA membranes to overcome this drawback.

Generally, the alternative membranes can be classified into three groups according to different methods of membrane fabrications. One is to attach charged units to a conventional polymer [50]. Most attentions to this type of alternative membranes have been paid to sulfonated polymer membranes and their composites [51]. Another group of membranes is named inorganic-organic composites or hybrid, which are fabricated by incorporating a polymer matrix with inorganic compounds [50]. Modified PFSA membranes [51], especially modified Nafion membranes [52, 53], are highly recommended due to competitive advantages of PFSA membranes in PEM fuel cell applications over others. Modifications of PFSA membrane are mainly focused on the proton conductivity at higher temperature, water uptake and retain at higher temperature, low humidification operations, and mechanical stability at higher temperature.

The third group of alternative membranes is acid-base polymer membranes. This type of membranes is complexes fabricated by doping strong acids or polymeric acids in conventional polymers [50]. So far, phosphoric acid (H_3PO_4, abbreviated to PA) doped polybenzimidazole (PBI) has been found to be one of the most commercially promising materials for PEM fuel cell operating at temperature higher than 100°C. The advantages of PBI over other polymers, including low cost [54], high glass transition temperature [55], excellent textile fiber properties [56], and great thermal stability [57], promised it to be an excellent polymer for membrane fabrications. One of the most significant advantages of a PA doped PBI membrane over a PFSA based membrane is that its conductivity no longer relies on the water content due to its unique proton conduction mechanism [50, 58-60], but strongly depends on the PA doping level [50, 61-63] and the operating temperature [50, 63, 64].

4.2.2. Impedance study

There are only limited studies of EIS applications to high temperature PEM fuel cells. Impedance measurement was pioneered to study the electrical conductivity of PBI-based films at the end of 1990s. Fontanella and his co-workers [65] utilized impedance measurement to study the conductivity of PA doped PBI films at the temperatures below 100°C at a pressure up to 0.25 GPa. Soon after that, Bouchet and Siebert [61] published their work of measuring the conductivity of acid doped PBI films with the help of impedance measurement. However, in these works, impedance measurement was utilized only as an auxiliary method.

EIS was not applied to a membrane electrode assembly (MEA) or a fully constructed PEM fuel cell based on high temperature membrane until 2005. Xu [66] employed impedance analysis to study the effect of relative humidity (RH) on oxygen reduction reaction (ORR) kinetics for a high temperature PEM fuel cell manufactured from Nafion-Teflon-phosphotungstic acid (NTPA) membranes. Almost at the same time, Ramani [67] published their impedance measurement to a PEM fuel cell based on PA doped Nafion membrane at 120°C and 35% RH.

Several EIS studies on PA-PBI based high temperature PEM fuel cells emerged from 2006 [68-72]. Jalani and his co-workers [68] published their impedance analysis of a single cell assembly, named Celtec®-P series 1000 MEA (BASF Fuel Cell). Qi and his group [72] applied EIS to study the performance and degradation of a PA-PBI based PEM fuel cell at 180°C under a current density of 0.2 A cm^{-2}. However, both groups did not perform EC simulations of the cell impedance. At the same time, a more completed EIS application was published by Jingwei Hu and his co-workers [22, 69-71], which included a series work of impedance measurement and analysis, EC simulations, and degradation tests of cell performance.

The published applications of EIS on high temperature PEM fuel cells, although limited, present different measurement results one from another. Several different EC diagrams were proposed to interpret measured impedance data. An ohmic resistance, introduced by cell components (electrodes, membranes, gas diffusion layers, and other supporting plates) and connections, and a wiring inductance were involved in impedance analysis of all published works. Main differences exist in the analysis and interpretation of polarization impedance. Cells tested with different configurations and operating conditions perform differently from each other; however, they should behave certain characteristics in common, especially the cells with the same type of membrane. The EIS applications to high temperature PEM fuel cells based on PA doped PBI membranes will be summarized and discussed in the following section. The emphases are placed on EC simulations and data interpretations.

4.2.3. EC simulation and data interpretation

So far, up to three arcs have been reported in Nyquist plots of measured impedance spectra, but they were not well separated in all cases. One arc overlapped with its neighboring ones when they shifted to a similar frequency range with the change of cell operating conditions. In some circumstances, one arc decreased to be negligible. The processes involved in data interpretation mainly include charge transfer process, mass transfer process, and gas diffusion process. However, some groups observing two impedance arcs in Nyquist plot preferred to classify the impedance arcs into anodic and cathodic processes [73-78].

1. Ohmic resistance

Ohmic resistance, R_Ω, is the sum of resistances of membranes, electrodes, catalyst layers, and gas diffusion layers, contact resistance, and any other resistances introduced by the hardware connected to the measuring circuit, such as wires, heaters, blowers, and controller boards. Although it is difficult to break down the total ohmic resistance according to different contributions, its change with varying operating conditions of cells mainly reflects the conductivity of membranes. Thus, the proton conductivity mechanism of PA doped PBI membranes can be studied with the help of ohmic resistance.

R_Ω was reported as a set of scattered numbers with increasing current density in Jalani's work [68]; however, a trend of decline was still observable from the published data. Later, it was confirmed in many other experiments that R_Ω decreases with increasing current density [77, 79-82]. Following Zhang's theory [79], it was accepted that the decrease of R_Ω was resulted from an increase in the proton conductivity of the membrane due to higher water productivity at higher current density. However, Zhang [79] also expected a constant level of R_Ω at a current density larger than 1.0 A cm^{-2} because of a constant water uptake of the membrane balanced between produced and purged water. This constant level was observed in Andreasen's experiments both on a 1 kW cell stack [78] and a single cell MEA [83] even at a current density lower than 1.0 A cm^{-2}.

The effect of operating temperature on R_Ω is more complicated than current density. And lots of attentions were paid to it since the activation behavior closely relates to the proton conductivity mechanism of membranes. Some reported that R_Ω decreased with increasing temperature [77, 79, 83]. The proton hopping mechanism proposed by Bouchet [60, 63] was applied to explain this thermal effect [79]. The membrane conductivity σ based on this mechanism is expressed as [50, 61]:

$$\sigma = \sigma^0 exp\left(\frac{-E_a}{RT}\right) = \frac{A}{T} exp\left(\frac{-E_a}{RT}\right) \qquad (9)$$

where σ^0 and A are pre-exponential factors, E_a is the activation energy, and R is the ideal gas constant.

But, inconsistent with these results, many other researchers reported an increase of R_Ω with increasing temperature when the temperature went higher than around 140°C [73, 80, 82].

2.   High frequency (HF) impedance arc

This arc appears right after the wiring inductive loop as frequency decreases in Nyquist plot. And generally it is quite a small semi-circle comparing to the following impedance arcs. It dominates the region of frequency from above 100 Hz up to 1000 Hz [68, 78, 79, 81-84], or even higher. The effects of temperature [79, 82] and current density [81, 82] on this impedance arc are observable but not as significant as on other impedance arcs dominating lower frequency regions. The resistance of this HF impedance arc decreases with increasing current densities, and its time constant decreases with increasing temperature. The latter can be observed in Nyquist plot as the impedance arc shifts toward higher frequency, or "shrinks". Kinetically, this phenomenon can be explained as the process occurring faster at higher temperature.

It was validated and discussed in many published works that this HF impedance arc is contributed by charge transfer processes. Its impedance is generally simulated by the ($R_{ct}C_d$) sub-circuit in EC models. The R_{ct} refers to charge transfer resistance and the C_d refers to double-layer capacitance introduced by the charge accumulation and separation in the interface of electrode-electrolyte. In some cases, the double layer capacitor may be substituted by a CPE to reflect the non-ideal characteristics of the interface. Many researchers preferred to ascribe this HF arc to the charge transfer process occurring on the anode, that was the charge transfer process involved in hydrogen oxidation reaction (HOR) [73, 77, 78, 80]. Some other groups stated that the charge transfer processes of both HOR and ORR contributed to this HF impedance arc; however, the contribution of HOR was negligible at low current density [79, 82].

3.   Middle frequency (MF) impedance arc

This arc is contributed by the most dominating process occurring in the cell. It usually appears as the largest semi-circle in Nyquist plot and spans from 100 Hz to 1 Hz [68, 78, 79, 81-84]. A consistent interpretation proposed for this arc were widely accepted in published cases that an activation process related to ORR contributed to this polarization loss.

In some cases, only one impedance arc was observed in Nyquist plot [22, 69-71, 85, 86]. The low frequency (LF) arc does not perform significant contribution to the total impedance at all circumstances (discussed later in the part of "LF impedance arc"). And sometimes, as mentioned in the discussion of HF arc, the HF arc shrinks to be hardly noticed. Thus, the only arc observed can be explained as mass transfer processes. In this case, the HF arc actually merges with the MF arc and the total impedance appears as one arc.

4.   Low frequency (LF) impedance arc

This arc only appears in some certain cases when the contributions of concentration processes to cell impedance are comparably significant. Generally, it dominates frequency region below 1 Hz down to around 0.1 Hz [68, 78, 79, 81-84]. The impedance of this arc strongly depends on the compositions of cathode inlet gas, generally consisting of air, oxygen, or a mixture of them. The LF arc enlarges with the increase of current loads and dominates the total impedance at high current loads instead of the MF arc. Studies on oxygen stoichiometry can provided further information for the study of diffusion processes.

5. Conclusions

This chapter emphasizes the existing necessities for the improvement and development of energy storage and conversion systems and highlights the competence of Electrochemical Impedance Spectroscopy (EIS) to dynamically characterize electrochemical systems. As the background, this chapter briefly provided fundamental knowledge of impedance measurement and data interpretation. The EIS applications are mainly focused on Pb-acid batteries, Ni-MH batteries, conventional proton exchange membrane (PEM) fuel cells, and high temperature PEM fuel cells based on phosphoric acid (H_3PO_4, abbreviated to PA) doped polybenzimidazole (PBI) membranes. For the rechargeable batteries and the conventional PEM fuel cells, investigation examples are presented with discussions on research challenges and further development. The high temperature PEM fuel cell is a freshly emergent research area. There are obvious differences between the behavior and variation tendency of impedance data collected from different systems. Several equivalent circuit (EC) diagrams and their physical interpretations were proposed for different high temperature PEM fuel cells. However, more experiments and impedance data are required to develop a consistent, validated, and generally accepted theory. Developed for over a hundred years, EIS will continuously contribute to the characterization, diagnosis, quality control, and further advanced areas of energy storage and conversion systems for energy economic considerations.

Author details

Ying Zhu, Wenhua H. Zhu and Bruce J. Tatarchuk*
Center for Microfibrous Materials
Department of Chemical Engineering
Auburn University, Auburn, AL, 36849, USA

Acknowledgement

This work was performed under a U.S. Army contract at Auburn University (W56HZV-05-C0686) administered through TARDEC. The authors would like to thank members from the Center for Microfibrous Materials for their assistance and contribution to this work.

6. References

[1] Tatarchuk BJ, Yang H, Kalluri R, Cahela DR, inventors. Microfibrous Media for Optimizing and Controlling Highly Exothermic and Highly Endothermic Reactions/Processes. 2011; Patent No: WO2011/057150A1.

[2] Sheng M, Cahela DR, Yang H, Tatarchuk BJ. Application of Microfiber Entrapped Catalyst in Fischer-Tropsch Synthesis. AIChE Annual Meeting, 2009; Nashville, TN, United States.

---

* Corresponding Author

[3] Sheng M, Yang H, Cahela DR, Tatarchuk BJ. Novel Catalyst Structures with Enhanced Heat Transfer Characteristics. Journal of Catalysis 2011; 281(2):254-262.

[4] Chew TL, Bhatia S. Catalytic Processes towards the Production of Biofuels in a Palm Oil and Oil Palm Biomass-Based Biorefinery. Bioresource Technology 2008;99(17):7911-7922.

[5] Chen Y. Development and Application of Co-Culture for Ethanol Production by Co-Fermentation of Glucose and Xylose: A Systematic Review. Journal of Industrial Microbiology & Biotechnology 2011;38(5):581-597.

[6] Demirbas A. Hydrogen Production from Carbonaceous Solid Wastes by Steam Reforming. Energy Sources Part A - Recovery Utilization and Environmental Effects. 2008;30(10):924-931.

[7] Yuan L-X, Wang Z-X, Dong T, Kan T, Zhu X-F, Li Q-X. Hydrogen and Liquid Bio-fuel Generated from Biomass. Journal of University of Science and Technology of China. 2008;38(6):6.

[8] Mantell CL. Batteries and Energy Systems: McGraw-Hill; 1983.

[9] Prokhorov DV. Toyota Prius HEV Neurocontrol and Diagnostics. Neural Networks. 2008;21(2-3):458-465.

[10] Heaviside O. The Induction of Currents in Cores. Electrical Papers. 1. New York: Macmillan and Co.; 1894. p. 353-415.

[11] Heaviside O. Electromagnetic Induction and Its Propagation (2nd Half). Electrical Papers. 2. New York: Macmillan and Co.; 1894. p. 39-154.

[12] Kennelly AE. Impedance. The 76th meeting of the American Institute of Electrical Engineers, 1893; New York, NY, United States; p. 172-232.

[13] Macdonald JR. Impedance Spectroscopy. Annals of Biomedical Engineering. 1992;20(3):289-305.

[14] Orazem ME, Tribollet B. Electrochemical Impedance Spectroscopy. Hoboken, New Jersey: John Wiley & Sons, Inc.; 2008.

[15] O'Hayre R, Cha S-W, Colella W, Prinz FB. Fuel Cell Fundamentals. Hoboken, New Jersey: John Wiley & Sons, New York; 2006.

[16] Barsoukov E, Macdonald JR. Impedance Spectroscopy: Theory, Experiment, and Applications: Wiley-Interscience; 2005.

[17] Wojcik PT, Agarwal P, Orazem ME. A method for maintaining a constant potential variation during galvanostatic regulation of electrochemical impedance measurements. Electrochimica Acta. 1996;41(7-8):977-983.

[18] Gamry Instruments. Gamry Instruments Product Brochure: FC350 Fuel Cell Monitor. http://www.gamry.com/assets/Uploads/FC350-Brochure.pdf.

[19] Zhu WH, Payne RU, Nelms RM, Tatarchuk BJ. Equivalent Circuit Elements for PSpice Simulation of PEM Stacks at Pulse Load. Journal of Power Sources. 2008;178(1):197-206.

[20] Shinners SM. Modern Control System Theory and Design: John Wiley & Sons; 1998.

[21] Gabrielli C. Identification of Electrochemical Processes by Frequency Response Analysis. Solatron Analytical AMETEK, Inc. 1998; Technical Report No. 004/83.

[22] Hu JW, Zhang HM, Gang L. Diffusion-Convection/Electrochemical Model Studies on Polybenzimidazole (PBI) Fuel Cell Based On AC Impedance Technique. Energy Conversion and Management. 200;49(5):1019-1027.

[23] Macdonald DD, Urquidimacdonald M. Application of Kramers-Kronig Transforms in the Analysis of Electrochemical Systems 1. Polarization Resistance. Journal of the Electrochemical Society. 1985;132(10):2316-2319.

[24] Boukamp BA. A Linear Kronig-Kramers Transform Test for Immittance Data Validation. Journal of the Electrochemical Society. 1995;142(6):1885-1894.

[25] Macdonald JR. Impedance Spectroscopy - Old Problems and New Developments. Electrochimica Acta. 1990;35(10):1483-1492.

[26] Boukamp BA. A Nonlinear Least-Squares Fit Procedure for Analysis of Immittance Data of Electrochemical Systems. Solid State Ionics. 1986;20(1):31-44.

[27] Vladikova D. The Technique of the Differential Impedance Analysis Part I: Basics of the Impedance Spectroscopy. International Workshop "Advanced Techniques for Energy Sources Investigation and Testing"; 2004; Sofia, Bulgaria.

[28] Research Solutions & Resources LLC. Electrochemistry Resources: Electrochemical Impedance: Diffusion and EIS: Warburg.
http://www.consultrsr.com/resources/eis/diffusion.htm

[29] Bard AJ, Faulkner LR. Electrochemical Methods: Fundamentals and Applications: Wiley; 2001.

[30] Payne RU, Zhu Y, Zhu WH, Timper MS, Elangovan S, Tatarchuk BJ. Diffusion and Gas Conversion Analysis of Solid Oxide Fuel Cells at Loads via AC Impedance. International Journal of Electrochemistry. 2011. DOI:10.4061/2011/465452.

[31] Zhu WH, Payne RU, Tatarchuk BJ. PEM Stack Test and Analysis in a Power System at Operational Load via AC Impedance. Journal of Power Sources. 2007;168(1):211-217.

[32] Keithley JF. The Story of Electrical and Magnetic Measurements: From 500 BC to the 1940s: Wiley; 1999.

[33] Berg H. Johann Wilhelm Ritter - The Founder of Scientific Electrochemistry. Review of Polarography. 2008;54(2):99-103.

[34] Nelson RF. Power Requirements for Batteries in Hybrid Electric Vehicles. Journal of Power Sources. 2000;91(1):2-26.

[35] Karden E, Buller S, De Doncker RW. A Method for Measurement and Interpretation of Impedance Spectra for Industrial Batteries. Journal of Power Sources. 2000;85(1):72-78.

[36] Hejabi M, Oweisi A, Gharib N. Modeling of Kinetic Behavior of the Lead Dioxide Electrode in a Lead-Acid Battery by Means of Electrochemical Impedance Spectroscopy. Journal of Power Sources. 2006;158(2):944-948.

[37] Furukawa J, Takada T, Kanou T, Monma D, Lam LT, Haigh NP, et al. Development of UltraBattery. CSIRO Energy Technology, Australia Furukawa Battery CO. Ltd., Japan, 2006.

[38] Lam LT, Louey R. Development of Ultra-Battery for Hybrid-Electric Vehicle Applications. Journal of Power Sources. 2006;158(2):1140-1148.

[39] Lam LT, Louey R, Haigh NP, Lim OV, Vella DG, Phyland CG, et al. VRLA Ultrabattery for High-Rate Partial-State-of-Charge Operation. Journal of Power Sources. 2007;174(1):16-29.

[40] Extreme Racing Owner's Manual. EnerSys Energy Products Inc., 2010.

[41] Zhu WH, Zhu Y, Tatarchuk BJ. A Simplified Equivalent Circuit Model for Simulation of Pb-Acid Batteries at Load for Energy Storage Application. Energy Conversion and Management. 2011;52(8-9):2794-2799.

[42] Sakai T, Uehara I, Ishikawa H. R&D on Metal Hydride Materials and Ni-MH Batteries In Japan. Journal of Alloys and Compounds. 1999;293:762-769.

[43] Shukla AK, Venugopalan S, Hariprakash B. Nickel-Based Rechargeable Batteries. Journal of Power Sources. 2001;100(1-2):125-148.

[44] Hariprakash B, Shukla AK, Venugoplan S. Editor-in-Chief: Jürgen G. Secondary Batteries - Nickel Systems | Nickel-Metal Hydride: Overview. Encyclopedia of Electrochemical Power Sources. Amsterdam: Elsevier; 2009. p. 494-501.

[45] Larminie J, Dicks A. Fuel Cell Systems Explained: J. Wiley; 2003.

[46] Mond L, Langer C, editors. A New Form of Gas Battery. Proceedings of the Royal Society of London; 1889; London: Harrison and Sons, St. Martin's Lane.

[47] Hoogers G. Fuel Cell Technology Handbook: Taylor & Francis; 2002.

[48] Zhu WH, Payne RU, Cahela DR, Nelms RM, Tatarchuk BJ. Performance Analysis of PEM Stacks at Operational Loads Using Equivalent Circuit Models. Proceedings of the 42nd Power Sources Conference; 2006; Philadelphia, PA.

[49] Costamagna P, Srinivasan S. Quantum Jumps in the PEMFC Science And Technology from the 1960s to the Year 2000 Part I. Fundamental Scientific Aspects. Journal of Power Sources. 2001;102(1-2):242-252.

[50] He RH, Li QF, Xiao G, Bjerrum NJ. Proton Conductivity of Phosphoric Acid Doped Polybenzimidazole and its Composites with Inorganic Proton Conductors. Journal of Membrane Science. 2003;226(1-2):169-184.

[51] Li QF, He RH, Jensen JO, Bjerrum NJ. Approaches and Recent Development of Polymer Electrolyte Membranes for Fuel Cells Operating above 100 Degrees C. Chemistry of Materials. 2003;15(26):4896-4915.

[52] Peighambardoust SJ, Rowshanzamir S, Amjadi M. Review of the Proton Exchange Membranes for Fuel Cell Applications. International Journal of Hydrogen Energy. 2010;35(17):9349-9384.

[53] Zhang JL, Tang YH, Song CJ, Xia ZT, Li H, Wang HJ, et al. PEM Fuel Cell Relative Humidity (RH) and its Effect on Performance at High Temperatures. Electrochimica Acta. 2008;53(16):5315-5321.

[54] Wainright JS, Wang JT, Weng D, Savinell RF, Litt M. Acid-Doped Polybenzimidazoles - A New Polymer Electrolyte. Journal Of The Electrochemical Society. 1995;142(7):L121-L123.

[55] Musto P, Karasz FE, Macknight WJ. Fourier-Transform Infrared-Spectroscopy on the Thermooxidative Degradation of Polybenzimidazole and of a Polybenzimidazole Polyetherimide Blend. Polymer. 1993;34(14):2934-2945.

[56] Chung TS. A Critical Review of Polybenzimidazoles: Historical Development and Future R&D. Journal of Macromolecular Science-Reviews In Macromolecular Chemistry and Physics. 1997;C37(2):277-301.

[57] Samms SR, Wasmus S, Savinell RF. Thermal Stability of Proton Conducting Acid Doped Polybenzimidazole in Simulated Fuel Cell Environments. Journal of the Electrochemical Society. 1996;143(4):1225-1232.

[58] Steininger H, Schuster M, Kreuer KD, Kaltbeitzel A, Bingol B, Meyer WH, et al. Intermediate Temperature Proton Conductors for PEM Fuel Cells Based on Phosphonic Acid as Protogenic Group: A Progress Report. Physical Chemistry Chemical Physics. 2007;9(15):1764-1773.

[59] Kreuer KD, Paddison SJ, Spohr E, Schuster M. Transport in Proton Conductors for Fuel-Cell Applications: Simulations, Elementary Reactions, and Phenomenology. Chemical Reviews. 2004;104(10):4637-4678.

[60] Schuster MFH, Meyer WH, Schuster M, Kreuer KD. Toward a New Type of Anhydrous Organic Proton Conductor Based on Immobilized Imidazole. Chemistry of Materials. 2004;16(2):329-337.

[61] Bouchet R, Siebert E. Proton Conduction in Acid Doped Polybenzimidazole. Solid State Ionics. 1999;118(3-4):287-299.

[62] Li QF, He RH, Berg RW, Hjuler HA, Bjerrum NJ. Water Uptake and Acid Doping of Polybenzimidazoles as Electrolyte Membranes for Fuel Cells. Solid State Ionics. 2004;168(1-2):177-185.

[63] Wannek C, Lehnert W, Mergel J. Membrane Electrode Assemblies for High-Temperature Polymer Electrolyte Fuel Cells Based on Poly(2,5-Benzimidazole) Membranes with Phosphoric Acid Impregnation via the Catalyst Layers. Journal of Power Sources. 2009;192(2):258-266.

[64] Bouchet R, Miller S, Duclot M, Souquet JL. A Thermodynamic Approach to Proton Conductivity in Acid-Doped Polybenzimidazole. Solid State Ionics. 2001;145(1-4):69-78.

[65] Fontanella JJ, Wintersgill MC, Wainright JS, Savinell RF, Litt M. High Pressure Electrical Conductivity Studies of Acid Doped Polybenzimidazole. Electrochimica Acta. 1998;43(10-11):1289-1294.

[66] Xu H, Song Y, Kunz HR, Fenton JM. Effect of Elevated Temperature and Reduced Relative Humidity on ORR Kinetics for PEM Fuel Cells. Journal of the Electrochemical Society. 2005;152(9):A1828-A1836.

[67] Ramani V, Kunz HR, Fenton JM. Stabilized Composite Membranes and Membrane Electrode Assemblies for Elevated Temperature/Low Relative Humidity PEFC Operation. Journal of Power Sources. 2005;152(1):182-188.

[68] Jalani NH, Ramani M, Ohlsson K, Buelte S, Pacifico G, Pollard R, et al. Performance Analysis and Impedance Spectral Signatures of High Temperature PBI-Phosphoric Acid Gel Membrane Fuel Cells. Journal of Power Sources. 2006;160(2):1096-103.

[69] Hu JW, Zhang HM, Hu J, Zhai YF, Yi BL. Two Dimensional Modeling Study of PBI/H_3PO_4 High Temperature PEMFCs Based on Electrochemical Methods. Journal of Power Sources. 2006;160(2):1026-1034.

[70] Hu JW, Zhang HM, Zhai YF, Liu G, Yi BL. 500h Continuous Aging Life Test on PBI/H(3)PO(4) High-Temperature PEMFC. International Journal of Hydrogen Energy. 2006;31(13):1855-1862.

[71] Hu JW, Zhang HM, Zhai YF, Liu G, Hu J, Yi BL. Performance Degradation Studies on PBI/H₃PO₄ High Temperature PEMFC and One-Dimensional Numerical Analysis. Electrochimica Acta. 2006;52(2):394-401.

[72] Qi ZG, Buelte S. Effect of Open Circuit Voltage on Performance and Degradation of High Temperature PBI-H₃PO₄ Fuel Cells. Journal of Power Sources. 2006;161(2):1126-1132.

[73] Lobato J, Canizares P, Rodrigo MA, Linares JJ. PBI-Based Polymer Electrolyte Membranes Fuel Cells - Temperature Effects on Cell Performance and Catalyst Stability. Electrochimica Acta. 2007;52(12):3910-3920.

[74] Lobato J, Canizares P, Rodrigo MA, Linares JJ, Ubeda D, Pinar FJ. Study of the Catalytic Layer in Polybenzimidazole-Based High Temperature PEMFC: Effect of Platinum Content on the Carbon Support. Fuel Cells. 2010;10(2):312-319.

[75] Lobato J, Canizares P, Rodrigo MA, Linares JJ, Pinar FJ. Study of the Influence of the Amount of PBI-H(3)PO(4) in the Catalytic Layer of a High Temperature PEMFC. International Journal of Hydrogen Energy. 2010;35(3):1347-1355.

[76] Lobato J, Canizares P, Rodrigo MA, Pinar FJ, Ubeda D. Study of Flow Channel Geometry Using Current Distribution Measurement in a High Temperature Polymer Electrolyte Membrane Fuel Cell. Journal of Power Sources. 201;196(9):4209-4217.

[77] Boaventura M, Mendes A. Activation Procedures Characterization of MEA Based on Phosphoric Acid Doped PBI Membranes. International Journal of Hydrogen Energy. 2010;35(20):11649-11660.

[78] Andreasen SJ, Jespersen JL, Schaltz E, Kær SK. Characterisation and Modelling of a High Temperature PEM Fuel Cell Stack Using Electrochemical Impedance Spectroscopy. Fuel Cells. 2009;9(4):463-473.

[79] Zhang JL, Tang YH, Song CJ, Zhang JJ. Polybenzimidazole-Membrane-Based PEM Fuel Cell in the Temperature Range of 120-200 Degrees C. Journal of Power Sources. 2007;172(1):163-171.

[80] Chen CY, Lai WH. Effects of Temperature and Humidity on the Cell Performance and Resistance of a Phosphoric Acid Doped Polybenzimidazole Fuel Cell. Journal of Power Sources. 2010;195(21):7152-7159.

[81] Mamlouk M, Scott K. Analysis of High Temperature Polymer Electrolyte Membrane Fuel Cell Electrodes Using Electrochemical Impedance Spectroscopy. Electrochimica Acta. 2011;56(16):5493-5512.

[82] Jespersen JL, Schaltz E, Kaer SK. Electrochemical Characterization of a Polybenzimidazole-Based High Temperature Proton Exchange Membrane Unit Cell. Journal of Power Sources. 2009;191(2):289-296.

[83] Andreasen SJ, Vang JR, Kær SK. High Temperature PEM Fuel Cell Performance Characterisation with CO and CO(2) Using Electrochemical Impedance Spectroscopy. International Journal f Hydrogen Energy. 2011;36(16):9815-9830.

[84] Oono Y, Sounai A, Hori M. Influence of the Phosphoric Acid-Doping Level in a Polybenzimidazole Membrane on the Cell Performance of High-Temperature Proton Exchange Membrane Fuel Cells. Journal of Power Sources. 2009;189(2):943-949.

[85] Modestov AD, Tarasevich MR, Filimonov VY, Zagudaeva NM. Degradation of High Temperature MEA with PBI-H(3)PO(4) Membrane in a Life Test. Electrochimica Acta. 2009;54(27):7121-7127.

[86] Lin HL, Hsieh YS, Chiu CW, Yu TL, Chen LC. Durability and Stability Test of Proton Exchange Membrane Fuel Cells Prepared From Polybenzimidazole/Poly(Tetrafluoro Ethylene) Composite Membrane. Journal of Power Sources. 2009;193(1):170-174.

Permissions

The contributors of this book come from diverse backgrounds, making this book a truly international effort. This book will bring forth new frontiers with its revolutionizing research information and detailed analysis of the nascent developments around the world.

We would like to thank Ahmed Faheem Zobaa, for lending his expertise to make the book truly unique. He has played a crucial role in the development of this book. Without his invaluable contribution this book wouldn't have been possible. He has made vital efforts to compile up to date information on the varied aspects of this subject to make this book a valuable addition to the collection of many professionals and students.

This book was conceptualized with the vision of imparting up-to-date information and advanced data in this field. To ensure the same, a matchless editorial board was set up. Every individual on the board went through rigorous rounds of assessment to prove their worth. After which they invested a large part of their time researching and compiling the most relevant data for our readers. Conferences and sessions were held from time to time between the editorial board and the contributing authors to present the data in the most comprehensible form. The editorial team has worked tirelessly to provide valuable and valid information to help people across the globe.

Every chapter published in this book has been scrutinized by our experts. Their significance has been extensively debated. The topics covered herein carry significant findings which will fuel the growth of the discipline. They may even be implemented as practical applications or may be referred to as a beginning point for another development. Chapters in this book were first published by InTech; hereby published with permission under the Creative Commons Attribution License or equivalent.

The editorial board has been involved in producing this book since its inception. They have spent rigorous hours researching and exploring the diverse topics which have resulted in the successful publishing of this book. They have passed on their knowledge of decades through this book. To expedite this challenging task, the publisher supported the team at every step. A small team of assistant editors was also appointed to further simplify the editing procedure and attain best results for the readers.

Our editorial team has been hand-picked from every corner of the world. Their multi-ethnicity adds dynamic inputs to the discussions which result in innovative outcomes. These outcomes are then further discussed with the researchers and contributors who give their valuable feedback and opinion regarding the same. The feedback is then collaborated with the researches and they are edited in a comprehensive manner to aid the understanding of the subject.

Apart from the editorial board, the designing team has also invested a significant amount of their time in understanding the subject and creating the most relevant covers. They scrutinized every image to scout for the most suitable representation of the subject and create an appropriate cover for the book.

The publishing team has been involved in this book since its early stages. They were actively engaged in every process, be it collecting the data, connecting with the contributors or procuring relevant information. The team has been an ardent support to the editorial, designing and production team. Their endless efforts to recruit the best for this project, has resulted in the accomplishment of this book. They are a veteran in the field of academics and their pool of knowledge is as vast as their experience in printing. Their expertise and guidance has proved useful at every step. Their uncompromising quality standards have made this book an exceptional effort. Their encouragement from time to time has been an inspiration for everyone.

The publisher and the editorial board hope that this book will prove to be a valuable piece of knowledge for researchers, students, practitioners and scholars across the globe.

List of Contributors

Mohammad Taufiqul Arif and Amanullah M. T. Oo
Power Engineering Research Group, Faculty of Sciences, Engineering & Health, Central Queensland University, Bruce Highway, Rockhampton, QLD 4702, Australia

A. B. M. Shawkat Ali
School of Computing Sciences, Faculty of Arts, Business, Informatics & Education, Central Queensland University, Bruce Highway, Rockhampton, QLD 4702, Australia

Haisheng Chen, Xinjing Zhang, Jinchao Liu and Chunqing Tan
Institute of Engineering Thermophysics, Chinese Academy of Sciences, Beijing, 100190, China

Luca Petricca, Per Ohlckers and Xuyuan Chen
Vestfold University College, Norway

Petr Krivik and Petr Baca
The Faculty of Electrical Engineering and Communication, Brno University of Technology, Czech Republic

Hussein Ibrahim
TechnoCentre éolien, Gaspé, QC, Canada

Adrian Ilinca
Université du Québec à Rimouski, Rimouski, QC, Canada

Yong Xiao, Xiaoyu Ge and Zhe Zheng
College of Information Engineering, Shenyang University of Chemical Technology, China

Masatoshi Uno
Japan Aerospace Exploration Agency, Japan

George Cristian Lazaroiu
Department of Electrical Engineering, University Politehnica from Bucharest, Bucharest, Romania

Sonia Leva
Department of Energy of the Politecnico di Milano, Milan, Italy

Yu Zhang and Jinliang Liu
College of Opto-Electronic Science and Engineering, National University of Defense Technology, Changsha, China

Ruddy Blonbou, Stéphanie Monjoly and Jean-Louis Bernard
Geosciences and Energy Research Laboratory, Université des Antilles et de la Guyane, Guadeloupe, France

Antonio Ernesto Sarasua
Instituto de Energía Eléctrica, Universidad Nacional de San Juan, Argentina

Marcelo Gustavo Molina and Pedro Enrique Mercado
CONICET, Instituto de Energía Eléctrica, Universidad Nacional de San Juan, Argentina

Ying Zhu, Wenhua H. Zhu and Bruce J. Tatarchuk
Center for Microfibrous Materials, Department of Chemical Engineering, Auburn University, Auburn, AL, 36849, USA